The Patrick Moore Practica

For further volumes:
http://www.springer.com/series/3192

Choosing and Using Astronomical Eyepieces

William Paolini

William Paolini
Vienna, VA, USA

ISSN 1431-9756
ISBN 978-1-4614-7722-8 ISBN 978-1-4614-7723-5 (eBook)
DOI 10.1007/978-1-4614-7723-5
Springer New York Heidelberg Dordrecht London

Library of Congress Control Number: 2013939732

© Springer Science+Business Media New York 2013
This work is subject to copyright. All rights are reserved by the Publisher, whether the whole or part of the material is concerned, specifically the rights of translation, reprinting, reuse of illustrations, recitation, broadcasting, reproduction on microfilms or in any other physical way, and transmission or information storage and retrieval, electronic adaptation, computer software, or by similar or dissimilar methodology now known or hereafter developed. Exempted from this legal reservation are brief excerpts in connection with reviews or scholarly analysis or material supplied specifically for the purpose of being entered and executed on a computer system, for exclusive use by the purchaser of the work. Duplication of this publication or parts thereof is permitted only under the provisions of the Copyright Law of the Publisher's location, in its current version, and permission for use must always be obtained from Springer. Permissions for use may be obtained through RightsLink at the Copyright Clearance Center. Violations are liable to prosecution under the respective Copyright Law.
The use of general descriptive names, registered names, trademarks, service marks, etc. in this publication does not imply, even in the absence of a specific statement, that such names are exempt from the relevant protective laws and regulations and therefore free for general use.
While the advice and information in this book are believed to be true and accurate at the date of publication, neither the authors nor the editors nor the publisher can accept any legal responsibility for any errors or omissions that may be made. The publisher makes no warranty, express or implied, with respect to the material contained herein.

Printed on acid-free paper

Springer is part of Springer Science+Business Media (www.springer.com)

This book is dedicated...
To my wife, whose love is more inspiring than any promise of celestial discovery, and whose beauty is more alluring than any star-filled sky...
To my parents, whose love and guidance is a beacon as sure as the Northern Star... To my siblings, whose companionship in life brings comfort like the sight of a familiar constellation... To my children, whose loving hands hold the shining stars that are the hopes, dreams, and achievements of the future.

Бул китерти...
Менин жубайыма, сүйүүсү, ааламдын берген ар түрлүү үмүттөрүнөн да артык, дем берген, суктандырган сулуулугу асмандагы жайнаган жылдыздардай болгон ардагыма арнайм...
Мени жөлөп-таяп, жылуу сүйүүсү менен таңкы жылдыздай колдогон ата-энелериме арнайм... Өз колдору менен, мага тааныш кошоктошкон жылдыздардай эле ылайыктуу шарт түзүп беришкен менин биртуугандарыма арнайм... Менин сүйүктүү балдарыма, ушул жаркыраган жылдыздарды - келечек, үмүт жана бакыт-таалай катары тапшырам.

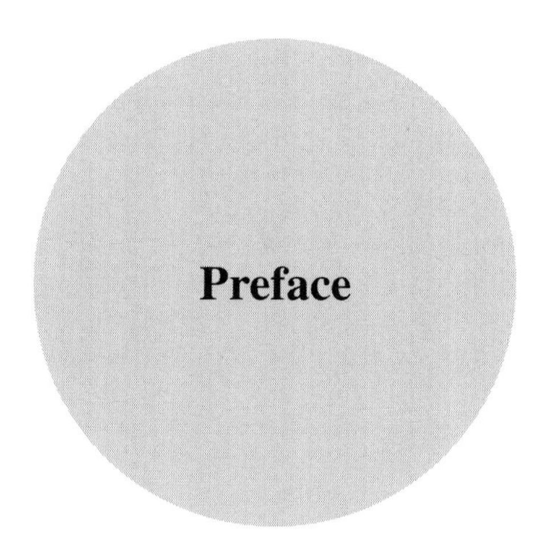

Preface

During the mid-1960s, I received my very first telescope. The prospect of what it would reveal was exciting to say the least, and with my brother's help, I took no time in setting it up so I could begin the adventure with some daytime observing. Back then, the attention was all about the "telescope," and eyepieces commanded no fascination of their own. In fact, eyepieces were just another "thing" that came in the box—they were simply a part of the telescope, like the Moon filter and the solar projection screen. Today of course, this has all changed and both the beginner and seasoned amateur astronomer maintain a keen interest in the eyepiece and the many special attributes it brings to the table that are separate and distinct from what the telescope can provide. With the many advances in eyepiece technology, it is also quite easy to assemble a set of eyepieces that individually contain more optical material than even the observer's telescope might have! And taken all together, a small collection of eyepieces can actually cost much more than the telescope itself. For some observers today, they even form fond attachments to their favorite eyepieces, and as they change and upgrade telescopes over time, their eyepieces remain. Because of all of this, the eyepiece of today has taken on a unique life of its own, as something very special and distinct, and as a much valued and discussed component in the observer's optical arsenal. This new standing for the eyepiece I feel is quite deserved, because it is not the telescope that brought us that experience which the popular idiom coined by Tele Vue Optics so wonderfully describes ("Space Walk Effect"), but it is the eyepiece alone that has accomplished this transformation for the amateur astronomer community. In many ways, the eyepiece of the twenty-first century has grown up and is no longer just another "thing" in the box with the telescope. Instead, the eyepiece is for many a well-deserved center of attention in the consumer astronomy market.

In this book, *Choosing and Using Astronomical Eyepieces*, I strove to present the eyepiece in five distinct ways: introductorily for those not too familiar with its basic form and function (Chap. 1), more in depth for those wanting to deepen their knowledge and learn a myriad of tricks-of-the-trade when using an eyepiece (Chap. 2), from the perspective of the eyepiece's Apparent Field of View capability as this is now a popular way of thinking about an eyepiece and has become the basis by many for how they choose and use eyepieces (Chap. 3), from the perspective of other amateur astronomers from around the world to give a broader perspective that goes beyond the technical and into the heart and soul of the eyepiece (Chap. 4), and, finally, from the all-important "gear-head" perspective where, in the final desk reference section of the book, the reader can peruse and refer to the many statistics and photographs on the over 200 popular new and used eyepiece brands and lines detailed (Chap. 5).

The astronomical eyepiece, once a little considered accessory, is today a much revered, discussed, and transformative component of a telescopic system. No longer are our explorations restricted to "porthole" views with eyes held tight to optics; instead the eyepiece has been the liberator, taking us to ever new heights in observing where the bounds of Earth can seemingly slip away as we immerse our way through space and among the stars.

Vienna, VA, USA William Paolini Jr.

Acknowledgments

My sincere gratitude goes out to the many individuals and organizations for their contributions and support. Without their selfless contributions of expertise, advice, essays, and photographs, this book would not have been possible. Thanks to Agena Astro Products, Al Nagler, Alexander Kupco, Alpine Astronomical , Andreas Braun, Andy Howie, Baader Planetarium, Blake Andrews, Bob Luffel, Bob Ryan, Brendan Cuddihee, Brent Kikawa, Carlos Hernandez, Carol Anderson, Celestron, Charles Brault, Chris Mohr, Christoph Bosshard, the CloudyNights.com online community, D. Regan, Daniel Mounsey, David Elosser, David Ittner, David Nagler, Denkmeier Optical, Don Pensack, Doug Bailey, Doug Richter, Ed Moreno, Ed Ting, Elmira Amanbekova, Eric Sheperd, Erik Bakker, Erika Rix, Eyepieces Etc., Fred Lamothe, Fukuura Masato, Gary Chiang, Gary Hand, Gary Russell, Geoff Chester, Glen Moulton, Hands On Optics, Hernando Bautista, I. R. Poyser Brass Telescopes, Ian Poyser, Irvin Vann, James Curry, James Spriesterbach, Jamie Crona, Janet Deis, Jeremy K., Jiang Shi Hua, Jim Barnett, Jim Rosenstock, John Levine, John W., John Watson, Judson Mitchell, Keith Howlett, Konstantinos Kokkolis, Kson Optics-Electronics Company, Larry Eastwood, Long Perng Optics, Lyra Optic, Malcolm Neo, Marisa Dominello, Mark Liu, Marty Stevens, Maury Solomon, Meade Instruments Corporation, Michel Guévin, Michelle Meskill, Mike Bacanin, Mike H., Mike Hankey, Mike Ratcliff, Mike Rowles, Mike Sutherland, Mike Wooldridge, Neil English, Neville Edwards, Orion Telescopes & Binoculars, Patrick O'Neil, Paul Surowiec, Paul Webb, Peter Sursock, Phil Piburn, Raisa Kydyralieva, Roman Seestakov, Russ Lederman, Russell Optics, Samuel de Roa, Sherry Hand, Sophia Tu, Springer Science and Business Media, Stephen Bueltman, Stephen Chen, Steve Couture, Steve G., Steve Stapf, Steven Cotton, Surplus Shed, TALTELEOPTICS, Tamiji Homma, Tele Vue Optics, Thomas McCague, Tim Wetherell, Tom Morris, Tony Miller, U.S. Naval Observatory, Uwe Pilz, Vixen Optics, Wade Wheeler, and William Rose. If I have inadvertently omitted a contributor to this effort, please accept my sincerest apologies and know that I am forever appreciative of the help and support you so graciously provided.

I also wish to thank my family and friends for their unwavering encouragement and support throughout this process: Aaron Paolini, Anthony Paolini, Jr., Ayanna Amanbekov, Aydar Shaildayev, Christina Paolini, Christopher Paolini, Claire Paolini, Dastan Sadykov, David Paolini, Dinara Amanbekova, Emily Straub, Frank DiLuzio, Helen Paolini, Jacqueline Diluzio, Janice Kupersmith, Jeanne Ayivorh, John Condia, Laritta Paolini, Louis Paolini, Michael Mencer, Murat Amanbekov, NiiBen Ayivorh, Patricia Amanbekova, Pepper, Raisa Kydyralieva, Russell Kupersmith, Sharon Condia, Ted Straub, Tursun Sadykov, Ulan Amanbekov, Virginia Paolini, and William Paolini, Sr.

Contents

Part I Background

1 Introducing the Astronomical Eyepiece 3
Historical Beginnings 3
Basic Function 7
Physical Construction 8
Optical Construction 12
Optical Design Characteristics 15
Focal Length 16
Apparent Field of View (AFOV) 18
Eye Relief 20
Exit Pupil Behavior 22
Internal Reflections and "Ghosting" 24
Aberrations (Including Distortions) 25

2 Choosing Eyepieces and Observing Strategies 33
Viewing Comfort and Usability Considerations 33
Eye Relief 34
Construction and Mechanical Features 35
Size and Weight 41
Visual Impact Considerations 43
Apparent Field of View (AFOV) 44
True Field of View (TFOV) 45
Magnification, Brightness, and Contrast 53
Aberration Control (Telescope Dependencies) 55
Observing Strategies 56
Focal Length Choices 56

Calculating Exit Pupil ... 60
High Magnification ... 66

3 **Eyepieces and Accessories for Celestial Targets** ... 73
Celestial Targets ... 74
The Moon ... 74
Stars and Clusters ... 76
Nebulae and Galaxies ... 77
The Planets ... 79
Comparing One Eyepiece to Another ... 83
Eyepiece Accessory Considerations ... 84
Barlows and Amplifiers ... 85
Filters ... 94
Astro Zoom Zoomset (www.astrozoom.de) ... 103
Care and Cleaning of Eyepieces ... 104

4 **Popular Eyepieces by AFOV Class** ... 109
The 40° (and Less) AFOV Class ... 109
The 50° AFOV Class ... 118
The 60° AFOV Class ... 127
The 70° AFOV Class ... 131
The 80°–90° AFOV Class ... 135
The 100° AFOV Class ... 144
The 120° AFOV Class ... 147

5 **Advice from the Amateur Astronomer Community** ... 149
20 Years of Personal Eyepiece History (1992–2012) ... 150
Achieving Focus with Long Focal Length-Rated Eyepieces ... 150
Astronomical Eyepieces: Objects of Pleasure ... 150
Benefits of Wider AFOV Eyepieces in Undriven 'Scopes ... 151
Binoviewing on a Budget ... 152
28 mm Edmund Scientific RKE.. ... 153
24 mm Brandons ... 153
Edmund Plössls ... 153
19 mm Smart Astronomy EF's.... ... 154
Choosing a Low Magnification Wide-Field Eyepiece ... 154
Choosing a Zoom Eyepiece ... 154
Couture Ball Eyepieces (The Do-It-Yourself Planetary) ... 155
Eyepiece Basics I ... 161
Eyepiece Basics II ... 162
Eyepiece Designs for the Critical Lunar and Planetary Observer ... 163
Eyepieces: Less Is More ... 164
General Comments on Eyepieces ... 165
Jolly Small Eyepieces ... 166
Learning to Use Hyper-Wide AFOV Eyepieces ... 167
Learning What Eyepieces Satisfy Your Needs ... 167
Microscope Eyepieces on Your Telescope ... 168

My Thoughts on Eyepieces for General Observing ... 169
Observing Comets and Nebulae: Brilliance and Wide Field Is What We Need First ... 170
Planetary Observing: Resolution and Contrast Is What We Need First ... 171
Quick Tip for New Telescope Owners ... 172
Sample Eyepiece Sets for Typical Amateur Telescopes ... 173
Shallow Sky Sketching Eyepieces ... 173
Should I Buy Complete Sets of Eyepieces or Mix-and-Match? ... 174
Some Tips for Viewing the Universe Through an Eyepiece ... 175
The DIY Eyepiece ... 175
The TV Ethos Series of Eyepieces: Virtues Beyond Just 100° of Panorama ... 176
Try Before You Buy ... 180
Using Zoom Eyepieces Successfully ... 180
What Is the Best Eyepiece? ... 181
What Makes a "Favorite" Eyepiece? ... 183
What to Do with Your Old, Inexpensive Eyepieces ... 183
Young People Can Own Nice Eyepieces, Too! ... 184

Part II Desk Reference of Astronomical Eyepieces

6 How to Use This Guide ... 187
Market Overview ... 187

7 Agena to Docter ... 191
Agena ED (*Discontinued*) ... 191
Agena: Enhanced Wide Angle (EWA) ... 192
Agena: Mega Wide Angle (MWA) ... 193
Agena: Super Wide Angle ... 194
Agena: Ultra Wide Angle (UWA) ... 195
Agena: Wide Angle ... 196
Antares: Classic Erfle ... 197
Antares: Elite Plössl ... 198
Antares: Ortho (*Discontinued*) ... 198
Antares: Plössl ... 199
Antares: Speers-WALER Series (Wide Angle Long Eye Relief) ... 201
Antares: W70 ... 202
APM: UWA Planetary ... 203
Apogee: Super Abbe Orthoscopic (*Discontinued*) ... 204
Apogee: Widescan III (*Discontinued*) ... 205
Astrobuffet: 1 RPD ... 205
Astro-Physics: Super Planetary AP-SPL (*Discontinued*) ... 206
Astro-Professional: EF Flatfield ... 207
Astro-Professional: Long Eye Relief Planetary ... 207
Astro-Professional: Plössl ... 208

Astro-Professional: SWA 208
Astro-Professional: UWA 209
Astro-Tech: AF Series 70° Field 209
Astro-Tech: Flat Field 210
Astro-Tech: High Grade Plössl 211
Astro-Tech: Long Eye Relief 212
Astro-Tech: Paradigm Dual ED 213
Astro-Tech: Series 6 Economy Wide Field 214
Astro-Tech: Titan Type II ED Premium 2″ Wide Field 215
Astro-Tech: Value Line Plössl 216
Astro-Tech: Wide Field 217
Baader Planetarium: Classic Ortho/Plössl 218
Baader Planetarium: Eudiascopic 219
Baader Planetarium: Genuine Abbe Ortho (*Discontinued*) 220
Baader Planetarium: Hyperion/Hyperion Aspheric 221
Standard Hyperions 222
Aspheric Hyperions 222
BST: Explorer ED 222
BST: Flat Field 223
Bresser: 52° Super Plössl 223
Bresser: 60° Plössl 224
Burgess: Planetary (*Discontinued*) 225
Burgess: Wide Angle (*Discontinued*) 226
BW Optic: Ultrawide (*Discontinued*) 226
Carton: Plössl (*Discontinued*) 227
Cave: Orthostar Orthoscopic (*Discontinued*) 228
Celestron: Axiom (*Discontinued*) 229
Celestron: Axiom LX (*Discontinued*) 230
Celestron: E-Lux (*Discontinued*) 231
Celestron: Erfle (*Discontinued*) 232
Celestron: Kellner (*Discontinued*) 234
Celestron: Luminos 235
Celestron: Omni 235
Celestron: Ortho (*Discontinued*) 237
Celestron: Silvertop Plössl (*Discontinued*) 238
Celestron: Ultima (*Discontinued*) 241
Celestron: Ultima LX 242
Celestron: X-Cel (*Discontinued*) 243
Celestron: X-Cel LX 243
Clavé: Plössl (*Discontinued*) 244
Coronado: CeMax 246
Couture: Ball Singlet 247
Criterion: Ortho/Kellner/A.R. (Achromatic Ramsden) (*Discontinued*) 248
Denkmeier: D21/D14 249
Docter: UWA 250

8 Edmund Scientific to Nikon ... 253
Edmund Scientific: Ortho ... 253
Edmund Scientific: Plössl ... 254
Edmund Scientific: RKE ... 255
Explore Scientific: 68° Nitrogen Purged (ES68 N2) ... 258
Explore Scientific: 82° Nitrogen Purged (ES82 N2) ... 259
Explore Scientific: 100° Nitrogen Purged (ES100 N2) ... 260
Explore Scientific: 120° Nitrogen Purged (ES100 N2) ... 261
Galland/Gailand/Galoc: Ortho/Erfle/König (*Discontinued*) ... 262
Garrett Optical: Orthoscopic (*Discontinued*) ... 263
Garrett Optical: Plössl ... 263
Garrett Optical: SuperWide Angle ... 264
GSO (Guan Sheng Optical): Kellner ... 264
GSO (Guan Sheng Optical): Super Plössl ... 265
GSO (Guan Sheng Optical): Superview ... 266
GTO: Plössl ... 267
GTO: Proxima ... 268
GTO: Wide Field ... 268
I.R. Poyser: Plössl and Adapted Military ... 269
Kasai: Astroplan (AP) ... 270
Kokusai Kohki: Abbe Orthos (*Discontinued*) ... 271
Kokusai Kohki: Erfle (*Discontinued*) ... 273
Kokusai Kohki: Kellner (*Discontinued*) ... 274
Kokusai Kohki: Widescan III (*Discontinued*) ... 275
Kson: Super Abbe Orthoscopic ... 275
Leitz: Ultra Wide (30 mm 88°) ... 276
Long Perng: 68° Wide Angle ... 277
Long Perng: Long Eye Relief ... 278
Long Perng: Plössl ... 279
Masuyama: Masuyama (*Discontinued*) ... 280
Masuyama: Orthoscopic ... 281
Meade: Research Grade Ortho and Wide Field (*Discontinued*) ... 282
Meade: Series II Modified Achromatic MA (*Discontinued*) ... 284
Meade: Series II Orthoscopic (*Discontinued*) ... 285
Meade: Series 3000 Plössl (*Discontinued*) ... 286
Meade: Series 4000 QX Wide Angle (*Discontinued*) ... 287
Meade: Series 4000 Super Plössl (Four-Element and Five-Element) ... 288
Meade: Series 4000 Super Wide Angle (SWA) (*Discontinued*) ... 290
Meade: Series 4000 Ultra Wide Angle (UWA) (*Discontinued*) ... 291
Meade: Series 5000 HD-60 ... 293
Meade: Series 5000 Plössl ... 294
Meade: Series 5000 Super Wide Angle (SWA) ... 295
Meade: Series 5000 Ultra Wide Angle ... 296
Meade: XWA ... 297
Moonfish: Ultrawide ... 297
Nikon: NAV-HW (Hyper Wide) ... 298

Nikon: NAV-SW (Super Wide) 299
Nikon: Ortho (*Discontinued*) 300

9 Olivon to Surplus Shed 301
Olivon: 60° ED Wide Angle 301
Olivon: 70° Wide Angle 302
Olivon: 70° Ultra Wide Angle 302
Olivon: 80° Ultra Wide Angle 303
Olivon: Plössl 303
Olivon: Wide Angled Plössl 304
Opt: Plössl 305
Opt: Super View 305
Orion: Deep View 306
Orion: E-Series 307
Orion: Edge-On Flat-Field 308
Orion: Edge-On Planetary 309
Orion: Epic ED II (*Discontinued*) 310
Orion: Expanse Wide-Field 310
Orion: GiantView 100° UltraWide 311
Orion: HighLight Plössl 312
Orion: Lanthanum Superwide (*Discontinued*) 313
Orion: Long Eye Relief 313
Orion: MegaView Ultra-Wide 314
Orion: Optiluxe (*Discontinued*) 315
Orion: Premium 68° Long Eye Relief 316
Orion: Q70 Super Wide-Field 317
Orion: Sirius Plössl 318
Orion: Stratus Wide-Field 319
Orion: Ultrascopic (*Discontinued*) 320
Owl Astronomy: Advanced Wide Angle 321
Owl Astronomy: Black Knight Super Plössl 321
Owl Astronomy: Enhanced Superwide 322
Owl Astronomy: High Resolution Planetary 324
Owl Astronomy: Knight Owl Ultrawide Angle 324
Parks: Gold Series 325
Parks: Silver Series 326
Pentax: SMC Ortho (*Discontinued*) 327
Pentax: XF 329
Pentax: XL (*Discontinued*) 330
Pentax: XO (*Discontinued*) 331
Pentax: XP (*Discontinued*) 332
Pentax: XW 332
Rini: Various Eyepieces 334
Russell Optics: 1.25 in. Series 335
Russell Optics: 2 in. Series 336
Siebert Optics: MonoCentricID 337

Siebert Optics: Observatory 338
Siebert Optics: Performance Series 340
Siebert Optics: Planisphere 341
Siebert Optics: Star Splitter/Super Star Splitter 342
Siebert Optics: Ultra 343
Sky-Watcher: AERO 344
Sky-Watcher: Extra Flat 345
Sky-Watcher: Kellner 345
Sky-Watcher: LET/Long Eye Relief (LER) 346
Sky-Watcher: Nirvana UWA 346
Sky-Watcher: PanaView 347
Sky-Watcher: Sky Panorama 347
Sky-Watcher: SP-Series Super Plössl 348
Sky-Watcher: Super-MA Series 348
Sky-Watcher: Ultra Wide Angle 349
Smart Astronomy: Extra Flat Field (*Discontinued*) 349
Smart Astronomy: SA's Solar System Long Eye Relief (*Discontinued*) 350
Smart Astronomy: Sterling Plössl 351
Stellarvue: Planetary 352
Surplus Shed: Erfles 353
Surplus Shed: Wollensak 354

10 Takahashi to Zooming Eyepieces 357
Takahashi: LE 357
Takahashi: Ortho (*Discontinued*) 358
Takahashi: UW 359
TAL: Super Wide Angle 360
TAL: Symmetrical (Super Plössl) 361
TAL: Ultra Wide Angle 362
Telescope Service: Edge-On Flat Field 363
Telescope Service: Expanse ED 363
Telescope Service: NED "ED" Flat Field 364
Telescope Service: Ortho 364
Telescope Service: Paragon ED 365
Telescope Service: Planetary HR 365
Telescope Service: RK 366
Telescope Service: Plössl 367
Telescope Service: Super Plössl 368
Telescope Service: SWM Wide Angle Eyepieces 368
Telescope-Service: Wide Angle (WA) 369
Tele Vue: Delos 370
Tele Vue: Ethos 371
Tele Vue: Nagler 372
Tele Vue: Panoptic 374
Tele Vue: Plössl 375

Tele Vue: Radian (*Partially Discontinued*) 377
Tele Vue: Wide Field (*Discontinued*) 378
TMB: 100 379
TMB: Aspheric Ortho (*Discontinued*) 380
TMB: Paragon (*Discontinued*) 381
TMB: Planetary II 382
TMB: Supermonocentric (*Partially Discontinued*) 383
Unitron: Kellner/Ortho/Symmetrical Achromat (*Discontinued*) 384
University Optics: 70° 385
University Optics: 80° 386
University Optics: Abbe HD Orthoscopic (*Discontinued*) 386
University Optics: Abbe Volcano Orthoscopic (*Discontinued*) 387
University Optics: König/König II/MK-70/MK-80 (*Discontinued*) 388
University Optics: O.P.S. Orthoscopic Planetary Series (*Discontinued*) 390
University Optics: Super Abbe Orthoscopic 391
University Optics: Super Erfle (*Discontinued*) 392
University Optics: Widescan II/III (*Discontinued*) 393
VERNONscope: Brandon 394
Vixen: Lanthanum (LV) (*Discontinued*) 396
Vixen: Lanthanum Superwide (LVW) 397
Vixen: NLV 398
Vixen: NPL 399
William Optics: SPL (Super Planetary Long Eye Relief) 400
William Optics: SWAN 400
William Optics: WA 66° 401
Williams Optics: UWAN 402
Zeiss: CZJ Ortho (*Discontinued*) 403
Zeiss: ZAO I/ZAO II (*Discontinued*) 404
ZAO-I 405
ZAO-II 405
Zhumell: Z100 406
Zhumell: Z Series Planetary 406
Zooming Eyepieces 407

Appendix 1 Formulas and Optical Design Data 413

Appendix 2 Eyepiece Performance Classes 419

Appendix 3 Glossary of Terms 423

About the Author 429

Index 431

Part I

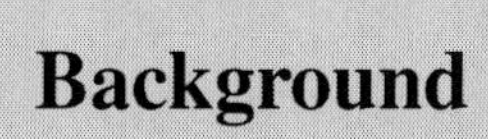

Chapter 1

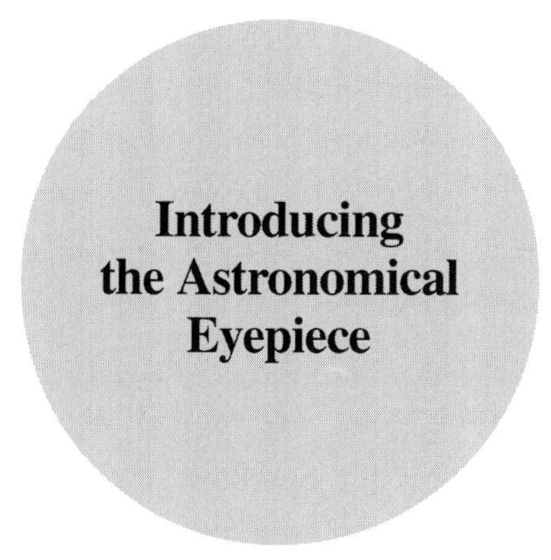

This chapter will introduce you to the astronomical eyepiece—its historical beginnings, basic function, physical construction, optical construction, and optical design characteristics. You will also gain a deeper understanding of some of the more technical aspects of eyepiece performance parameters related to focal length, apparent field of view, eye relief, exit pupil, stray light control, and optical aberrations.

Historical Beginnings

In the early 1980s amateur astronomers were treated to a new kind of eyepiece that would eventually revolutionize the way observers think about both eyepieces and observing. This bold new design was the Tele Vue Optics Nagler. For the first time amateur astronomers had access to an eyepiece where the view was so wide, and so well corrected, that the term "spacewalk" was coined for the views it provided. Since then, the amateur astronomer community's attention on the eyepiece effectively skyrocketed, and the eyepiece remains today one of the most actively discussed topics among observers.

As we examine many of today's truly exciting marvels of eyepiece technology, it is easy to overlook the eyepiece's very humble beginnings. When was the very first eyepiece conceived, and who was the first to use one? How many lenses did the first eyepieces use and how sharp were the views they provided? If we take a journey through time, and examine the archeological discoveries related to optics, we find many lens-like objects that existed as far back as 2500 B.C., more than 4,000 years before the first documented telescope!

You can begin your journey with a visit to the Louvre Museum in Paris or the Egyptian Museum in Cairo and see these ancient optic-like artifacts as polished

W. Paolini, *Choosing and Using Astronomical Eyepieces*, The Patrick Moore Practical Astronomy Series, DOI 10.1007/978-1-4614-7723-5_1,
© Springer Science+Business Media New York 2013

convex crystal lenses used for the eyes in Egyptian statues. Move forward to the period from 500 B.C. to 700 B.C. and you will find what is today called the Nimrud lens. This artifact appears to be an actual plano-convex lens discovered in an area of the world that was ancient Assyria. Then you can find other clues from history that may have, or did have, optical applications:

- A polished magnifying crystal found on Mt. Ida in Crete that can magnify up to 7× clearly and up to 20× with distortions from 500 B.C.
- Egyptian hieroglyphs depicting the use of glass lenses in 500 B.C.
- The Roman philosophers Seneca and Pliny who wrote about magnifiers and "burning glasses" during the first century A.D.
- The Persian scientist Ibn al-Haytham publishing a seven-volume book dedicated to optics called *Kitāb al-Manā ir* (The Book of Optics) between A.D. 1011 and 1021.
- The English friar Roger Bacon writing of the magnification properties of lenses and their possible use as corrective lenses in his *Opus Majus* in 1262.
- Venetian glass makers producing "disks for the eyes" in the 1400s.
- Nicholas of Cusa from Germany using concave lenses to correct near-sightedness in 1451.

As can be seen, the actual history of making and grinding optics, including the convex and concave lenses that are the basis of the telescope, dates back much further than the earliest recorded use of telescopes in the time of Galileo. However, although all these discoveries of optic-like lenses and crystals, burning glasses, and magnifying lenses may lead the imagination to the possibilities that telescope-like devices could have existed long before we believe, none of them actually points firmly to use of a combination of optics to produce an actual refractive telescopic system with both objective and eyepiece.

As intriguing as all these historical clues may be to imagining someone stumbling upon a secret ancient telescope, it was not until the early 1600s that we have the first confirmed use of optics for a telescope, invented by a Dutch-German optician named Hans Lippershey. Once this new invention was revealed, the speed at which copies of this device made it to all parts of the globe, enthralling culture after culture with its almost magical capabilities, is testimony that this was most likely the real genesis of both the eyepiece and the telescope.

This very first telescope used a singlet convex lens with a focal length of approximately 600 mm for an objective, and a singlet concave lens of approximately 200 mm focal length for an eyepiece. Together they produced a telescope with only 3× magnification. Then, in 1608, having perfected his invention, Lippershey applied for a patent calling it the Dutch Perspective Glass.

History does not record if Hans Lippershey ever turned his invention towards the heavens, but we do know that another individual did just that when, in 1609, this exciting new technology fell into the hands of the Italian physicist Galileo Galilei. Galileo enthusiastically embraced the new technology, carefully scrutinizing its operation, and then worked feverishly to improve the design and exploit its power for discovery. His efforts resulted in a telescope with a 37 mm diameter plano-convex objective that had an extended focal length of 980 mm, and the use of a 22 mm diameter plano-concave eyepiece with a shorter 50 mm focal length.

Together these provided a six-fold boost in performance over the original telescope design, increasing its magnification from 3× to almost 20×.

From this point forward, the eyepiece and telescope revealed the universe as never seen before, and over the next four centuries the eyepiece has blossomed from a single lens mounted in a paper or parchment tube to our current modern marvels of technology using highly complex multi-element designs, exotic rare earth glasses, coatings that have their layers deposited atom by atom, and in some cases even the incorporation of image-intensifying electronics. Compared to its humble beginnings, the eyepieces of today truly bear little resemblance to the simple single-element designs that dominated the first half century of its existence.

Moving from Galileo with his first telescopic discoveries using the simple Lippershey/Galilean plano-convex lens eyepiece through the golden era of astronomical discovery (1600s–1800s) we find that the many discoveries of the period were all made with the most simple of eyepieces. The Lippershey/Galilean was of course the very first eyepiece and used extensively by Galileo. Within a few years, the Kepler eyepiece was invented; again only a singlet lens but having a double-convex design.

Then, in 1671, a major advancement was made with the development of the Huygen eyepiece. This design used two optical elements and provided a much improved color-corrected field of view. It was this eyepiece design that remained in use throughout all of the major discoveries made by the great astronomers of the classical period of astronomical discovery. The next advancement in eyepiece design, the two-element Ramsden, was invented near the end of this classical era of visual astronomy and therefore could only have participated in discoveries near the end of the period.

Today, many amateur astronomers sometimes refer to the time-honored designs such as the Huygen or the Ramsden as "junk" eyepieces. This could not be further from the truth, as these eyepiece designs, especially the Huygen, were what was used for a vast majority of the major visual discoveries in astronomy, a distinction none of the modern designs can claim. The following is an accounting of prominent astronomers who made major visual discoveries during the times when the Lippershey/Galilean, Kepler, and Huygen designs were the most technically advanced available:

- Thomas Harriot, an English astronomer and mathematician, using a telescope in 1609 makes the first drawing of the Moon.
- Galileo, starting in 1610, discovers topographical features on the Moon, that Venus appears in phases like the Moon, that the Milky Way is composed of individual stars instead of being a nebula, moons around Jupiter, markings on both Mars and Jupiter, that the Sun rotates, and unusual bodies close to Saturn (e.g., its rings).
- Giovanni Cassini, an Italian astronomer who, in 1665, discovered the oblate aspect of Jupiter and in 1675 discovered the division in Saturn's ring system that now bares his name.
- Christiaan Huygens, a Dutch astronomer who, in 1671, discovered the first of Saturn's moons and also invented the first compound eyepiece, the two-element Huygen eyepiece that greatly improved color correction compared to previous designs.

- Edmund Halley, an English astronomer who, in 1705, accurately predicted the comet of 1682 that now bears his name would return in 76 years.
- Charles Messier, a French astronomer who, in 1774, published a catalog of over 100 deep sky objects.
- William Herschel, an English astronomer who discovered the planet Uranus in 1781, then in 1781–1821 published catalogs of over 800 binary star systems and over 2,400 deep sky objects.
- Johann Gottfried Galle, a German astronomer who, in 1846, made the first visual confirmation of the existence of the planet Neptune (Galileo noted it as a star).
- Asaph Hall, an American astronomer who, in 1877, discovered the two moons of Mars.

As can be seen from the list above, even as late as 1877 when Asaph Hall discovered the moons of Mars using the 26″ Clark refractor currently operating at the U. S. Naval Observatory in Washington, D.C., the Huygen design eyepiece was the standard. So when today's amateur astronomers use their modern ultra-high technology eyepieces to view the many celestial objects from such famous lists as the Messier catalog or the Hershel catalog, they need to realize that all these wondrous celestial objects were discovered and cataloged for future generations using eyepieces no more advanced than that of the humble two-element Huygen.

Fig. 1.1 Huygen eyepiece from the U. S. Naval Observatory's 26 in. Clark refractor (the telescope used by Asaph Hall in 1877 to discover the moons of Mars). Inset of 1.25″ Meade 7 mm RG Ortho is for reference and to scale (Clark eyepiece from the U. S. Naval Observatory collection. Image by the author)

Basic Function

At its heart, just what is an astronomical eyepiece? One can think of an eyepiece as really nothing more than a specialized magnifying glass. The main objective of any telescope, whether it be the primary mirror of a Newtonian or the large glass lens at the front of a refractor, focuses the image the telescope produces at what is called the focal plane of the telescope. Then, when you insert the eyepiece into the focuser of a telescope, the eyepiece's job is to magnify this image produced by the telescope at that focal plane. At its simplest, the eyepiece is merely used as a specialized magnifying glass to observe the image produced by the telescope at the main objective's focal plane. So the eyepiece's function is not to improve the image a telescope can produce in any way, but instead its function is to magnify the telescope's image as best it can with the least amount of aberration and distortion possible.

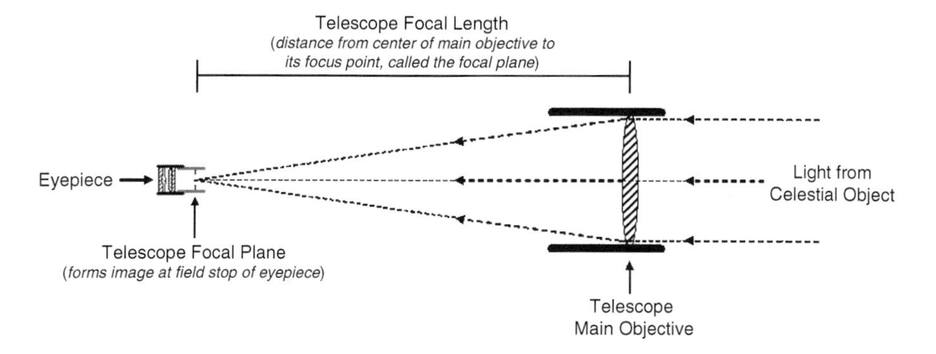

Fig. 1.2 The eyepiece's relation to the telescope's main objective (Illustration by the author)

Beyond the simple, however, the astronomical eyepiece is in reality very much more. For the visual astronomer, the astronomical eyepiece is nothing less significant than the "user interface" of the telescopic system. The eyepiece is therefore the most critical component for the visual observer because it enables the observer to connect to the celestial objects the telescope reveals in a thoroughly personal and engaging way. As such, its importance can't be understated, bringing an array of functions and capabilities to the telescope that no other part of the system can accomplish. As testimony to this, all one has to do is to re-visit the 1980s when that new small company called Tele Vue Optics introduced the first high-quality mass-produced ultra-wide field eyepiece. This one offering transformed the telescope experience from a mostly porthole view of the universe into something much more exciting. As with the Dobsonian revolution that brought large aperture (aperture is the diameter of the main lens or mirror of the telescope) telescopes within reach of the amateur, with the advent of the Nagler technological breakthrough the amateur astronomer community similarly exploded with an enthusiasm that more than a quarter of a century later feverishly continues.

Physical Construction

With our perspective on the history and basic function of the eyepiece, it's time to examine the major components of the eyepiece's construction. Examining any modern eyepiece one can see that it is made of several distinctly different sections. Each of these sections is important to the eyepiece and serves a distinct function for the eyepiece. The major components of any eyepiece are:

- The housing or mount
- The barrel
- The shoulder
- The optics
- The field stop
- The eyeguard (optional)

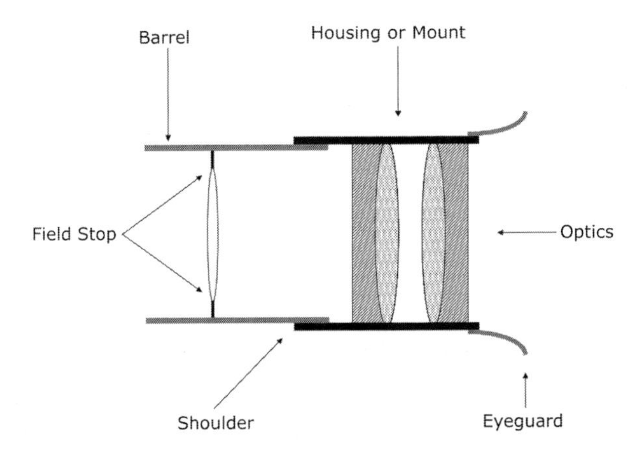

Fig. 1.3 Major components of an eyepiece (Illustration by the author)

Of the major components, the two most obvious parts are the housing (sometimes called the mount) and the barrel (or the top and bottom of the eyepiece). The housing is typically composed of metal, Delrin, or some other convenient hard polymer/plastic material. The brand name and focal length of the eyepiece are typically imprinted or engraved into the housing, and sometimes the coatings used are indicated on the housing as well (e.g., fully coated, multi-coated, or fully multi-coated). On the outside of the housing, some eyepiece manufacturers place rubberized panels or engrave a diamond pattern into the surface to allow secure gripping of the eyepiece. Also attached to the eyepiece housing, or integrated as part of the housing, is some type of eyeguard. The function of the eyeguard is to shield the observer's eye from stray light that may be in the distance around the observer. Eyeguards are typically a rigid or foldable rubber shield, and in some cases they can be a mechanical feature that is raised or lowered as the observer needs.

The second most obvious part of the eyepiece, the barrel, is typically chromed or nickel-plated brass, polished or anodized aluminum, or some other base metal. In some rare instances the barrel is stainless steel or, in the case of vintage eyepieces, an uncoated brass. Older eyepieces typically have a barrel that is entirely smooth, whereas on more modern eyepieces manufacturers alter them with a feature called an undercut or taper. This safety feature is designed so the focuser can "catch" the eyepiece should it not be tightened securely in place or accidentally positioned where it could fall out of the focuser.

The edges of the inset can be sharp (called a full undercut) or have a gentle bevel (called a beveled or tapered undercut). The purpose of the beveled design is to reduce the likelihood of the undercut getting stuck on the focuser's set screw or compression ring when inserting or removing the eyepiece from the telescope. A further improvement to this safety design is the tapered barrel, where there is no machined section but instead the entire barrel gently angles inward so that the part of the barrel closest to the housing is smaller in diameter than the bottom of the barrel. Amateur astronomers generally have mixed feelings about this small feature and can sometimes have very passionate opinions about it. The lines are generally divided between those liking the feature and others strongly disliking it, as it often makes the eyepiece difficult to remove from the focuser.

Fig. 1.4 Barrel features left to right: smooth, undercut, beveled undercut, tapered (Image by the author)

Inside the eyepiece housing (and sometimes in the barrel) are the optics of the eyepiece. These optics, sometimes referred to as the optical assembly, are a combination of lenses of different sizes with surfaces of different curvatures that are grouped and spaced to the specific design of the optician. The individual lens types that any eyepiece uses have a generic name based on the direction of their surface curves and overall shape. The illustration below shows the basic lens types. In the vast majority of all eyepieces, the curves on the lens surfaces are spherical, with their shape being the radius of the circle that their curve defines. Although there is sometimes mention of non-spherical shapes for lens surfaces (called aspherical), this is very rare and usually adds cost to the design due to the complexity in grinding a lens without a uniform shape. (Note that the Leica Vario ASPH Zoom and the TMB Aspheric Ortho are two examples of eyepieces using at least one lens with an aspheric surface.)

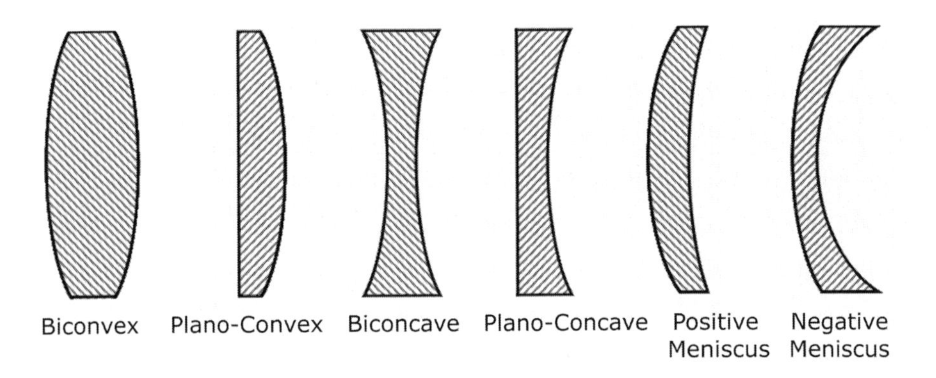

Fig. 1.5 Typical lens types in an eyepiece's optical assembly (Illustration by the author)

Within the eyepiece's optical assembly, the lens closest to the eye is commonly referred to as the "eye lens," and the lens furthest from the eye is referred to as the "field lens" (see illustration below). Also inside the eyepiece, either in the housing or in the barrel, depending on the optical design, is a fixed circular opening or diaphragm that is located at the eyepiece's focal plane. This is called the "field stop" of the eyepiece. Ideally, the field stop is located at or very near the shoulder of the eyepiece. The shoulder is where the barrel meets the housing of the eyepiece. This shoulder is also where the eyepiece comes to rest when it is inserted into the focuser of the telescope.

Fig. 1.6 The apparent field of view (AFOV) of an eyepiece (Illustration by the author. Lunar astrograph courtesy of Mike Hankey, Freeland, MD, USA—www.mikesastrophotos.com)

The purpose of the field stop is to limit light rays that are outside of the design parameters of the eyepiece from entering the field of view. The field stop also provides the distinct circular outline where the image in the field of view ends when observing through the eyepiece (called the apparent field of view, or AFOV).

Limiting the AFOV of the eyepiece with a field stop is a critical part of its design parameters. If the field stop is removed to allow additional light rays and widen the AFOV, then this additional view is likely to be less sharp and show significant levels of aberration in this extended portion of the AFOV. So while some ambitious amateurs may want to widen their eyepiece's field stop to attain a larger AFOV, doing this will most likely result in poor image quality near the edge of the field of view.

Since the field stop is located at the eyepiece's focal plane, when an observer holds his or her eyepiece up to a light and looks through it, he or she will see a bright field of view bordered with a sharply defined edge. This edge of the view appears sharp because the field stop within the eyepiece is at the focus point of the eyepiece. So when one looks through an eyepiece, it is like looking through a magnifying glass that is focused on the area between the circle of the field stop within the eyepiece. If there are any imperfections in the edge of the physical field stop, then the observer will clearly see those imperfections at the edge of the field of view when observing. It is therefore wise to be careful when handling or inspecting the inside of the eyepiece so as to not damage the sharp edge of the field stop.

Since the eyepiece's function is to magnify whatever is located at its field stop, when the eyepiece is inserted into the telescope the function of the focuser becomes to move the eyepiece so the image produced at the focal plane of the telescope is precisely positioned at the field stop of the eyepiece. When this happens, then the eyepiece can clearly magnify the image the telescope is forming at that location. Eyepieces should ideally locate their field stop at their shoulder (e.g., where the barrel meets the housing of the eyepiece, which is also where the eyepiece comes to rest when it is inserted into the focuser of a telescope). This is considered the "ideal" location because the magnification factor of a Barlow lens accessory assumes that the field stop is located at the shoulder of the eyepiece. If it is not, then the magnification factor will change slightly.

When the field stop is physically located at the same position on different eyepieces, then these eyepieces are said to be "parfocal." This means that when one eyepiece is placed in the focuser and the image is brought into focus, then all the other eyepieces will automatically be in focus when they are placed in the focuser as well. Again, at its simplest, this demonstrates how the eyepiece is a specialized magnifying glass that is magnifying the image projected by the telescope when this image is positioned at the field stop of the eyepiece.

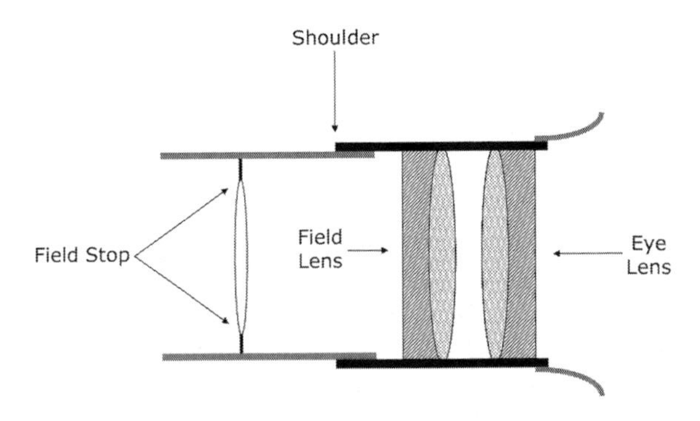

Fig. 1.7 Eye lens, field lens, field stop, and shoulder locations on the eyepiece (Illustration by the author)

Optical Construction

The optical construction, or optical assembly, is the heart of any eyepiece. This assembly typically contains multiple lenses, arranged and grouped into a specific optical prescription, more commonly called the optical "design" of the eyepiece (e.g., Kellner, Ortho, Plössl, König, Erfle, Nagler). This design then directly controls how the eyepiece magnifies the image produced by the telescope and influences such things as how far your eye needs to be from the eyepiece to see the view, how wide or narrow the view appears, how sharply stars appear in different parts of the view, and other characteristics as well.

Given the impact the optical design has upon the perceptions of the observer, there have been hundreds of optical designs developed over time to address many different goals and needs. However, the predominant designs offered commercially to observers are more limited and generally fall into a much smaller number of design types. These optical designs are also sometimes referred to by the name of their inventor, and all of them evolved from the very first astronomical eyepiece in the early 1600s that was the Lippershey/Galilean design (a negative concave singlet lens) used by Hans Lippershey and Galileo.

The table and illustration that follows represents a broad overview of these major optical designs. The "Elements" column of the table indicates the number of individual glass elements within the eyepiece. Since multiple elements can be cemented together to form a single group composed of multiple elements of glass, the "Groups" column indicates how many groupings the individual elements are arranged into. Finally, the "Arrangement" column indicates the layout of the groups. As an example, a 1-3-2 arrangement means that the first group in the eyepiece, or the eye lens, is a single element of glass as the first group (e.g., the first "1" in the 1-3-2 sequence). The second group in the eyepiece (e.g., the middle "3" in the 1-3-2 sequence) is a group composed of three lenses cemented into a triplet group. Finally, the last group, which is the field lens (e.g., the last "2" in the 1-3-2 sequence), is two lenses cemented into a doublet group.

Besides the numbers and groupings of lenses, numerous other factors in an eyepiece's optical design produce its performance characteristics. The number of glass elements, their groupings and arrangements provide the basics to help identify an eyepiece's design. However, many of the designs developed throughout history have similar or the same elements, groups, and arrangements. The other defining characteristics that influence the optical characteristics of an eyepiece are the radius of the curves on each surface of a lens element, the type of glass used for each lens and its index or refraction, and the spacing between the groups. All these factors blend together to make each eyepiece's optical design, and performance characteristics, unique.

Eyepiece design	Invented	Lens elements	Lens groups	Lens arrangement (Eye-to-field lens)
Lippershey/Galilean	1608	1	1	1
Kepler	1611	1	1	1
Huygen	1662	2	2	1-1
Ramsden	1782	2	2	1-1
Kellner	1849	3	2	2-1
Plössl/symmetrical	1860	4	2	2-2
Abbe (Ortho)	1880	4	2	1-3
Monocentric	≈1883	3	1	3
König	1915	3–5	2–3	1-2-1/1-3/1-2
Erfle	1921	5	3	2-1-2
Kaspereit	≈1923	6	3	2-2-2
Brandon	1949	4	2	2-2
RKE	1977	3	2	1-2
Tele Vue Nagler	1979	6–7	4–5	2-2-1-2
LE (Takahashi)	1980s	5	3	2-1-2
Tele Vue Wide Field	1982	6	4	2-1-1-2
Tele Vue Panoptic	1992	6	4	2-1-1-2
Pentax XL/XW	1996/2003	7	5	1-2-1-1-1-1 (for 10 mm)
Meade 4000 SWA/UWA	1997/2003	6–8	4–5	1-2-1-2/2-1-2-1-2
Williams Optics UWAN	2005–2006	7	4	–
Tele Vue Ethos	2007	≈7–9	≈4–5	2-1-2-1-3 (est. for 13 mm)
Nikon NAV-SW/HW	2009–2010	7–12	5–8	–
Takahashi-UW	2011	8–10	5–6	–
Tele Vue Delos	2011	–	–	–
Explore Scientific 120°	2012	12	8	–

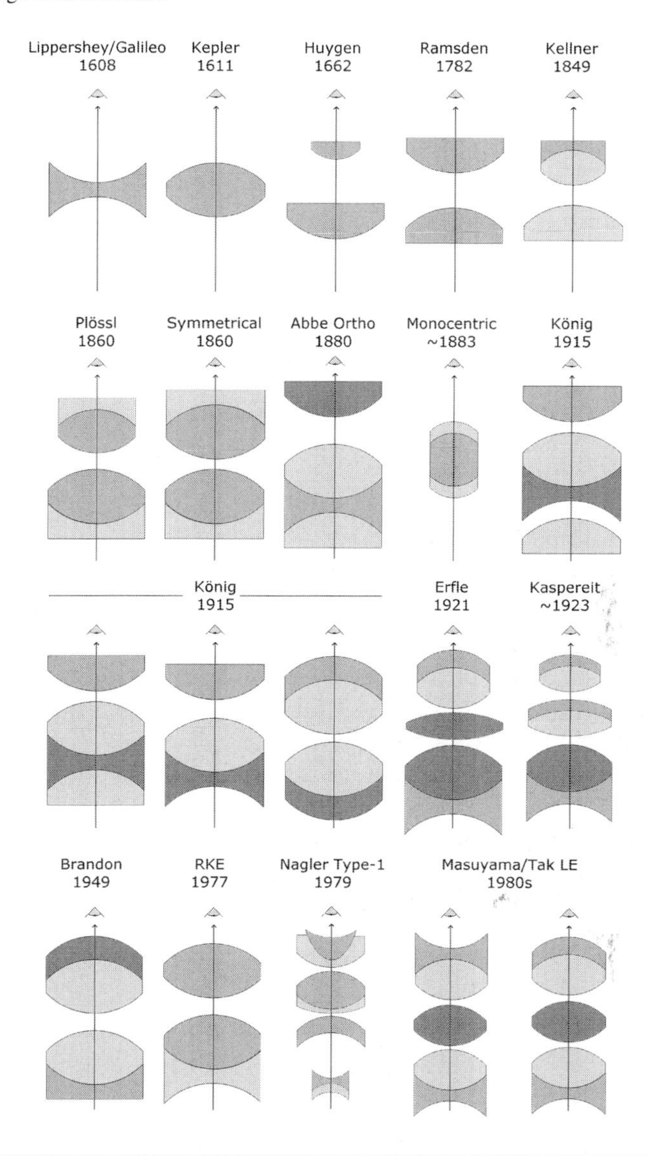

Fig. 1.8 Historical eyepiece designs. For illustrative purposes of lens types used, curves/sizes/spacing not to scale (Illustration by the author)

Optical Design Characteristics

When reviewing the performance of any eyepiece, the lens prescription of the optical design directly controls seven major characteristics important to observers. When opticians create or modify an optical design, they will optimize the subset of characteristics they are most interested in, while balancing the others to provide the best mix of characteristics to meet their design goals for the eyepiece. The primary characteristics that observers should understand about any eyepiece they intend on using are:

- Focal length
- Apparent field of view (AFOV)
- Eye relief
- Exit pupil behavior (i.e., eye position sensitivities)
- Internal reflections and "ghosting" (controlled with internal light baffles and antireflection surfaces)
- Aberrations (includes distortions)

Focal Length

The focal length is probably the first characteristic that one looks at when considering an eyepiece. The reason this is so important is that the focal length of the eyepiece will determine the magnification it will produce when placed in a telescope. The magnification the eyepiece produces is unique to each telescope because it is dependent on the focal length of the telescope. The way it is calculated is to simply divide the focal length of the telescope by the focal length of the eyepiece. As an example, if the telescope has a focal length of 1,000 mm and the eyepiece has a focal length of 20 mm, then the magnification of this eyepiece with this telescope is $1{,}000 \div 20 = 50\times$.

For a line of eyepieces to provide a good range of magnifications in any telescope, they must be available in a range of focal lengths. A full range is usually considered to have the longest focal length be between 32 and 40 mm, and the shortest focal length near 4 mm, with multiple others available between those two end points. For a line of eyepieces with 1.25 in. barrels, it is typical to see the following focal lengths being available: 40 mm, 32 mm, 24 mm, 16 mm, 12 mm, 9 mm, 7 mm, 5 mm, 4 mm. For eyepieces with 2 in. barrels, they typically only focus on the longer focal lengths as the larger barrel can accommodate long focal lengths with wider AFOVs. Therefore, 2 in. eyepieces are usually found in focal lengths as long as 56 mm, with some rarer ones having up to 100 mm focal lengths.

Some eyepiece lines, however, are not made in a full range of focal lengths, and many amateur astronomers often wonder why a manufacturer does not extend the line to a full range. What these observers do not realize is that the available range of focal lengths that an eyepiece line can accommodate is often directly related to the optical design type used by that eyepiece line. As an example, the time-honored Erfle design, which observers have been using for many decades, is rarely ever seen in a focal length shorter than 16 mm. Similarly, the very popular Tele Vue Panoptic line of 68° AFOV eyepieces is only available in focal lengths as short as 15 mm. The reason that shorter focal lengths are not produced for these designs is that at focal lengths shorter than approximately 15 mm, the resulting eye relief of these designs becomes too short to be effectively used by an observer. Every optical design is therefore not able to support all the focal lengths that may be desired by the amateur astronomer.

Because an eyepiece line may not have available eyepieces in a shorter focal length that may be needed by an observer, this does not mean that the observer needs to look for another brand or line of eyepieces. To attain shorter focal lengths using designs such as the Erfle, the Panoptic, and others that are not available in shorter focal lengths, all an observer need do is to simply use a Barlow lens with those eyepieces. When an observer finds they enjoy a particular line of eyepieces not made in shorter focal lengths, with the incorporation of a quality 2× or 3× Barlow lens he or she can easily attain these shorter focal lengths (and without shortening the eye relief of the eyepiece).

To illustrate, the Tele Vue Panoptic has 15 mm as its shortest available focal length; however this eyepiece operates at an effective focal length of 7.5 mm when used with a 2× Barlow, and operates at an effective focal length of 5 mm with a 3× Barlow. Incorporating a Barlow can therefore greatly extend the focal lengths of any eyepiece line. The illustration below shows how a single Barlow of the proper magnification factor can eliminate the need to purchase three additional eyepieces. Depending on the expense of the eyepieces, using a Barlow can save the observer from $100 to more than a $1,000 when expensive ultra-wide field eyepieces are being considered.

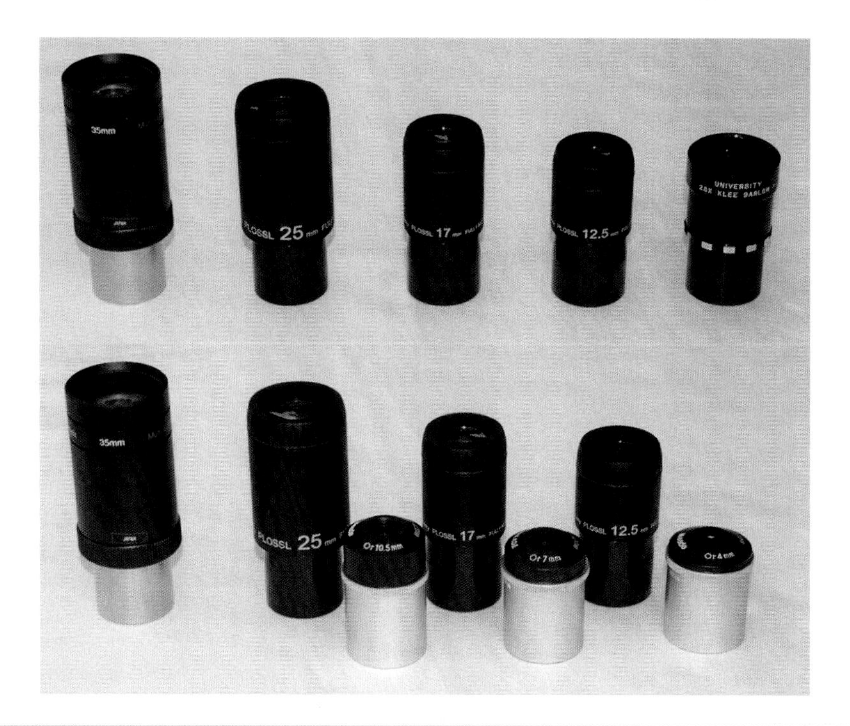

Fig. 1.9 Comparison of two complete eyepiece sets with similar range of focal lengths. Using a 2.8× Barlow (*top right*) allows the four eyepieces shown with it to produce a similar range of focal lengths as shown by seven eyepieces (*bottom*). Using a Barlow can therefore reduce the number of eyepieces needed and save money (Image by the author)

Apparent Field of View (AFOV)

The apparent field of view (AFOV) of an eyepiece is probably the most talked about aspect of the eyepiece. AFOV is how wide the eyepiece's field of view appears as you observe through the eyepiece. In effect, it is how large the "porthole" looks as you view. It is measured as the angle of the view from the furthest left to the furthest right of the field of view. Since the popularization of wide-field eyepieces with the introduction of the Nagler design by Tele Vue Optics in the 1980s, eyepieces with the widest AFOVs possible have become the rage among the vast majority of observers.

The eyepiece's optical design is what controls how large the AFOV of an eyepiece can be while maintaining an acceptably good image from center to edge. The eyepiece then uses the field stop as the physical device to limit or "stop" the AFOV to the specification of the optical design. The sharply defined edge to the field of view you observe through the eyepiece is actually this physical field stop device. Therefore, any damage to the field stop, or a field stop that is not properly formed to a smooth knife edge, will be immediately noticeable as an irregularity in the outer circle of the field of view. Extreme care should therefore always be exercised when inspecting or cleaning the field stop of an eyepiece. The table below lists the AFOV that is common for the major eyepiece designs.

Eyepiece design	AFOV
Lippershey/Galilean	10°
Kepler	10°
Huygens	40°
Ramsden	35°
Kellner	45°
Plössl/symmetrical	50°
Abbe (Ortho)	45°
Monocentric	25°–30°
König	55°–65°
Erfle	60°–70°
Kaspereit	68°
Brandon	50°
RKE	45°
Tele Vue Nagler	82°
LE (Takahashi)	52°
Tele Vue Wide Field	65°
Tele Vue Panoptic	68°
Pentax XL/XW	65°–70°
Meade 4000 SWA/UWA	65°–82°
Williams Optics UWAN	20°

(continued)

(continued)

Eyepiece design	AFOV
Tele Vue Ethos	100°–113°
Nikon NAV-SW/HW	72°/102°
Takahashi-UW	90°
Tele Vue Delos	72°
Explore Scientific 120°	120°

Today, larger AFOVs are generally considered more desirable by most observers, as they provide a more natural view, like that of the unaided eye, which is approximately 140° at its widest. However, some observers still feel quite comfortable with, and even prefer, AFOVs that are more constrained in size, as smaller AFOVs can reduce observing strain. Smaller AFOVs can make it easier to take in the entire view at a glance, reducing the effort needed to "look around" when the AFOV is large. Additionally, smaller AFOV eyepieces are many times less prone to aberrations and distortions off-axis, since their field of view is constrained. The 50° Plössl is an excellent example of such an eyepiece—easy to produce at high quality for a modest price with performance that is very good in all telescope designs and focal ratios.

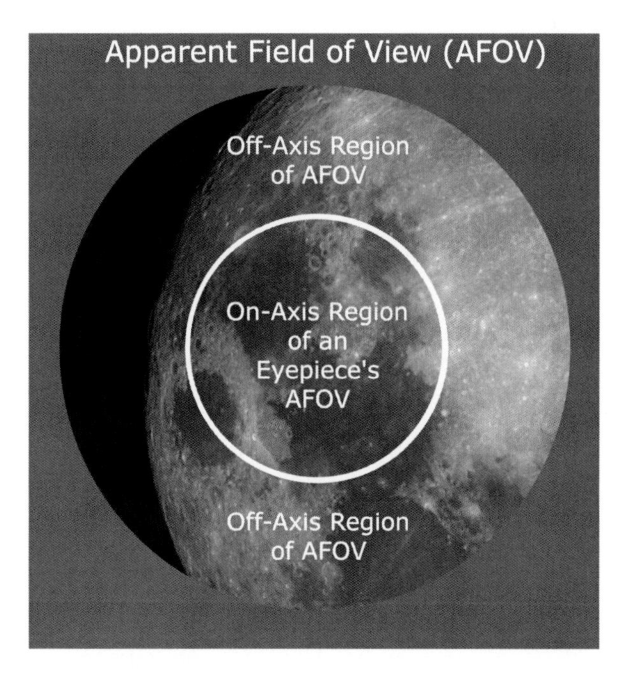

Fig. 1.10 The on-axis and off-axis regions of the field of view (Illustration by the author. Lunar astrograph courtesy of Mike Hankey, Freeland, MD, USA—www.mikesastrophotos.com)

Although the AFOV of an eyepiece is a hot topic among amateur astronomers and often the basis for recommendations, it remains a completely personal preference, unique to the likes and dislikes of each observer. New observers should therefore always experiment for themselves at a local astronomy club or organized evening star party to determine their AFOV size preferences for an eyepiece before they commit to any purchases.

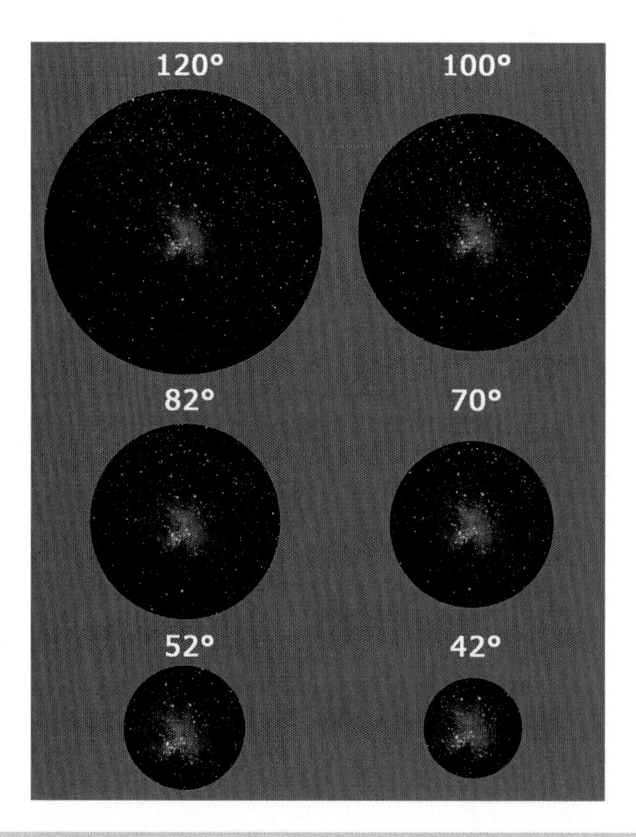

Fig. 1.11 Comparison of common AFOV sizes and how AFOV can affect the view: 120° – 100° – 82° – 70° – 52° – 42° (Eagle Nebula (M16) astrograph courtesy of Mike Hankey, Freeland, MD, USA—www.mikesastrophotos.com. Illustration by the author)

Eye Relief

Eye relief is controlled by the optical design and is the measure of the distance from the center of the top of the eye lens of the eyepiece to the point above the eye lens where the eyepiece magnifies and focuses the image. Eye relief of an eyepiece is a very important factor to consider, as it affects comfort and ease of use. If an eyepiece has an advertised eye relief of 10 mm, then the observer will need to place his or her eye 10 mm above the center of the eyepiece's eye lens surface to see the image magnified by the eyepiece. When the eye relief is short, some observers find prolonged observing becomes uncomfortable. If the eye relief is less than about 15 mm, then for observers who wear eyeglasses while observing may find it impossible to see the entire AFOV of the eyepiece. Eye relief is therefore a critical consideration when both choosing and using an eyepiece.

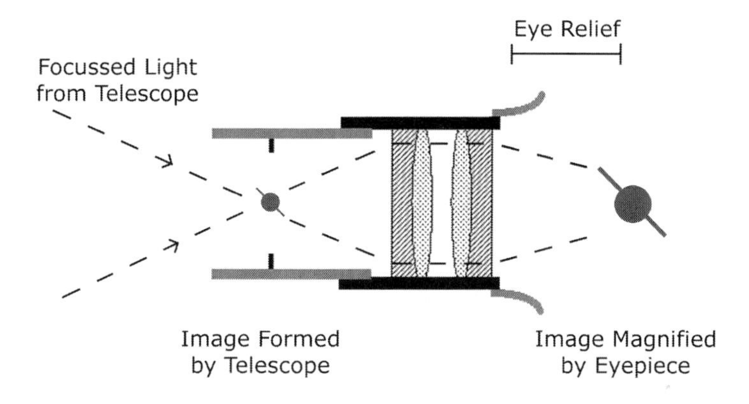

Fig. 1.12 Eyepiece eye relief (Illustration by the author)

For the majority of optical designs, the eye relief of the eyepiece varies as a function of the focal length (FL) of the eyepiece. In a few instances, however, opticians created their designs so the eye relief is fixed and does not vary regardless of the focal length of the eyepiece. The table that follows illustrates both types of approaches to eye relief for many of the available eyepiece designs:

Eyepiece design	Eye relief
Lippershey/Galilean	–
Kepler	≈ (0.90 × FL)
Huygens	≈ (0.10 × FL)
Ramsden	≈ (0.00 × FL)
Kellner	≈ (0.45 × FL)
Plössl/symmetrical	≈ (0.68 × FL)
Abbe (Ortho)	≈ (0.80 × FL)
Monocentric	≈ (0.80 × FL)
König	≈ (0.92 × FL)
Erfle	≈ (0.60 × FL)
Kaspereit	≈ (0.30 × FL)
Brandon	≈ (0.80 × FL)
RKE	≈ (0.90 × FL)

(continued)

(continued)

Eyepiece design	Eye relief
LE (Takahashi)	≈ (0.73 × FL)
Tele Vue Nagler	≈ (0.6–1.4 × FL)
Tele Vue Wide Field	–
Tele Vue Panoptic	≈ (0.68 × FL)
Pentax XL/XW	Fixed at 20 mm
Meade 4000 SWA/UWA	≈ (0.7 × FL)
Williams Optics UWAN	Fixed at 12 mm
Tele Vue Ethos	Fixed at 15 mm
Nikon NAV-SW/HW	Fixed at 16–19 mm
Takahashi-UW	Fixed at 12 mm
Tele Vue Delos	Fixed at 20 mm
Explore Scientific 120°	Fixed at 13 mm

Under the "Eye Relief" column in the chart above note those with formulas, e.g., 0.68 × FL. For these eyepiece designs the eye relief can be calculated. As an example, referencing the table for the Plössl design, it shows the eye relief as 0.68 × FL (the focal length of the eyepiece). This formula can be used to predict the approximate eye relief of all eyepieces using the Plössl design. Therefore, a 20 mm Plössl can be expected to have an optical eye relief of 0.68 × 20 mm = 13.6 mm. Note that the "usable" eye relief of the 20 mm Plössl may be less, as the top surface of the eye lens can be recessed into the housing. The eye relief for any given named design may also vary slightly from these formulas as manufacturers often alter the exact design prescription to accommodate manufacturing efficiencies or to improve performance from the standard design. Regardless of any deviations, these formulas provide an excellent guideline to predict approximate eye relief for an eyepiece if it is not specified by the manufacturer and the approximate optical design is known.

Exit Pupil Behavior

The ease of placing your eye over the eyepiece to view the image is affected not only by "where" the image is formed (eye relief), but it is also affected by "how" that image is formed by the eyepiece. The area above the eye lens where the eyepiece produces the image is called the exit pupil of the eyepiece. The exit pupil naturally occurs at the eye relief point of the eyepiece. Some designs form that image in a flat plane of space above the eye lens, while others form the image on a curved surface of space above the eye lens. Each eyepiece design type (i.e., Abbe Ortho, Plössls, Wide Fields, etc.) forms the image at the exit pupil differently. Depending on how the exit pupil is formed, the image may—or may not—be very easy for your eye to see without being sensitive to small head movements.

For some eyepieces, any small movements of the observer's head can cause the image to momentary black out, to form blank spots in the field of view that are shaped like kidney beans, or even to cause the different spectra of light to separate and show portions of the image with separated colors. The most common phenomenon to occur is for the image to vanish or "black out." When this happens it is typically not because of the exit pupil but simply because the long eye relief of the eyepiece makes it difficult for the observer to hold his or her eye in the proper position. Usually sitting while observing and using an adjustable eyeguard on the eyepiece will greatly reduce this issue.

When only a portion of the field vanishes, and this portion is shaped like an oval or a kidney bean, this is more likely a phenomenon caused by a technical design issue referred to as spherical aberration of the exit pupil. There is no way to solve this eye positioning sensitivity issue except to gain skill in maintaining as steady a position as possible while viewing. Since the exit pupil design is not specified by manufacturers for their various eyepiece lines, the experiences and advice from other observers is invaluable in determining if an eyepiece is more or less likely to have any of these eye position sensitivity issues. And when considering a long eye relief eyepiece with 16 mm or more of eye relief, it is best if the eyepiece has an adjustable eyeguard to alleviate any potential blackout issues.

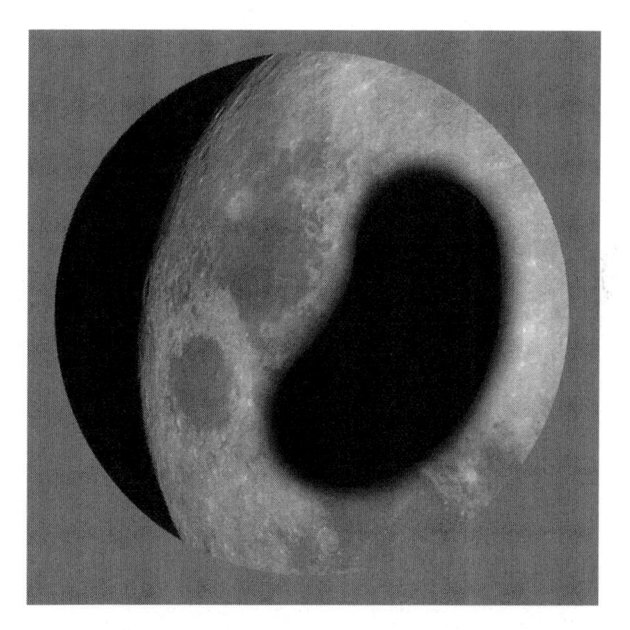

Fig. 1.13 Kidney-beaning in the field of view (Underlying lunar astrograph courtesy of Mike Hankey, Freeland, MD, USA—www.mikesastrophotos.com. Illustration by the author)

Internal Reflections and "Ghosting"

Finally, each eyepiece's optical design has a unique set of lenses of different shapes separated by distances that are unique to the design, and these unique properties can lead to unwanted reflections. As a light ray enters the eyepiece, a very small percentage of the light will be reflected off the surface of each lens surface. Depending on the curve of the lens' surface, the reflected light could then bounce off a reflective part of the internal housing or back into the light path of the observer. When this happens, this stray light ray may appear either as a dim ghost of the object being observed or an unwanted beam of light or bright area in the field of view (flare).

Eyepiece manufacturers use a variety of techniques to suppress these reflections, which include the use of anti-reflection coatings on the lens surfaces to reduce the reflections, internal baffles, applying non-reflective flat black coatings to the interior of the eyepiece, or even adjusting the lens curves and/or spacing as necessary to reduce the possibility of reflections. Generally, an eyepiece design that has the flat surfaces of two lenses closely adjacent to each other will be prone to these unwanted reflections. Similarly, if an observer places two filters onto the bottom of an eyepiece's barrel these two flat surfaces can induce unwanted reflections and ghosting as well. If ghosting, flare, or glare occurs in an eyepiece it is generally out of the control of the average observer to effectively remedy it, as it often requires disassembly of the eyepiece, constructing and installing additional baffles (e.g., small diaphragms inside the eyepiece to block stray light rays) and/or painting internal reflective surfaces.

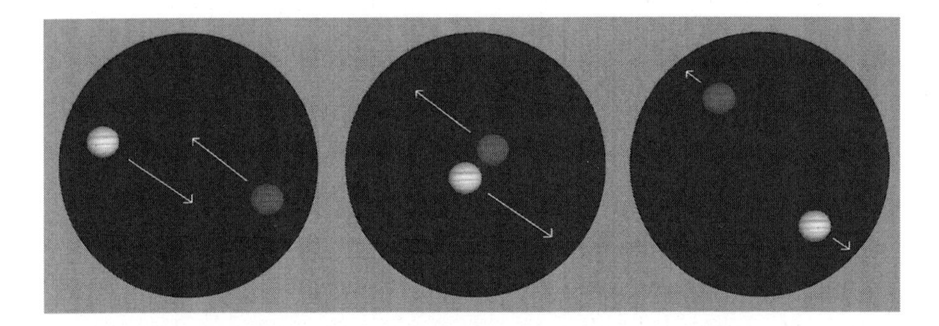

Fig. 1.14 Example of a lens ghost of Jupiter as it drifts through the field of view (Jupiter astrograph courtesy of Mike Hankey, Freeland, MD, USA—www.mikesastrophotos.com. Illustration by the author)

Aberrations (Including Distortions)

Each eyepiece design corrects for aberrations differently than any other design, whether the design is as simple as a common Plössl or as complex as a name brand ultra-wide field. The degree to which these aberrations will be corrected may also vary with the focal ratio of the telescope. Therefore, understanding the strengths and weaknesses of an eyepiece's optical design when used in different telescopes enables the experienced observer to make better decisions on appropriate eyepieces for appropriate telescopes. In the end, there is no good or bad eyepiece, as any eyepiece will work to their design specifications when used in a proper instrument.

The eight optical aberrations that are most commonly encountered and can be easily distinguished in eyepieces are:

- Astigmatism
- Axial chromatic aberration
- Coma
- Field curvature
- Spherical aberration
- Angular magnification distortion
- Radial or rectilinear distortion
- Lateral or transverse chromatic aberration

Of these, axial chromatic aberration and coma are not at all common in modern eyepieces, even inexpensive ones. However, depending on the optical design of the eyepiece and the focal ratio of the telescope, astigmatism and field curvature appear much more readily. Eyepieces with wider AFOVs will always show some level of distortion to the experienced eye; in particular they will always show some degree of rectilinear distortion or angular magnification distortion. This is true because it is impossible to correct the field of view for both rectilinear distortion and angular magnification distortion simultaneously, and when the AFOV is greater than approximately 57.3° (e.g., one radian in angular width), then these distortions can become apparent. So wide field eyepieces will have at least some visible combination of these two distortions, balanced to give the best view possible. For the other common aberrations, particularly astigmatism and field curvature, these can become more severe when the telescope's focal ratio is faster (for astigmatism) or its focal length is short (for field curvature).

Astigmatism: This causes a star point to not come to focus properly. As you adjust focus, the star will generally form a line with a central bulge instead of a perfectly round point. If the in-focus star image is defocused (e.g., moving the focuser inward), then the line will change into an extended oval instead of a circle. When the in-focus star image is defocused in the opposite direction (e.g., moving the focuser outward) the oval pattern will change direction by 90°. This aberration is primarily seen in the off-axis of the field of view of the eyepiece and typically appears when an eyepiece is used in a telescope of a shorter focal ratio than the eyepiece design allows.

Generally, modern eyepieces do not show astigmatism on-axis. Astigmatism is considered a severe aberration, as it alters the image to the point where details are lost and/or become blurry. The illustration below simulates the view of a star point with and without astigmatism, and shows how the defocused star image will shift by 90°, which is the classic telltale sign of this aberration. The top left shows a field of view with no astigmatism, all stars are points. The top right shows a field of view with astigmatism; off-axis stars are not points but are more line-shaped. The bottom left shows a focuser racked "outward" of focus, and field of view shows off-axis stars as large ovals instead of large circles when out of focus. The bottom right shows the focuser racked "inward" of focus, and field of view shows off-axis stars as large ovals that are now rotated 90° from how they appeared when the focuser was racked outward instead of inward.

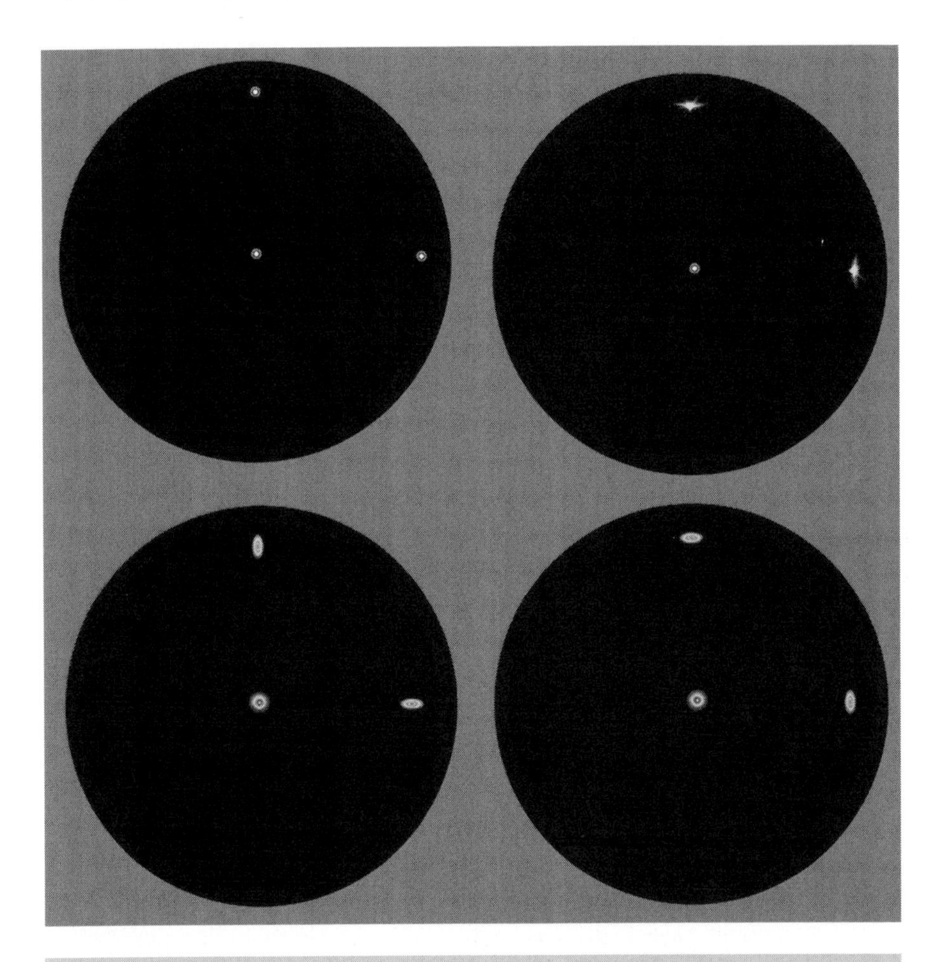

Fig. 1.15 Off-axis astigmatism of a star point. No astigmatism (*top left*), with astigmatism (*top right*), defocused with out-focus (*bottom left*), defocused with in-focus (*bottom right*) (Illustration by the author)

Axial (or Longitudinal) Chromatic Aberration: This causes an in-focus star point to separate into a small spectrum of colors around the fringe of the star point. The color fringing from this aberration, when present, will occur throughout the field of view. If any color fringing is seen only when the bright star, or limb of the Moon or planet is positioned in the off-axis of the field of view, then this is not axial chromatic aberration but is instead a distortion called lateral or transverse chromatic aberration, or simply lateral color (see the following discussion on distortions). Axial chromatic aberration is rarely encountered in modern eyepieces.

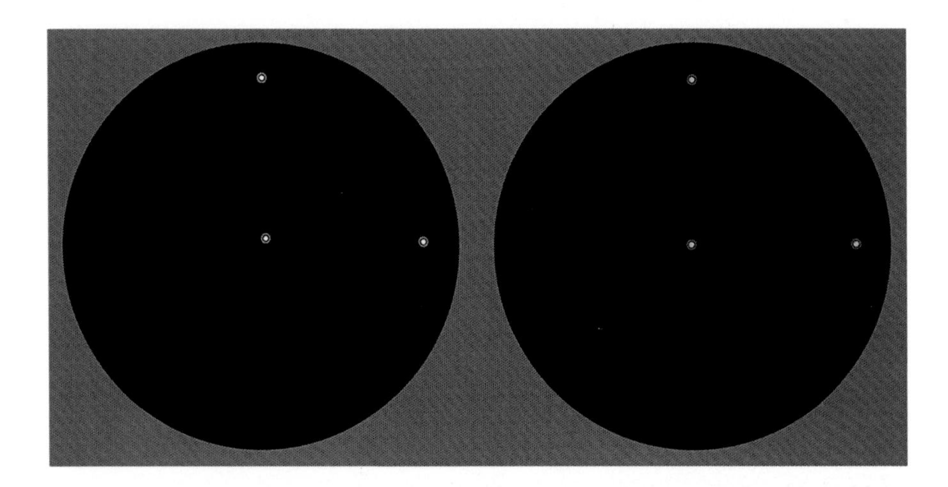

Fig. 1.16 Field of view without axial chromatic aberration (*left*) and with (*right*). Note that regardless of position in the field of view the stars show color separation (Illustration by the author)

Coma: This aberration occurs in the off-axis of the field of view of an eyepiece and makes a star appear as a small comet shape, with the tail always pointing directly away from the center of the field of view. In modern eyepieces this aberration is rarely due to the optics of the eyepiece. When it is seen viewing through a modern eyepiece, it is typically due to the misalignment of the optics of the telescope, or due to the particular design of the telescope. As an example, coma is a property of telescopes using parabolic mirrors, like Newtonian reflectors, where the shorter focal ratio of a Newtonian telescope will produce more visible coma and require the use of a coma corrector attachment to suppress or eliminate this coma.

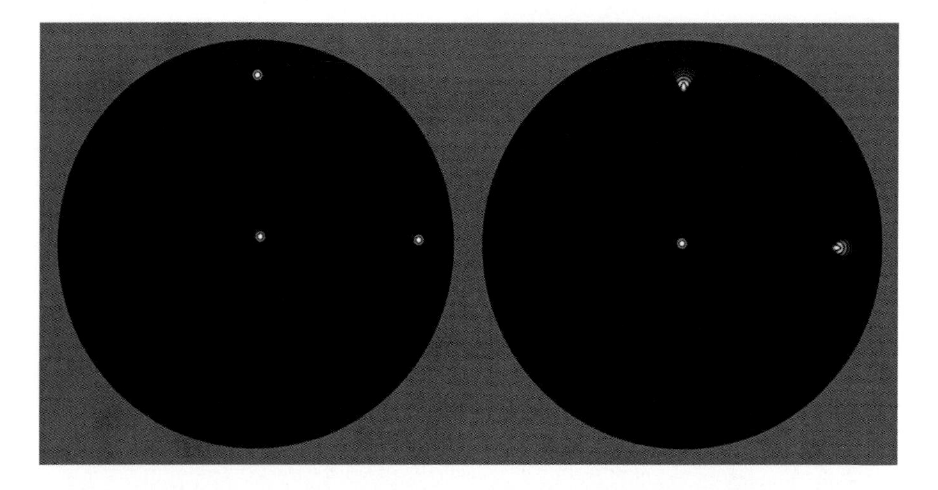

Fig. 1.17 Field of view without coma (*left*) and with coma (*right*). Note that the coma appears in the off-axis as stars in the center remain points (Illustration by the author)

Field Curvature: This aberration is when the image comes to focus on a curved surface instead of a flat plane. It appears in the eyepiece as a focused image in the center of the eyepiece's field of view and an out of focus image in the off-axis of the field of view. A small turn of the telescope's focuser will bring the image in the off-axis into sharp focus, but doing so will cause the on-axis image to go out of focus. When this aberration is seen when observing, it may come from either the eyepiece or from the telescope, as many telescope designs present their image at the telescope's focal plane on a curved surface instead in a flat plane. When it is encountered, it is often impossible to determine if this aberration is coming from the eyepiece or from the telescope. If field curvature is seen, and the field curvature is not too great, an observer can many times choose a focus point that is in the middle of the range of the field curvature. The human eye then has some ability to "accommodate" the difference, and the image will actually be perceived as sharp across the field of view. The ability of the human eye to accommodate different levels of field curvature varies by individual, and this ability is generally diminished with age.

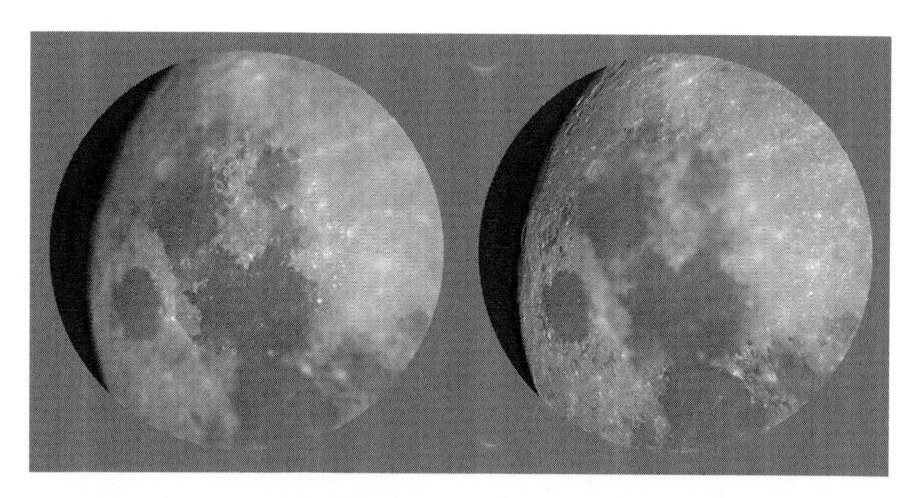

Fig. 1.18 Field curvature of the Moon. Center is focused and off-axis becomes blurry (*left*). Off-axis is focused and center becomes blurry (*right*) (Lunar astrograph courtesy of Mike Hankey, Freeland, Maryland—www.mikesastrophotos.com. Illustration by the author)

SPHERICAL ABERRATION: This aberration occurs when the optics of the eyepiece (or telescope) are not focusing all the light rays into a single point. In general, when this happens the image will appear less distinct or blurry, depending on the severity of the spherical aberration. For star points, they appear as larger, less focused points with more prominent diffraction rings if the eyepiece has any spherical aberration.

Like most other aberrations, spherical aberration can also come from the main objective of the telescope and is not solely a characteristic of the eyepiece. Several eyepiece optical designs have an inherent amount of spherical aberration, and this aberration is reduced when used in telescopes of longer focal ratios. As an example, the Ramsden design has less inherent spherical aberration than the Huygens design; it is therefore a better choice than a Huygens eyepiece, especially for higher magnifications and probably why many of the vintage .965″ barrel eyepieces use the Ramsden design for the shortest focal lengths.

ANGULAR MAGNIFICATION DISTORTION: This distortion occurs when the eyepiece's focal length is not constant across the field of view but varies slightly. As the focal length varies, this then results in the magnification produced by the eyepiece with the telescope to also vary, either increasing slightly or decreasing slightly as the object being observed moves across the field of view. The primary impacts are when scanning star fields, observing double stars, or observing the Moon and planets. Since the magnification varies, when panning the telescope across a star field the stars will appear to trace a path as if they were being viewed through a fish-bowl or through a fish-eye lens. This can sometimes cause a nauseated or queasy feeling for some observers. For double stars, if they are allowed to drift through the field of view of the eyepiece and the magnification is changing, then the double star will either no longer show itself as being split or the distance it is split will vary depending on its location in the field of view. Finally, for the Moon and planets, if this distortion is present then they will change shape from round to either a partially squashed or partially elongated shape, when they approach the field stop in wide-field eyepieces. If the magnification is increasing towards the field stop, then a portion of the Moon or planet will appear to reach out for the field stop; if the magnification is decreasing towards the field stop then a portion of the Moon or planet will appear to shy away from the field stop as it approaches the edge of the field of view.

Fig. 1.19 Moon without angular magnification distortion (*left*) and with this distortion increasing the magnification towards the edge of field (*right*) (Lunar astrograph courtesy of Mike Hankey, Freeland, MD, USA—www.mikesastrophotos.com. Illustration by the author)

Radial or Rectilinear Distortion: This distortion causes straight lines to appear bowed/bent inward or outward. It is typically seen in the off-axis of the field of view of the eyepiece and appears stronger as you move closer to the field stop.

The three sub-types of this rectilinear distortion are (1) pincushion distortion (a positive rectilinear distortion where straight lines are bowed inward toward the center of the field of view), (2) barrel distortion (a negative rectilinear distortion where straight lines bowed outward), and (3) mustache distortion (mixed positive and negative rectilinear distortion where straight lines appear wavy—very rare).

Rectilinear distortion is also tied to angular magnification. As a result, these two cannot be completely eliminated simultaneously. If one is completely corrected, then the other becomes uncorrected, and vice versa. Because of this relationship between rectilinear distortion and angular magnification, optical designers correct for one more than the other depending on the intended use of the optic. Generally, most optical designers choose to better correct for angular magnification rather than rectilinear distortion with astronomical optics, since angular magnification adversely impacts many observing situations in visual astronomy. Conversely, if the intended use is for daytime terrestrial observing, the general preference is to better correct for rectilinear distortion so daytime viewing of familiar straight or linear objects, like telephone poles and fence posts, remain natural looking. Regardless of the preference of which distortion to better correct, both distortions have visible impacts whether one is observing terrestrial objects during the daytime or observing astronomical objects during nighttime.

In astronomical observing, the primary impacts of rectilinear distortion are seen during panning of the telescope and in lunar observing. During panning star paths will be seen taking a curved path through the field of view. In lunar observing straight features on the Moon such as rilles, crater ejecta rays, and the lunar terminator will change shape and bend as they are moved closer to the field stop in the field of view of the eyepiece. Shapes, such as the full Moon and planet disks can also be impacted by rectilinear distortion when they are placed far off axis. Generally the shape will be slightly elongated radially from the center of the field of view. However, this is purely radial elongation is not often noticed as the planet disk subtends too small of a portion of the field of view for the impact to be apparent. Some observers strongly dislike this distortion, so again the best advice for new observers is to borrow eyepieces from their fellow hobbyists or to try eyepieces that other observers may be sharing for viewing at their telescopes at organized star parties to assess for themselves how this distortion common to wide-field eyepieces may or may not affect them.

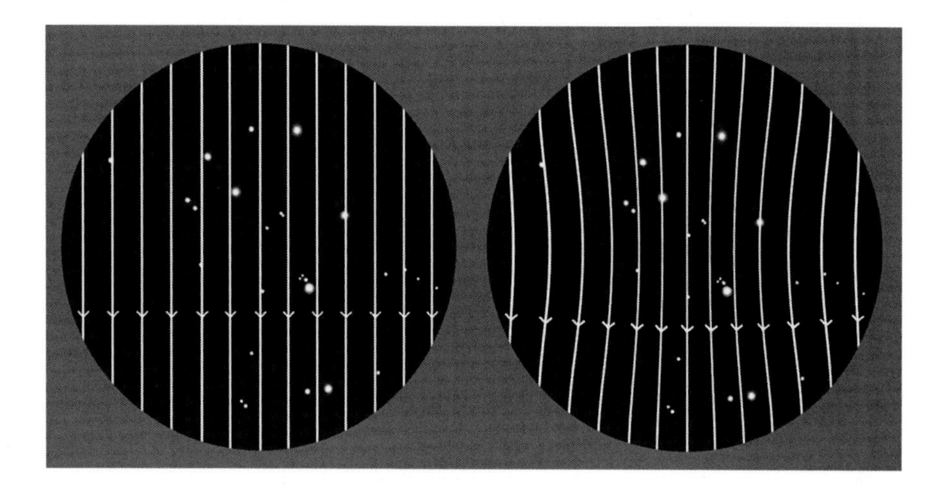

Fig. 1.20 Star paths in an undistorted field (*left*) and in a rectilinear distorted field (*right*). As with angular magnification distortion, stars will also be positioned incorrectly relative to each other with rectilinear distortion (Illustration by the author)

Lateral or Transverse Chromatic Aberration: This is commonly referred to as lateral color. It is primarily seen in the off-axis region of the eyepiece's field of view as a blue, red, green, or yellow fringe on the limb of the Moon, planet, or star point. Observers often strongly disagree on the impact this aberration has when observing. Some feel it is inconsequential when present and others strongly dislike the aberration. Most eyepiece lines will only show minor amounts of this aberration when present.

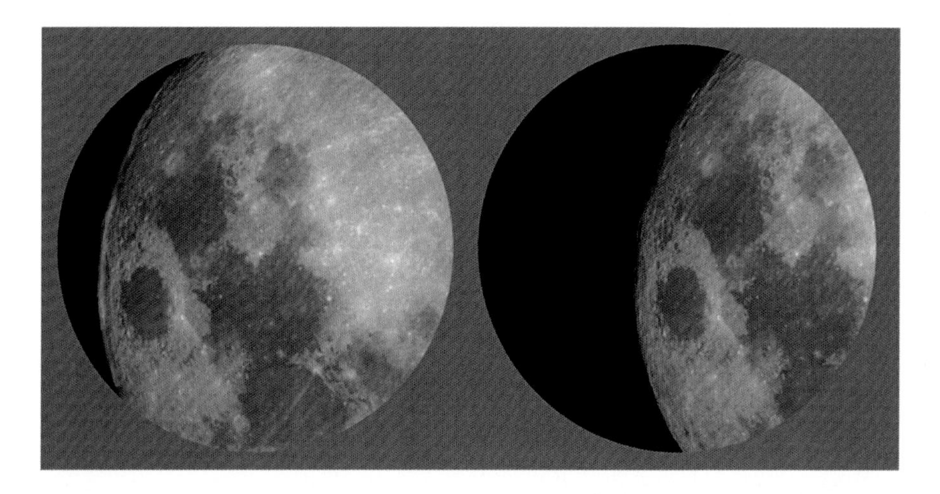

Fig. 1.21 Eyepiece with lateral color shows the aberration on the lunar rim only when off-axis. Note that there is usually a color shift when moved to opposite side of the FOV (Lunar astrograph courtesy of Mike Hankey, Freeland, MD, USA—www.mikesastrophotos.com. Illustration by the author)

Chapter 2

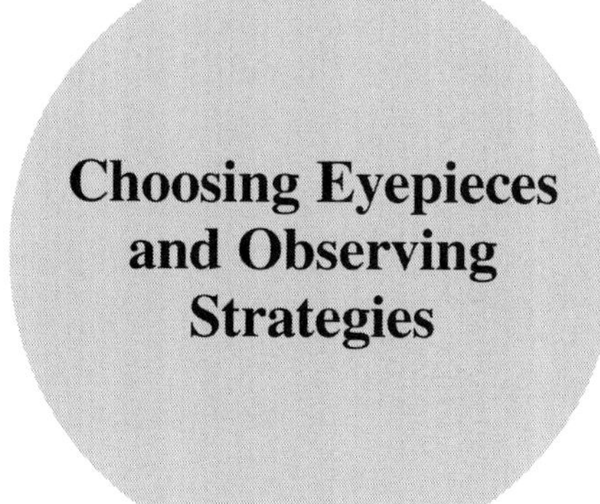

In the marketplace today, there are literally many hundreds of eyepiece lines from which to choose. How can one possibly navigate such a large marketplace and ever hope to make a good decision? The answer is, of course, to arm yourself first with knowledge. When considering the choice and use of an eyepiece, it helps to first consider:

- Viewing comfort and usability considerations
- Visual impact considerations
- Observing strategies

Knowing these will provide the necessary insight to choosing and using eyepieces that will perform their best optically in your equipment for the celestial targets you most enjoy, provide you the ergonomic qualities that make them easiest to use, and give you strategies that will best leverage their full potential.

Viewing Comfort and Usability Considerations

The viewing comfort and usability of an eyepiece are primarily influenced by three characteristics, which when taken together are sometimes vital to the observer's impression of how much they "enjoyed" using the eyepiece:

- Eye relief
- Construction and mechanical features
- Size and weight

W. Paolini, *Choosing and Using Astronomical Eyepieces*, The Patrick Moore Practical Astronomy Series, DOI 10.1007/978-1-4614-7723-5_2,
© Springer Science+Business Media New York 2013

There are many eyepieces in the marketplace that may provide less than the best images yet be highly ergonomic and enjoyable to use, whereas other eyepieces may provide the best optical performance possible and yet be very difficult or awkward to use. Experienced observers often find that the correct balance of both optical and comfort/usability factors are critical for an eyepiece to be a successful choice for long-term use. The balance of factors is also often quite unique to each observer, so while one observer may find an eyepiece highly enjoyable and capable, another observer can find that same eyepiece just the opposite. The reason for this is that an eyepiece is a highly personal part of the telescopic system, and finding the correct eyepiece involves not simply how well its optics will perform but is always dependent on the personal likes, dislikes, needs, and desires of each observer.

Eye Relief

Eye relief is a critical factor influencing comfort when using an eyepiece. It is the distance from the center of the top surface of the eye lens of the eyepiece to where the image is formed by the eyepiece. Eye relief is therefore the distance above the eye lens where the observer needs to place the eye to view the image. Eyepieces with shorter eye relief, requiring the observer to place the eye close to the lens, are generally less comfortable to use than those with longer eye relief. However, when the eye relief becomes too long, the eyepiece may also become difficult because of the problem many observers encounter trying to keep their eye positioned properly so far above the eyepiece. For many observers, an eye relief of about 20 mm is the maximum distance where proper eye position can be maintained without losing the view.

If an observer must wear eyeglasses when observing, then the eye relief of an eyepiece becomes the most critical factor of an eyepiece's performance. Observers who wear eyeglasses typically report that an eye relief not less than 13–15 mm is needed to see the entire field of view of an eyepiece when wearing eyeglasses. Luckily, it is often not necessary to wear eyeglasses when observing, since the eyepiece can simply be refocused to accommodate the needs of an observer's vision. Where an observer will need to wear eyeglasses for observing is if the eyeglass prescription corrects any astigmatism. Any refocusing of the lens will not correct for astigmatism. Observers whose eyes have astigmatism must either use longer eye relief eyepieces with 15–20 mm of eye relief so they can wear their eyeglasses while observing, or find eyepieces that have optional add-on accessories to accommodate correction of astigmatism.

One such accessory is the Tele Vue Dioptrx. The Dioptrx is made to fit on top of compatible Tele Vue eyepieces and is available in various prescriptions to accommodate different amounts of astigmatism. Although the Tele Vue Dioptrx is designed for Tele Vue eyepieces, there are some other eyepiece brands of long eye relief where this accessory can fit (Vixen LVWs, Baader Hyperions, and a few others).

Fig. 2.1 The Tele Vue Dioptrx for observers who have astigmatism (Image © 2013 Tele Vue Optics—www.TeleVue.com)

Construction and Mechanical Features

Although the optical performance of an eyepiece should of course never be overlooked, other considerations can sometimes be just as important, depending on the observer's personal preferences. The first and most obvious aspect of any eyepiece will be its construction and its mechanical features, as these are what will provide one's initial impression. The construction of the housing and the barrel provide the first impression. Housing construction usually varies from metal to various hard polymers such as Delrin. If the housing is metal, then it is typically either anodized some color, or coated with a durable paint. As a result of these differences, how the housing will wear over time will then vary. For housings made of solid materials that are of uniform color throughout, such as Delrin, any wear over time will simply show as a depression. If the housing is made of a coated metal, however, time will wear these surface coatings away, and the bare metal can eventually show. Anodized coatings are typically more resistant than painted housings.

Fig. 2.2 Eyepiece housings: Delrin (*left*), powder coated (*center*), anodized (*right*) (Image by the author)

Next, how the graphics on the housing are applied can cause them to wear away quite quickly when subjected to constant handling. Graphics and lettering are typically applied as surface painting, surface silk screening, or are engraved into the housing then filled with paint. The longest wearing, and most easily refurbished, application is the engraved one that is paint filled. Given that the paint-filled engraving is recessed from the surface, the paint is less prone to touch and wear, making it the most long-wearing. If the paint-filled engraving ever does loosen or wear away from age or use, it is an easy task to fix at home by simply filling with a same color paint. Of the other two common applications, silk screening and surface painting, silk screening generally resists wear slightly better than surface painting.

Fig. 2.3 Eyepiece housing graphics: surface silk screen (*top*) and paint-filled engraved (*bottom*) (Image by the author)

There are also housing features designed to improve secure gripping of the eyepiece. Many feel housings with smooth sides are aesthetically the best. However a diamond etching in the surface or a rubberized band will greatly aid in maintaining a secure grip on the eyepiece, especially when handled in colder weather when cold hands may make one's grip less tight or secure. Some eyepiece brands also place a rubberized material completely around the housing, or construct the housing entirely out of a sure grip material as an approach to maximize the grip-ability of the eyepiece. These types provide the most secure handling characteristics and are especially good in cold weather, as they never feel overly cold to the touch.

Fig. 2.4 Housing grips: smooth (*left*), rubber strip (*center*), completely rubberized (*right*) (Image by the author)

The final common feature of the housing is the eyeguard. Most eyepieces come with some sort of eyeguard to block external light from entering your eye when you use the eyepiece. However, there are still some brands where no eyeguard is provided or integrated into the housing. For some of these it is actually impossible to offer an eyeguard due to the short eye relief of the eyepiece, which would make it impossible to place your eye close enough to view the entire field.

Eyeguard construction comes in a variety of applications, from fixed non-adjustable ones to ones made of soft rubber that can be folded back, offering in effect two positions, to ones with "wings" on one side (very useful when the eyepiece is used in binoviewers) and to adjustable height eyeguards. For foldable eyeguards, the thickness and pliability of the rubber is important, as those made with very thin rubber can wear and crack along the fold points if folded often over time. For adjustable height eyeguards, the mechanisms to accomplish this are typically push-pull types or rotating types to adjust the height. Weak points to look for in these designs are their ability to stay in any adjusted position when pressure is applied on top. Once an eyeguard is adjusted for optimum viewing position, it is not desirable for the position to easily move when contacted during observing, or when removed or inserted into the focuser. Rotating mechanisms are usually less prone to accidental movement when in use. However some push-pull types have locking mechanisms so once the height is chosen if can be firmly locked in position (e.g., the Tele Vue Delos line of eyepieces).

Fig. 2.5 Eyeguards: none (*left*), folding rubber (*center*), adjustable (*right*) (Image by the author)

After the housing, the barrel is the next major construction element of the eyepiece. Most eyepieces have barrels that are chromed/nickel plated brass, polished aluminum or other base metal, or black anodized aluminum or other base metal. The chromed/nickel coated brass types are typical of older classic eyepieces. Today, however, more vendors are moving to the polished or anodized aluminum or base metals. In rare instances, some eyepieces may even have stainless steel used for the barrels (e.g., the discontinued Astro-Physics Super Planetary line). The chromed/nickel coated brass barrels, although having a nice heft and beautiful appearance, will wear over time from constant insertion into the focuser. After many years of heavy use, the wear can show as an area where the gold color of the brass shows through. Fortunately, it generally takes many decades of use for this to happen.

Similarly, barrels that are anodized can show wear of the anodized coating, revealing the color of the underlying metal. Although these types of wear result in cosmetic changes to the barrel, rarely do they affect the function of the barrel in any way. Solid metal barrels that are uncoated and only polished offer the longest wear with fewer cosmetic issues over time.

Another aspect of the barrel that some eyepieces possess is a safety feature to help prevent accidental slipping out of the focuser. For barrels with smooth sides, if the retaining screw of the focuser is not firmly tightened to hold these eyepieces in place, or if it loosens for any reason, then these eyepiece can fall out of the focuser if the telescope is placed into a position where the eyepiece is inverted.

To reduce the risk of this possibility, manufacturers have developed three different styles of a safety feature on the eyepiece barrel: a full undercut barrel, a beveled or tapered undercut barrel, and a tapered barrel. The full undercut types have widely milled bands around the barrel. When the focuser's set-screw or pressure ring is tightened into these milled inset areas of the eyepiece barrel, if it loosens for any reason then as the set-screw or the compression ring contacts the limit of the inset area the eyepiece will be stopped, preventing it from falling out of the focuser. The advantage with this feature is that it reduces risk of the eyepiece slipping out of the focuser, falling, and breaking if the focuser is inverted. The disadvantage of this feature is that many amateurs complain that the eyepiece often gets stuck or hung on this feature when removing it from the focuser. To reduce this tendency of getting stuck, the undercut style of some eyepieces has a bevel in the inset instead of being a sharp corner; these are called beveled undercut. The tapered barrel design, however, is probably the best design for reducing any annoying tendency for the eyepiece to sometimes get stuck when being inserted or removed. As small of a design feature as the undercut is on an eyepiece, it is often a hot topic on online astronomy boards, as many observers tend to have strong feelings one way or the other related to this feature.

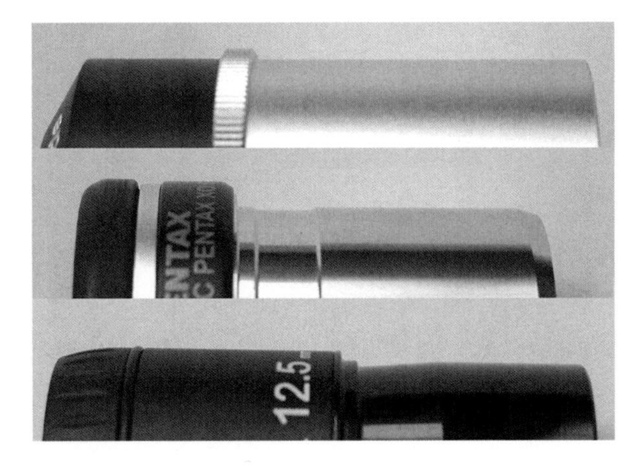

Fig. 2.6 Barrel safety features: none (*top*), undercut barrel (*center*), tapered barrel (*bottom*) (Image by the author)

The final construction element we will examine that is important to the best function of an eyepiece is how the eyepiece is baffled and if reflective surfaces are suppressed. Unfortunately, is not always feasible to inspect these without disassembling the eyepiece, which is not a recommended practice, as it may both void any warranty and cause damage if done improperly. What can be done, however, is to invert the eyepiece and look into its interior. When inspecting the interior of the barrel of an eyepiece, there should not be obvious highly reflective surfaces visible. All portions of the interior should be uniformly blackened with a flat anti-reflective paint or anodized flat black to minimize the possibility of stray light reflections entering the field of view. A well blackened interior is critical when observing brighter targets such as the Moon to eliminate flare within the field of view. Additionally, all visible parts in the interior should typically be made of metal and held in place mechanically with screwing threads rather than glue or pressure, as glued or the pressure-based method may not have the longevity of mechanical-based methods.

Fig. 2.7 Inside of barrels: well blackened (*left*), bare metal parts (*center*), plastic parts (*right*) (Image by the author)

Overall, the construction of an eyepiece should not be overlooked or deemed less important than the eyepiece's optical characteristics. Mechanical features and construction will often determine how well the eyepiece will wear and survive use over time. This is important not only for aesthetic reasons, but it can also be important for functional longevity of the eyepiece, as features such as rubberized grip panels and eyeguards can deteriorate over time if not robustly constructed, and the use of glue or plastics as construction elements can similarly deteriorate or be insufficiently robust.

Size and Weight

The physical size and weight of an eyepiece can also be important considerations, as these two characteristics affect many aspects of usability, including the balance of the telescope. Size, the most obvious feature, can be an issue for some observers when the eyepiece is too large or too small. Where this might also be an issue is related to ergonomics in use e.g., an eyepiece case of six or seven large eyepieces may not only require a large case to hold but may also be considered overly heavy if portability is a concern. Additionally, many eyepiece holders between the tripod legs of equatorial mounts will not accommodate more than one or two of the large format 2 in. eyepieces.

Fig. 2.8 Examples of size and weight variations of eyepieces. 40 mm Pentax XW (*left*) vs. 40 mm Meade 3000 Plössl (*right*) (Image by the author)

A usage area where the size of an eyepiece can be critical is in the width of the housing. This becomes important if the observer plans to choose eyepieces for use in a binoviewer. If the eyepieces are too wide, then an observer may not be able to move the two eyepieces close enough to each other in the binoviewer so that their eyes can be positioned over the center of the eye lenses. Therefore, if the intention is to use the eyepieces for binoviewing, then it is critical to know the distance between the pupils of one's eyes, called the interpupillary distance. Eyepieces for binoviewing must have a maximum housing width that is no greater (or slightly less preferred) than the interpupillary distance of the observer; otherwise the observer might not be able to effectively view through the eyepieces when used in the binoviewer.

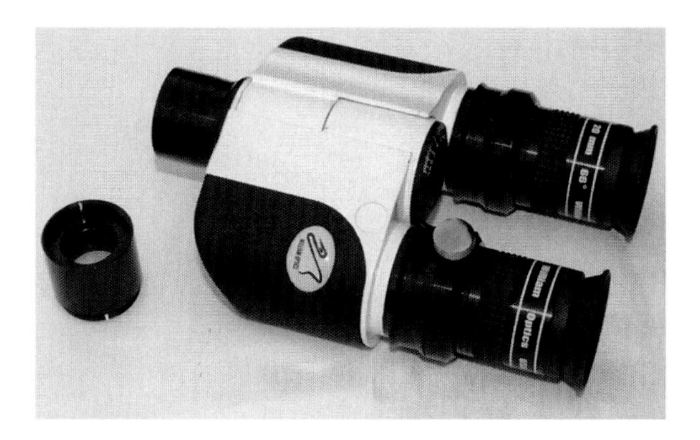

Fig. 2.9 William Optics binoviewer with optical corrector adapter and William Optics eyepieces (Image by the author)

Since many times large eyepieces may be of significant weight, this can also impact the balance of the telescope. As an example, many Dobsonian telescopes do not have a tension adjustment for the altitude movement of the telescope's tube. Instead, these telescopes rely on the tube being balanced as it sits in its mount. If a very heavy eyepiece is placed in the focuser, this may then require the addition of weights near the base of the telescope's tube so the telescope will remain balanced and pointing where it is positioned. Alternatively, for those telescopes that have motorized tracking mounts, a heavy eyepiece may similarly cause an out of balance situation where additional weights must be used or the tube must be repositioned in its holder so the tracking accuracy is maintained.

Although not extremely critical, when choosing an eyepiece the observer should always be cognizant of its listed dimensions and weight to determine if these characteristics will be a problem for the telescope's balance, accessory holders, binoviewers, or portability needs.

Visual Impact Considerations

When observing, there are numerous characteristics of the eyepiece, both exclusive to the eyepiece and in combination with the telescope, that will influence the impact of the view for the observer. The better understanding of how these various characteristics visually impact the view through the eyepiece, the better choices can be made to ensure the eyepiece provides the best visual experience possible.

The primary characteristics that provide a substantial influence over the visual impact an eyepiece will provide include:

- Apparent field of view (AFOV)
- True field of view (TFOV)
- Magnification, brightness, and contrast
- Aberration control

Apparent Field of View (AFOV)

As was introduced earlier, the AFOV of an eyepiece is probably the most talked about aspect of eyepieces today. The AFOV is how wide the eyepiece's field of view appears when observing through the eyepiece (e.g., how large the "porthole" looks as you observe through the eyepiece). AFOV is important when choosing an eyepiece because it conveys the "experience" of viewing differently depending whether the AFOV is smaller or larger. Eyepieces that have narrower AFOVs tend to not engage the observer as prominently as eyepieces with larger AFOVs. Since a single human eye's vision is approximately 140° from left to right (and from 160° to more than 200° degrees when using both eyes together), the larger the eyepiece's AFOV the more natural it appears compared to our normal unaided eye's vision. This makes AFOV an important consideration.

In addition to how engaging the AFOV makes the view through the eyepiece, the AFOV size also directly impacts the true field of view (TFOV) attained by the eyepiece with the telescope (e.g., how large of a patch of sky will be visible). The illustration below demonstrates how the Great Orion Nebula, M42, appears in a 250 mm f/4.7 Dobsonian telescope using a 24 mm eyepiece. The eyepiece used for the field of view shown on the left has an AFOV of 68°, and the one on the right has an AFOV of 44°. The magnifications are exactly the same, but the eyepiece

Fig. 2.10 AFOV-TFOV simulation: 24 mm 68° AFOV (*left*) and 24 mm 44° AFOV (*right*). Note how the larger AFOV eyepiece shows more TFOV around the object being observed (M42 astrograph courtesy of Mike Hankey, Freeland, MD, USA—www.mikesastrophotos.com)

with the larger AFOV allows one to see a more aesthetically pleasing larger patch of sky around M42 than does the eyepiece with the smaller AFOV.

Depending on the telescope used, especially for shorter focal ratio telescopes, the AFOV of the eyepiece is not only an aesthetic consideration, but is also a critical performance consideration. For some shorter focal ratio telescopes, the only way to obtain a large TFOV with the telescope is to use shorter focal length eyepieces that have the maximum AFOV size possible so the exit pupil of the eyepiece and telescope combination does not exceed the limit of the average human eye, which is approximately 7 mm (see the section on Calculating Exit Pupil under Observing Strategies later in this chapter for a more in-depth discussion of this limitation).

To illustrate, if you are using one of the new generation of ultra short focal ratio Dobsonian telescopes, such as the Webster 28 inch f/3.3 telescope, then a 23 mm eyepiece is the longest focal length that can be used without exceeding the 7 mm exit pupil limit of the typical human eye. Given this, it is preferable to use an eyepiece with the largest AFOV possible so the TFOV is maximized. In the case of our 28 inch f/3.3 telescope example, the 21 mm Tele Vue Ethos, with its 100° AFOV, would give us twice the TFOV compared to using a typical 20 mm Plössl. For short focal ratio telescopes a wider AFOV eyepiece is therefore sometimes necessary for TFOV capability. As a general rule, if the telescope being used has a focal ratio of f/5 or shorter, then at least one wide-field eyepiece with a 2 in. barrel will be required if one wants to maximize the TFOV capability of the telescope and keep the exit pupil smaller than the typical 7 mm maximum that the human eye can handle.

Although the AFOV of an eyepiece can be important for aesthetic reasons, or critical for performance reasons, this does not mean that eyepieces with larger AFOVs are necessarily better, or even preferable, to those with smaller AFOVs. For many observers, larger AFOV sizes can be more stressful or difficult to use, so they develop a preference for the smaller AFOV eyepieces. In some specialized observing tasks, such as double star or planetary observing, some observers prefer smaller AFOVs because a larger TFOV around these objects is generally of no advantage to the observation, either technically or aesthetically. Additionally, these smaller AFOV eyepieces typically need fewer glass elements for a well corrected view, can be made in smaller and easier to handle form factors, and are usually less expensive than the more complex wide-field optical designs. So although AFOV is an important consideration when choosing an eyepiece, individual observer preferences vary greatly, making larger AFOVs not necessarily better. The best advice is always to either borrow eyepieces of different AFOV sizes so observers can determine for themselves what works, or to participate in local astronomy club observing evenings to try different eyepieces that club members may be using with their telescopes.

True Field of View (TFOV)

The true field of view (TFOV), sometimes referred to as the actual field of view, is an angular measure, expressed in degrees, of the maximum amount of sky visible in the eyepiece from the furthest left to the furthest right of the field of view. It is useful because it gives the observer the ability to predict how well the selected eyepiece with telescope will frame the intended target. TFOV is not an attribute of

Fig. 2.11 The Moon as it would appear in a 24 mm Panoptic eyepiece in a 203 mm (8 in.) f/10 SCT (*left*) and with the same eyepiece in a 100 mm f/7.5 refractor (*right*) (Moon astrograph courtesy of Mike Hankey, Freeland, MD, USA—www.mikesastrophotos.com)

the eyepiece alone but varies depending on the focal length of the telescope, the focal length of the eyepiece, and the AFOV of the eyepiece used.

When considering the best eyepiece to use for a planned observing session for a particular target, it is often important to consider using an eyepiece that will provide a TFOV larger than the target itself to capture the surrounding context of the star field. For some targets, such as the Great Orion Nebula, Messier 42, the surrounding context of stars provides a richly rewarding view. The Great Orion Nebula M42 has an angular size of 0.6°, so an eyepiece that would show 0.6° of TFOV in the telescope would only "just" fit Messier 42 in the field of view. However, compare how differently the Great Orion Nebula appears in a TFOV that is over twice as large, at 1.3°. With M42's surrounding context of the entire Sword of Orion fully framed

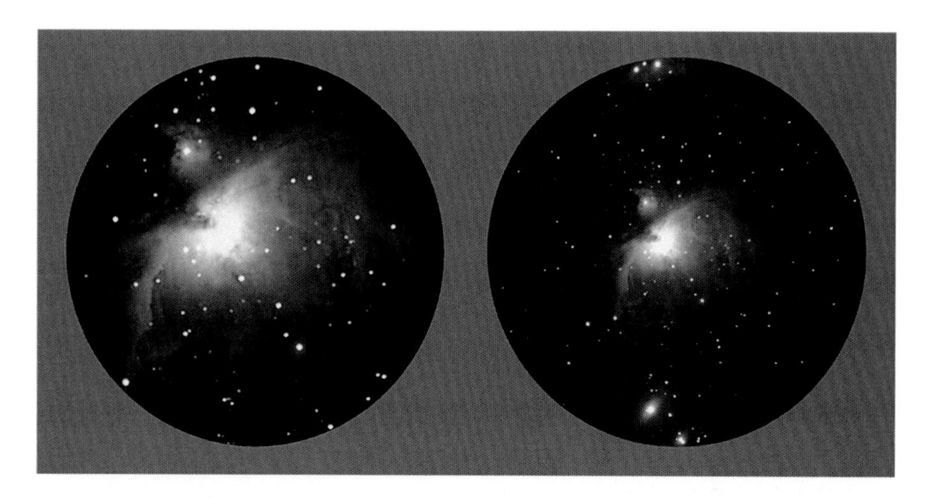

Fig. 2.12 M42 in a higher magnification 0.6° TFOV (*left*) and in a lower magnification 1.3° TFOV (*right*) (M42 astrograph courtesy of Mike Hankey, Freeland, MD, USA—www.mikesastrophotos.com)

by the eyepiece, the observing experience of this target is completely different, and for many observers, preferred.

To calculate the TFOV of your eyepiece and telescope combination, you can use any of the several formulas that follow. Some calculation methods are more precise than others; however, even the least accurate methods are typically only 5–10 % less accurate. In the list of formulas below, the first set is an approximation-based method that will generally yield an answer several percent less accurate than the most accurate methods.

For calculating TFOV (in degrees) based on manufacturer provided data:

1.
$$\text{TFOV} \approx \frac{\text{AFOV}}{\text{Magnification}}$$

Notes: AFOV is the apparent field of view of the eyepiece in degrees. Magnification is the magnification produced by the eyepiece in the telescope.

2.
$$\text{TFOV} \approx \frac{\text{AFOV}}{(\text{FL [telescope]} \div \text{FL [eyepiece]})}$$

Notes: AFOV means the apparent field of view of the eyepiece in degrees. FL[telescope] means the focal length of the telescope in millimeters. FL[eyepiece] means the focal length of the eyepiece in millimeters. This method is only approximate, as any distortions in the eyepiece's field of view will make the results inaccurate by as much as 10 %.

For calculating TFOV (in degrees) based on drift time observations of a star:

3. $$\text{TFOV} = \frac{\text{DT[sec]}}{239}$$

Notes: DT[sec] means drift time in seconds; this method is only accurate when the star is near the celestial equator.

4. $$\text{TFOV} = ABS(\text{DT[sec]} \times .0041781 \times COS(\text{DEC[star]} \div 57.3))$$

Notes: ABS means absolute value. DT[sec] means drift time across the entire field of view in seconds. COS means cosine. DEC[star] means the declination of the star in degrees. This method is accurate for any star chosen by drift time since the declination of the star from the celestial equator is taken into account.

For calculating TFOV (in degrees) based on field measures:

5. $$\text{TFOV} = \frac{\text{Eyepiece Field Stop Diameter}}{\text{FL[telescope]}} \times 57.3$$

Notes: Eyepiece field stop diameter is in millimeters. FL[telescope] means the focal length of the telescope in millimeters. COS means cosine; and DEC[star] means the declination of the star in degrees.

6. $$\text{TFOV} = \frac{(\text{Tape Measure[observed]} \times 57.3)}{\text{Distance[telescope-to-tape measure]}}$$

Notes: Tape Measure[observed] means the number of inches (or millimeters) of the tape measure that are observed through the eyepiece in the telescope. Distance[telescope-to-tape measure] means the distance in inches (or millimeters) from surface of the objective of the telescope to the wall where the tape measure is mounted.

Formulas 1 and 2 are the easiest to use since all the needed information is provided by the manufacturers of your eyepiece and telescope. However, they only provide results close to the actual TFOV that will be produced. You can expect calculations using these two formulas to be accurate to within about 5 % or so of the actual TFOV. The reason these may be not as precise is due to distortions that may exist in the far off-axis of the AFOV of the eyepiece, particularly in wider AFOV eyepieces that typically have rectilinear and/or angular magnification distortions of several percent. If the eyepiece has several percent of these distortions in the off-axis affecting the AFOV, then the results of this TFOV calculation method will also be off by several percent. However, even with this imprecision, these are still fairly accurate and usable formulas, providing a quick and easy method.

Formulas 3 and 4 provide a method of calculating the TFOV without knowing any information at all about the eyepiece or the telescope. These are called drift-time methods and involve pointing the eyepiece and telescope at a star, placing the star just outside the field stop, then timing it with a stopwatch as it comes into view until it goes out of view at the other end of the field of view. If the star chosen is on or close to the celestial equator, then these formulae get simpler, less confusing, and produce more accurate results. (*Note:* The celestial equator is visible from almost everywhere on Earth.) The difficultly with this method is ensuring that the star crosses the exact center of the field of view during its transit. Since it is often difficult to ensure this, observers generally take the timings for several drifts and then average them for the most accuracy.

The first drift formula (3) provided is less accurate than the second (4) since it does not account for the declination of the star being observed. If the star is offset far from the celestial equator, then the constant in the denominator of the equation will be different. So this quick and easy drift formula is only very accurate when you choose a star very close to the celestial equator. However, an easy solution to this issue is to choose a star that you can identify, look up its declination, then use this in formula 4 for maximum accuracy. Since the declination of a star can be a positive or negative number, this formula uses the absolute value of the result, since it is only the declination's offset from the celestial equator that is important and not its direction of offset.

In formula 5, the TFOV is measured using the field stop diameter of the eyepiece. Although this is an excellent method, many current day wide field designs have the field stop within the main housing of the eyepiece set between some of the glass elements. Eyepieces of this design are generically called positive–negative eyepieces because the elements in the barrel are a negative group, like a Barlow. With these designs the physical size of the field stop is of no use because of the elements in the barrel of the eyepiece that are in part acting like a Barlow. For positive–negative design eyepieces, however, some manufacturers provide the "equivalent" or "effective" field stop measurements for their positive/negative designs. If these are supplied, then they can be used with formula 5 for accurate results. Otherwise, the other TFOV formulas will work just fine if the manufacturer has not provided what is often called the effective or equivalent field stop size for these positive–negative type eyepieces.

If the eyepiece is a classic design, where the field stop is visible in the barrel of the eyepiece below the field lens, a direct measure of the field stop size to use formula 5 will result in very effective results. Since field stops are usually small, a caliper or similar precision instrument is needed as a measuring device to accurately measure to a thousandth of an inch or a fraction of a millimeter.

If taking a direct measure of an eyepiece's field stop, be careful not to put too much pressure on the edges of the field stop because any dent or inadvertent etch in the knife-edge of the field stop will be seen as a notch at the edge of the field of view of the eyepiece when observing. The shorter the focal length of the eyepiece the easier it will be to see any inadvertent damage to the field stop's edge because of the increased magnification of short focal length eyepieces, so extreme care is always advised.

Once the field stop is measured, then simply plug this value into formula 5 to determine the TFOV that will be observed using that eyepiece with each telescope. To illustrate, assume the telescope is a 100 mm f/8 (800 mm focal length) refractor and the eyepiece is a 20 mm Plössl eyepiece that has a 50° AFOV. When measuring the field stop of this eyepiece, the caliper reads its diameter as 17.5 mm. For accuracy, repeat the measurement at least two more times. The next two measurements are: 17.1 and 16.7 mm. Taking the average of the three measurements the result is an average of 17.1 mm. Since the three measurements varied by 0.8 mm (e.g., maximum minus the minimum or $17.5-16.7=0.8$) the accuracy of the measurements is half of the variance or +/−0.4 mm (or expressed as a percent it is variance/average $=0.4\div17.1$ +/−2.3 %). Taking the average measurement for the field stop, use formula 5 to calculate the TFOV this eyepiece for the example's 100 mm f/8 telescope with its 800 mm focal length. To illustrate this example:

$$\begin{aligned}\text{TFOV} &= (\text{Eyepiece Field Stop Diameter} \times 57.3) \div \text{Telescope Focal Length}\\ &= (17.1\ \text{mm} \times 57.3) \div 800\ \text{mm} = 1.22^{\circ}\end{aligned}$$

Although performing TFOV calculations for all eyepiece-telescope combinations can be tedious, the one place where performing TFOV calculations is the most important is to ensure that an eyepiece is available to adequately frame the largest celestial object you intend to observe. To illustrate, presume the intent for the evening's observation is to drive to a dark site with a 100 mm f/8 telescope, and that the largest object on the observing plan is Messier 45, the Pleiades Cluster. Assume that for this observing trip a 2 in. format eyepiece will not be taken to reduce the bulk of all eyepieces being taken for the observation. Will a 1.25 in. eyepiece that produces the most TFOV in a 1.25 in. barrel, such as a 32 mm 50° eyepiece, be sufficient to view Messier 45? Using the TFOV calculation methods just reviewed it is now possible to easily answer this question. Since Messier 45 is approximately 1.7° in size, to frame it better and get a little more context around the target the plan is to have an eyepiece that will produce about 1.5× more TFOV than the size of Messier 45, or $1.5\times1.7°=2.6°$. Will the planned 1.25 in. 32 mm 50° eyepiece provide the needed 2.6 in. of TFOV in the 100 mm f/8 telescope? Doing the calculations shows that this eyepiece only produces a TFOV of approximately 2.0°, much less than the 2.6° desired. To illustrate this example:

$$\begin{aligned}\text{TFOV} &\approx \text{AFOV} \div (\text{Telescope Focal Length} \div \text{Eyepiece Focal Length})\\ &\approx 50^{\circ} \div (800\ \text{mm} \div 32\ \text{mm}) \approx 2^{\circ}\end{aligned}$$

Although in this scenario the plan was to not bring 2 in. format eyepieces, taking a little time to verify TFOV calculations revealed that our selected 1.25 in. eyepiece could frame Messier 45 tightly, but that it falls short of the 2.6° desired for extra context around the target. Knowing how to perform the TFOV calculations for an observation planning, particularly one that will be at a remote location where the observer might not have access to all his or her equipment, can be highly beneficial. In our scenario, bringing a 2 in. wide-field that will produce more TFOV than our 32 mm Plössl would be best to gain the extra context that a larger 2.6° provides for Messier 45. Repeating the above calculations for a 2 in. Pentax 30 mm XW with a 70° AFOV shows it will produce the more desired 2.6° TFOV when used with the 100 mm f/8 telescope.

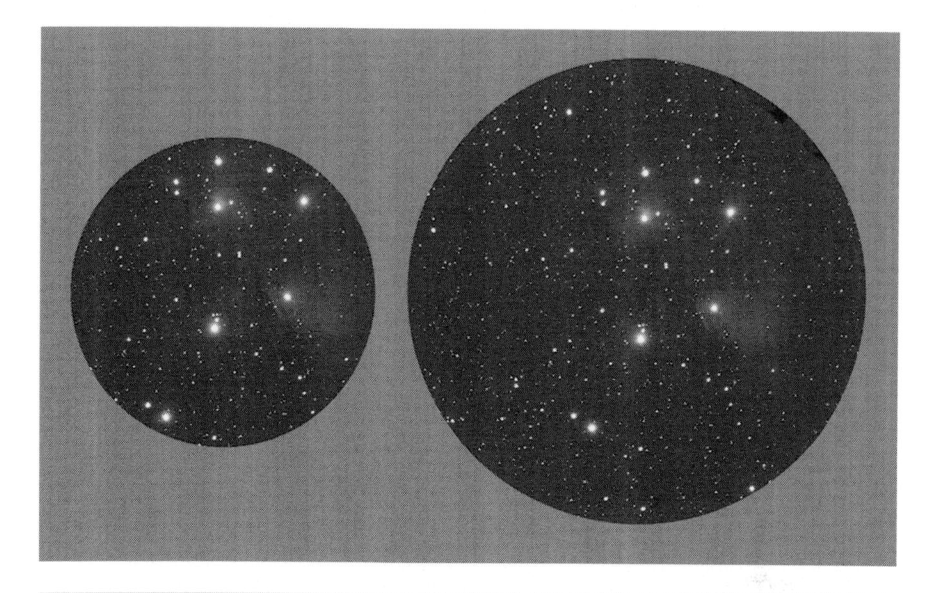

Fig. 2.13 M45, the Pleiades Cluster, in a 30 mm Plössl (*left*) and 30 mm Pentax XW (*right*) (M45 astrograph courtesy of Mike Hankey, Freeland, MD, USA—www.mikesastrophotos.com)

The final formula, 6, uses a method where the eyepiece and telescope are set up during the day, and direct measures of the size of the TFOV they produce are calculated based on measuring a target placed down field. This can be done outdoors or indoors, as long as your eyepiece and telescope can come to focus on the target. This method is highly accurate, as it is independent of any unmeasured or manufacturer provided data, such as the focal length of the eyepiece or telescope or the magnification being produced, etc. With this method, two direct measures are taken; then a simple trigonometric formula is used to calculate the TFOV.

To set up for this method, first mount a tape measure which is several feet long that has closely spaced measurement intervals (e.g., 1/16 in. tic marks) on a wall in the distance. Next place the eyepiece and telescope at a workable distance away so they can come to focus on the target, then measure the distance from the surface of the telescope's objective lens to the tape measure on the distant wall. Next, observe the tape measure in the field of view of the eyepiece and count the number of inches and fractions of an inch (or centimeters and millimeters) that is observed through the eyepiece from one end of the field of view to the other. Note that it is vital that the tape measure exactly bisects the field of view of the eyepiece so the measurement is taken from the widest part of the view from left to right. The unit of measure you choose is not relevant to the formula, as long as both measures are in the same unit of measure. Therefore, if you are counting the number of millimeters you can view on the tape measure across the field of view, then the distance from the telescope's main objective to the tape measure must also be in millimeters.

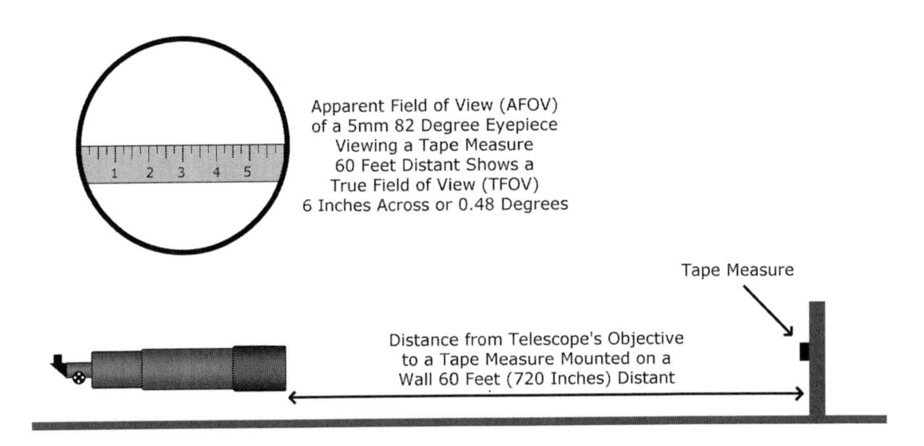

Fig. 2.14 Direct field measurements to determine the TFOV of an eyepiece and telescope (Illustration by the author)

To illustrate this method, assume the equipment used is a 100 f/8 telescope together with a 5 mm 82° AFOV eyepiece. The tape measure is placed on a wall that is measured as 60 ft (720 in.) from the front surface of the telescope's main objective. The tape measure is mounted on the wall from left to right so it is horizontal to the ground. Next place the 5 mm 82° AFOV eyepiece in the telescope and aim the telescope at the tape measure so the top of the tape measure is in the exact center (from top to bottom) of the eyepiece's field of view. Focus the telescope on the tape measure and assume for our example that a count of exactly 6 in. is visible from the furthest left of the field of view to the furthest right of the field of view.

Using the TFOV formula 6, TFOV = Observed Tape Measure Size × 57.3 ÷ Distance from Telescope Objective to Tape Measure, insert the measured and observed values into the formula. To illustrate this example:

$$\text{TFOV} = (6\ \text{inches} \times 57.3) \div 720\ \text{inches} = 0.48^\circ\ \text{of TFOV}$$

The accuracy of any of these formulas can only be as accurate as the measure taken by the observer. It is therefore recommended that several measurements be taken and the average of the multiple measurements is used. When this is done the resulting answers can be provided with a level of accuracy based on the multiple measurements taken. The formulas to calculate this simple accuracy method is:

$$\text{Results} = \frac{(Value1 + Value2 + ValueN)}{\text{N}}$$

Notes: When taking field measurements, the measurement should be repeated several times, then the average taken using the formula for averages above. Once the average measure is calculated, then the formula for accuracy below can be used to express the accuracy of the average.

$$\text{Accuracy} = \pm\frac{(MaximumValue - MinimumValue)}{2}$$

To illustrate the above formulas, if the above TFOV measurement was conducted three times, where each of the measures was repeated and the three results were 0.43°, 0.48°, and 0.53°, then the result's average that would be reported is = (0.43° + 0.48° + 0.53°) ÷ 3 = 0.48°. The accuracy that would then be reported would be: (0.53°–0.43°) ÷ 2 = ± 0.05°.

Magnification, Brightness, and Contrast

Most people both inside and outside of the astronomy hobby understand the relationship between magnification and brightness—which is the greater the magnification the larger the image appears and the dimmer it appears. Some even understand the more exact relationship of brightness to magnification, which is that if you double your magnification then the image becomes only one quarter as bright. A more accurate representation of what is being observed is called the inverse-square law. Light follows what is called the inverse-square law, which for our use in visual astronomy with a telescope we would represent as follows:

$$\text{Brightness} = \frac{1}{(\text{Magnification2} \div \text{Magnification1})^2}$$

To illustrate, when observing the Moon at 100×, if the magnification is then tripled for the second observation to 300×, then using the formula above the brightness of the Moon through your telescope for the second observation would appear only 1/9 as bright:

$$\text{Brightness} = 1 \div (300 \div 100)^2 = 1 \div (3)^2 = 1 \div 9 = 1/9\times$$

Going the other direction, observing the Moon at 300×, if the magnification is reduced to 100× for the second observation, then the brightness of the Moon through the telescope would appear 9× as bright for the second observation:

$$\text{Brightness} = 1 \div (100 \div 300)^2 = 1 \div (1/3)^2 = 1 \div 1/9 = 9\times$$

Even though light follows this inverse-square law, using it will not always predict how the brightness will change when viewing through the eyepiece and telescope in certain circumstances. In our example of observing the Moon, this target does follow the inverse-square law faithfully, since the Moon will many times completely fill the field of view of the eyepiece, and as we further magnify the Moon its image will grow and become dimmer as expected. However, when observing stars things may start behaving much differently. Stars generally do not follow our expectations for the inverse-square law because for the typical observing magnifications stars will not enlarge with more magnification and they will remain a point of light. Because of this, all of the light energy coming from that star remains in that one point in the eyepiece's field of view, not growing larger, and consequently will not appear to dim. In fact, as the magnification is increased the stars may even appear brighter! Why?

If observing from a location that has any light pollution, the background sky itself will have a small glow that is seen through eyepiece. The background in the field of view of the eyepiece will be dark, but not completely black when observing from light-polluted locations. This is especially evident when using lower magnifications. However, when magnification increases the background glow of the sky will be magnified and dim faithfully according to the inverse-square law, just as the Moon dims. So with more magnification when observing stars, the background field of view in the eyepiece will grow darker while the stars will appear to have the same brightness. This increase in contrast between a dimming and darkening background, and a constantly bright star point, makes the stars in the field appear more distinct and brighter to our perceptions. In addition, as the background field of view darkens more and more stars in the eyepiece's field of view will actually appear, as these dimmer stars will no longer be hidden by the bright background. This latter situation is why the limiting magnitude of stars visible to a telescope of a given aperture is also dependent on both the prevailing light pollution of the observing site as well as the magnification used to observe, and not just the aperture of the telescope.

The lesson of magnification in astronomical observing is therefore that for some targets, such as the Moon and planets, they will appear dimmer as magnification is increased. However, for other targets, such as stars, open clusters, globular clusters, and even nebulae and galaxies, these become easier to see as magnification is increased (up to a point), since the background field of view will dim and darken, allowing the stars to stand out more distinctly. Understanding the intricacies of these relationships and experimenting for yourself will aid greatly in an observer's ability to use eyepieces most effectively.

Aberration Control (Telescope Dependencies)

As discussed earlier, there are numerous optical designs, each of which has distinctive optical characteristics and performance capabilities. When choosing an eyepiece, optical designs become an important consideration, depending on the telescope that is used with the eyepiece. Where the optical design of the eyepiece becomes important is when the focal ratio of the telescope is short, particularly for telescope focal ratios of f/5 or less. As an example, if the intent is to use a wider field eyepiece, like one based upon the König design, then the majority of the off-axis of the field of view will have moderate to severe aberrations if the telescope's focal length is much shorter than f/8. So star points will show as aberrations in the off-axis, and the view will not appear pleasing to many observers. However, place that same König eyepiece in an f/10 telescope, and the majority of the field of view will show beautifully sharp and well defined.

As a general rule-of-thumb, simpler optical designs, those with five or fewer elements, should not be expected to perform well off-axis if they have AFOVs wider than 55° and are used in telescopes with focal ratios shorter than f/6 or f/7. This is not to say that it is impossible for these simpler designs to perform well off-axis in short focal ratio telescopes, but it should simply raise a warning flag that the observer is best advised to seek observers knowledgeable with the eyepiece in question prior to committing to any purchase. As observers become more familiar and experienced with the various eyepiece designs, they will begin to get a fairly accurate intuitive feel for which may perform better in the various telescope designs and focal ratios.

With the more modern and complex optical designs using specialized glass types and/or many more optical elements, there are some of these that have a distinct reputation for performing well in telescopes with f/6 and shorter focal ratios. Some of these include the Plössl, the Abbe Orthoscopic, the TMB Planetary, and the Tele Vue Radian for narrower AFOV eyepieces. For wide fields, check out designs branded as Tele Vue Ethos, Tele Vue Nagler, Tele Vue Panoptic, Tele Vue Delos, Explore Scientific 100/82/68 Series, Meade 4000 and 5000 SWA/UWA, Pentax XL/XW, and Nikon NAV.

Choosing an eyepiece for a short focal ratio telescope requires a considerably greater expenditure in time and research (as well as money) to ensure the eyepiece will perform up to expectations, particularly when the focal ratio is shorter than f/5. A best practice is to first borrow the eyepiece from a friend if feasible; otherwise conduct online research and ask fellow amateur astronomers to get as many opinions as possible. After this, make sure the return policy of the store is fully understood if expectations are not satisfied so the eyepiece can be returned.

In the end, since there are many factors that contribute to perceived aberrations, from the human eye to the eyepiece to the telescope and to the atmosphere, it is impossible to predict the exact performance of any eyepiece in a unique optical chain to a great degree of accuracy. Instead, only a ballpark prediction is usually feasible. Therefore, as in so many other circumstances, it is best to seek advice from others who have similar telescopes and the eyepiece in question to get their impressions, or to attend a local club's evening observing session to get first-hand experience.

Observing Strategies

The average observer approaches astronomical observing as taking a broad range of eyepiece focal lengths out to the telescope, then simply switching eyepieces in and out of the telescope until they see a view they consider good. Although there is nothing wrong with an approach to observing where an extensive range of eyepiece focal lengths are purchased and used in a trial-and-error process for all possible observing needs, there are a number of observing strategies that can also be employed instead of the trial-and-error method. Each of the strategies listed have their own distinct strengths and weaknesses, but they all provide the observer with more considered methods of choosing eyepieces to be used that can result in an improved view and less labor intensive method than the trial-and-error method. Some of these effective time-tested strategies are:

- Focal length choices (or magnification strategies)
- Exit pupil
- High magnification
- Intended targets
- Comparing one eyepiece to another

Focal Length Choices

Amateur astronomers use a variety of different strategies when choosing eyepiece focal lengths to build a range of needed capability with their telescope. A very time honored and popular strategy is to choose focal lengths that produce magnifications of 50×, 100×, and 150× in the telescope. Some amateurs extend this rule and say it should also include focal lengths to obtain 200×, and even 250× or 300×. This strategy is actually a highly practical approach, as observing conditions rarely allow magnifications above 300× for most locations. Another nice aspect of this is that the eyepiece collection can be fairly small and still give an excellent range of magnifications. Basically, three eyepieces and one 2× Barlow are all that are needed to implement this strategy and have a full range of magnifications.

With just three eyepieces that produce 50×, 100×, and 150×, plus a 2× Barlow to enable the eyepieces that produce 100× and 150× to generate 200× and 300×, this small collection can provide all the capability the average observer will need for the vast majority of situations. Minimalist sets such as this are often praised by observers as being "freeing," since it takes their focus away from the myriad of choices when there are too many eyepiece choices, and instead allows more concentration on the act of observing. For the beginner, this strategy excels because it requires the least amount of equipment and expense, and for the seasoned observer it is highly effective because it allows greater attention to observation versus the equipment.

A second popular strategy is to have the eyepieces provide jumps in magnification in approximately 1.4× or 1.5× increments. To illustrate, if the first eyepiece produces a magnification of 50×, then using the 1.4× rule the second selected would produce 1.4×50=70×, then the third would produce 1.5×75=98×, and so on. This method can be calculated using only the eyepiece focal length instead of magnifications and have the same results. To do this, divide the focal length of the eyepiece by 1.4× magnification jump factor instead. When using this method with eyepiece focal lengths instead of magnifications, one must necessarily start with the longest focal length eyepiece they desire. To illustrate, if the lowest magnification eyepiece desired is a 20 mm wide field, then the next eyepiece focal length needed is 20 mm ÷ 1.4=14 mm, followed by 14 mm ÷ 1.4=10 mm, and so on. As can be seen, these focal lengths actually represent what some manufacturers provide (e.g., the Pentax XW line's 1.25 in. eyepieces come in focal lengths that have 1.4× separations: 20 mm, 14 mm, 10 mm, 7 mm, 5 mm, and 3.5 mm). The advantage of thinking of this rule in terms of eyepiece focal lengths is that, regardless of the telescope used, these focal lengths would produce magnification jumps in the same 1.4× increments, and if it is not possible to find eyepieces in the exact focal lengths dictated by the rule, simply choose an eyepiece with a focal length that is possible.

Fig. 2.15 The Pentax 1.25 in. XWs made in 1.4× focal length increments (Image by the author)

A third strategy, one touted by amateur astronomer Don Pensack from Los Angeles, California, is the 1x/2x/3x rule. With this rule the value of "x" in the rule varies based on the aperture of the telescope; for 6–8 in. aperture x = 50, for 10 in. aperture x = 60; for 12.5 in. aperture x = 70, and for 18–20 in. x = 80. The uniqueness of this rule is that it defines a nice minimalist set of magnifications that are also tuned to the magnification potential of each aperture class of telescope. Using this method, with a 10 in. 'scope where x = 60, the three eyepieces that result provide the following highly useful range of magnifications:

$$1 \times (60) = 60\times$$

$$2 \times (60) = 120\times$$

$$3 \times (60) = 180\times$$

If the telescope was a 1,200 mm focal length 250 mm (10 in.) f/4.7 Dobsonian, like those offered by Orion Telescopes and others, then to calculate the focal lengths needed for this 1x/2x/3x rule simply divide the focal length of the telescope by the magnifications specified by the rule. Since the rules says that a 10 in. would use x = 60, the 1x/2x/3x magnifications are: 1 × 60 = 60x, 2 × 60 = 120x, and 3 × 60 = 180x. The eyepiece focal lengths required then become:

$$1200 \div 60\times = 20\text{ mm}$$

$$1200 \div 120\times = 10\text{ mm}$$

$$1200 \div 180\times = 6.7\text{ mm}$$

Since these strategies are more optimized to produce a minimal set of eyepieces to satisfy a broad range of observing, they work best when special attention is paid to the longest focal length eyepiece. In the above example, using the 250 mm (10 in.) f/4.7 Dobsonian telescope, if the 20 mm eyepiece chosen was a 50° AFOV Plössl, then much of the maximum TFOV potential of the telescope would be lost. A 250 mm (10 in.) f/4.7 telescope with a 2 in. focuser has a maximum TFOV potential of approximately 2.2°, whereas the example 20 mm 50° AFOV Plössl would only give 0.82° TFOV. So with these methods, the longest focal length should be a quality ultra wide-field eyepiece so the capability exists for observing larger celestial objects (e.g., a 20 mm 82° AFOV eyepiece would produce 1.3° TFOV and a 20 mm 120° AFOV eyepiece would produce 2°, close to the maximum potential of the telescope).

Another strategy to use in choosing eyepiece focal lengths is to select them based on the specific exit pupils that are generally considered optimum for the basic observing situations (the section that follows this continues with an in-depth discussion on exit pupil). An example of this method is to choose focal lengths that produce exit pupils within the following categories so at least one eyepiece is available that is optimized for that category of observing:

- Maximize image brightness—6 mm (light-polluted observing site) / 7 mm exit pupil (dark site)
- General observing exit pupil—4 mm to 5 mm exit pupil
- Optimum deep sky object (DSO) exit pupil—2 mm to 3 mm exit pupil
- Optimum planetary exit pupils—0.75 mm, 1 mm, and 1.5 mm exit pupils
- Filler focal lengths if a transitional magnification or TFOV capability is desired between the major categories

Fig. 2.16 Example of how strategically using a Barlow allows almost no focal length overlap. No Barlow—35 mm, 25 mm, 17 mm, and 12.5 mm. With 2.8× Barlow—12.5 mm, 8.9 mm, 6.1 mm, and 4.5 mm (Image by the author)

The final strategy that observers use is the trial-and-error method. This strategy is very simplistic, popular, and unfortunately the highest cost option. It involves simply purchasing as many focal lengths as possible, from long focal lengths for widest TFOV to a number of special high magnification focal lengths for planetary observing. When at the telescope, since the observer has a large selection of eyepieces in many focal lengths, the process is simply to experiment with the different focal lengths until one is found that best portrays the target being observed. The advantage of this method is that no pre-planning is required, making the observing session simple and intuitive. The disadvantage is that an optimum focal length eyepiece may not be available for the particular target, or even known since no planning was conducted to best exploit the capabilities of the telescope. A further disadvantage is that many eyepieces are needed that complicates set-up of the equipment, especially if it is at a remote location. However, even though this strategy is more of a "brute force" approach, it still is a highly successful approach that is used by many observers.

Regardless of the strategy one may choose, seasoned observers realize that it is not necessary to have a large number of eyepieces for enjoyable and productive observing. Some amateurs use as little as only two or three eyepieces with a Barlow and never find themselves wanting. Whatever strategy is selected, it is important to take into account the maximum practical magnification for the telescope, and also how much magnification the typical sky conditions at the observing locations will permit. Many suburban observing sites do not allow magnifications greater than 200× regardless of the aperture of the telescope, whereas other observing sites have typical sky conditions that are more stable, allowing magnifications of 400×, 500× and more.

For the seasoned observer, eyepiece focal length choices typically involve assessment of the following factors: (a) the telescope's limits, (b) the TFOVs needed to best frame the anticipated celestial targets, and (c) the prevalent sky conditions of the intended observing site. For beginners, however, until they gain several years of experience, choosing eyepiece focal lengths that will produce 50×, 100×, 150×, together with a 2× Barlow accessory to attain the less used 200× and 300× magnifications, may be the soundest approach.

Calculating Exit Pupil

In the previous section in the discussion of eyepiece focal length/magnification strategies, the concept of taking advantage of exit pupils for observing strategies was introduced. As was learned, the exit pupil is not a property of an eyepiece by itself but is actually a combined property of both the eyepiece and the telescope.

There are at least five common situations where calculating exit pupils can be a beneficial strategy for observing. These situations range from maximizing the potential of the telescope's capability to finding exit pupils that work best on various celestial objects:

1. Equalizing image brightness between telescopes
2. Planetary high magnification improvement
3. Matching exit pupil to the eye's dark-adapted state
4. Darkening the background sky under light-polluted skies
5. Optimizing the search for deep sky objects

1. Equalizing Image Brightness Between Telescopes
   When an eyepiece is in the telescope and the image is brought to focus, this image is formed within a small circle above the eye lens at the eye relief point of the eyepiece. All the light gathered from the telescope is then contained in the circle of this image. Therefore, when the diameter of the exit pupil is the same in two different telescope-eyepiece combinations, this indicates that the brightness of the image produced by both telescopes is the same. Even though the two telescopes may have completely different aperture sizes, when eyepieces are chosen for each

telescope that produce an identical exit pupil, this means the image in each telescope will appear just a brightly. The magnifications may be different due to the apertures and focal lengths of each telescope, but the brightness will be the same when each eyepiece produces the same size exit pupil. Since the exit pupil is a measure of brightness that is "independent" of the aperture of the telescope, it becomes a valuable tool that the seasoned observer can use to their advantage.

To calculate the exit pupil of an eyepiece and telescope combination, either of the following formulas will provide equally accurate results:

a. $$\text{Exit Pupil} = \frac{\text{FL[eyepiece]}}{\text{FR[telescope]}}$$

Notes: FL[eyepiece] means the focal length of the eyepiece in millimeters. FR[telescope] means the focal ratio of the telescope, which is the focal length of the telescope in millimeters divided by the aperture of the telescope in millimeters.

b. $$\text{Exit Pupil} = \frac{\text{APERTURE[telescope]}}{\text{Magnification}}$$

Notes: APERTURE[telescope] means the diameter of the main objective of the telescope in millimeters. Magnification is the magnification produced by the eyepiece in the telescope, which is the focal length of the telescope divided by the focal length of the eyepiece.

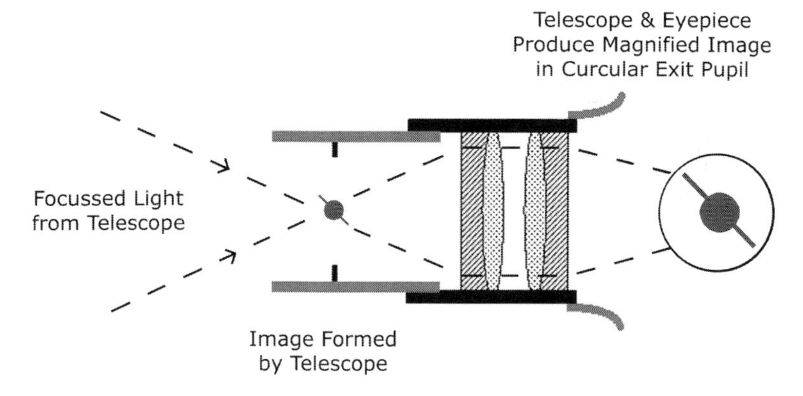

Fig. 2.17 Exit pupil circle (Illustration by the author)

To illustrate exit pupils in action, let's use formula 1 with two different eyepiece and telescope combinations. Assume that telescope "A" has an aperture of 200 mm, a focal ratio of f/6, a focal length of 1,200 mm, and has an eyepiece with a focal length of 12 mm. Using the formula to calculate magnification, this eyepiece will produce a magnification of 1,200 mm/12 mm or 100× in telescope "A." What will the exit pupil size be of the 12 mm eyepiece in our 200 mm f/6 telescope? Using formula 1 we would calculate it as the eyepiece focal length, or 12 mm, divided by the focal ratio of the telescope, which is f/6. So the exit pupil is 12 mm ÷ 6, or 2 mm. The 12 mm eyepiece in the 200 mm f/6 telescope will therefore produce a 2 mm exit pupil at a magnification of 100×.

If telescope "B" is a 100 mm f/8 telescope, what eyepiece is needed that would produce the same 2 mm exit pupil so its view will appear just as bright as it was in telescope "A"? Again, formula 1 can be used for this, but it needs to be rearranged so it solves for eyepiece focal length instead of exit pupil. To illustrate:

$$\text{Exit Pupil} = \text{Eyepiece Focal Length} \div \text{Focal Ratio of Telescope}$$

solved for eyepiece focal length becomes...

$$\text{Eyepiece Focal Length} = \text{Exit Pupil} \times \text{Focal Ratio of Telescope}$$

Inserting the values we have for our example's 100 mm f/8 telescope "B" in the above formula becomes:

$$\begin{aligned}\text{Eyepiece Focal Length} &= \text{Exit Pupil} \times \text{Focal Ratio of Telescope}\\ &= 2\text{ mm} \times f/8\\ &= 16\text{ mm}\end{aligned}$$

Performing the exit pupil calculations we have discovered that the image observed in the 200 mm f/6 telescope "A" with a 12 mm eyepiece will appear just as bright as the image observed in the 100 mm f/8 telescope "B" using a 16 mm eyepiece. The magnifications will be different, 100× in telescope "A" and 50× in telescope "B," but their brightness will appear the same. For observers who have multiple telescopes, or who wish to purchase new telescopes, the exit pupil calculations can be invaluable in gauging how the brightness of the view will compare for their favorite targets.

2. <u>Planetary High Magnification Improvement</u>

Another area where knowing the exit pupil becomes particularly important is in planetary observing. Observers typically find that when the exit pupil gets smaller than 0.75 mm, most of the primary planets such as Jupiter, Saturn, or Mars begin to appear dim and lack the high contrast necessary to show many of their fainter details. So regardless of the telescope's aperture, exit pupil calculations allow the observer to determine the optimum eyepiece and telescope combination to produce what will essentially be the highest magnification where brightness and contrast are at an optimum level for planetary observing.

To illustrate, assume that an observer has a 100 mm telescope and feels that once the magnification get above 150× planets generally appear too dim to the eye. So with a 100 mm telescope he or she prefers 150× as a maximum planetary magnification, where details are most pleasingly bright and shown in high contrast. After many years with the 100 mm telescope, a more powerful planetary telescope is wanted that could view planets at 225× and have the image appear just as bright and with just as much contrast as is currently enjoyed with the 100 mm telescope at 150×. What aperture telescope would provide this capability? To answer this, we use the exit pupil calculation.

What is required first is to calculate the exit pupil of the current 100 mm telescope when observing planets at maximum preferred magnification. Knowing the aperture of the telescope, 100 mm, and the magnification, 150×, using exit pupil formula 2 to calculate the exit pupil of this telescope at 150× becomes:

$$\text{Exit Pupil} = \text{Telescope Aperture} \div \text{Magnification}$$

$$= 100\text{ mm} \div 150\times$$

$$0.67\text{ mm exit pupil}$$

We now know that a 0.67 mm exit pupil is the observer's personal preference for maximum magnification when planetary observing with their 100 mm telescope. Using exit pupil calculations, what aperture telescope should observers upgrade to if their goal is to observe planets that will appear just as brightly at 225×? Again, exit pupil formula 2 can be rearranged to solve for the unknown aperture as follows:

$$\text{Exit Pupil} = \text{Telescope Aperture} \div \text{Magnification}$$

Solved for aperture the above formula becomes...

$$\text{Telescope Aperture} = \text{Exit Pupil} \times \text{Magnification}$$

$$= 0.67\text{ mm} \times 225\times$$

$$= 151\text{ mm aperture telescope}$$

Using exit pupil calculations gives the observer the flexibility to remove much of the guesswork from situations like these, and therefore permit more considered decisions. In the scenario just presented, understanding how to use exit pupil calculations allows the observer to predict outcomes without having to resort to trial and error experiments. Instead, using exit pupil calculations, they are able to reliably predict that a 151 mm aperture telescope will provide them their goal to view planets at 225×, instead of at 150×, and they will appear just as bright and high contrast as their current 100 mm telescope performs at 150×.

A note of caution, though. Exit pupil analyses such as these work best when the compared telescopes have the same overall transmission efficiency. That is, both telescopes are of the same optical designs, such as two doublet refractors or two Newtonians with both using similar anti-reflection technologies on their optics. A 151 mm SCT telescope at 225× producing a .67 mm exit pupil would not show its image as brightly as a 100 mm refractor at 100× producing the same .67 mm exit pupil. The reason for this is that mirrored surfaces do not transmit light as efficiently as glass surfaces. To adjust these exit pupil calculations to account for the different transmission efficiencies of typical telescope designs, consult a trusted resource on telescope optics, an online astronomy forum, or members of a local astronomy club for assistance.

3. Matching Exit Pupil to the Eye's Dark-Adapted State

   If an observer wishes to determine the eyepiece that will have an exit pupil with the lowest practical magnification for the telescope, the general rule of thumb is to keep the maximum exit pupil produced by the eyepiece-telescope combination at 7 mm or smaller to match the dark-adapted human eye. A 7 mm exit pupil is recommended, since popular wisdom advises that the average dark-adapted human eye can open to approximately 7 mm in diameter. This collective experience holds well with multiple studies that generally report that for a range of individuals with normal vision between 20 and 70 years of age, the average measured dark-adapted pupil dilation varies by age between 7.5 and 5.5 mm (*Note:* This was the average range; a very few individuals had a maximum opening as high as 9 mm and as low as 4 mm). Therefore, if the eyepiece and telescope are producing an exit pupil larger than 7 mm, then a portion of the light from the telescope is blocked by the eye's pupil if the observer is an average individual with a maximum pupil opening of 7 mm.

   To illustrate the issue with eyepiece-telescope combinations producing overly large exit pupils, if the eyepiece-telescope combination is producing an exit pupil of 8 mm, and the observer's eye can only open to 7 mm, then 1 mm of the light cone is being blocked, which represents a full 23 % loss in the available brightness of the image.

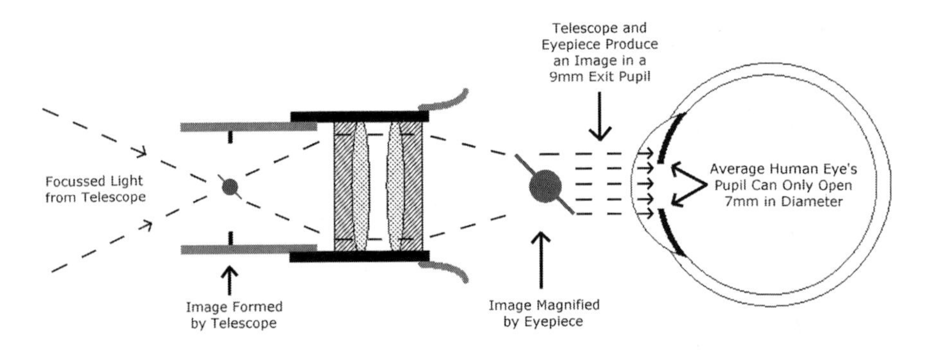

Fig. 2.18 Human eye blocking light when the eyepiece-telescope exit pupil is too large (Illustration by the author)

Another way to think about the potential light loss from excessively large exit pupils is to equate how this light loss would compare to telescope aperture. Since the area of a 7 mm diameter circle is approximately 23 % less than the area of an 8 mm diameter circle, this means that the observer is potentially losing 23 % of their telescope's light-gathering capability if they are using an eyepiece that produces an 8 mm exit pupil and their dark-adapted eye only opens to 7 mm. If they were using a 100 mm telescope, the brightness loss by using an eyepiece producing an 8 mm exit pupil would make it appear that the image was only as bright as what they would see in a telescope of approximately 88 mm. If they were observing with a 250 mm (10 in.) aperture telescope with an 8 mm exit pupil then it would be like their telescope was only 226 mm (8.9 in.) instead.

Because the pupil of the average dark-adapted human eye may only open to a maximum of 7 mm, does this mean larger exit pupils should never be used? The answer to this is "No," as using larger exit pupils than recommended only means that the observer should expect the image to appear dimmer than might be expected. If an observer desires to use an eyepiece-telescope combination that produces an exit pupil larger than 7 mm to obtain a view with wider TFOV, then this is a perfectly acceptable practice. It is actually a common practice by many observers to use exit pupils larger than 7 mm, as the brightness of star points are generally not perceived to dim nearly as much as with faint objects such as nebulae and galaxies.

Therefore, a large and bright cluster such as the Pleiades will show just as spectacularly using an 8 mm exit pupil as it does when using a 7 mm exit pupil. However, targets like the great Orion Nebula will appear much dimmer with less nebulosity visible at an 8 mm exit pupil than it does at a 7 mm exit pupil. So it is best to use an eyepiece-telescope combination that produces exit pupils larger than the eye can accommodate wisely; acceptable if the goal is to see the widest amount of sky possible, not recommended if the goal is to observe fainter objects, as they may appear dimmer than desired.

4. Darkening the Background Sky under Light-Polluted Skies

To determine the eyepiece-telescope combination that will have the largest "practical" exit pupil when observing under brighter light-polluted skies, many observers feel that keeping the largest exit pupil to 6 mm or even 5 mm is a best guideline. Since the background sky appears brighter with larger exit pupils, using an eyepiece that generates a 5 or 6 mm exit pupil will make the background sky in the eyepiece appear darker and richer. This darker background then allows the stars to show themselves with more contrast against the darker background than with an eyepiece that produces a larger exit pupil. Observers should experiment for themselves if they routinely observe from sites with light-polluted skies, as they may find that a shorter focal length eyepiece that has a wider AFOV and produces a 5–6 mm exit pupil proves an aesthetically improved view with little practical loss in TFOV if they choose the widest AFOV eyepiece available. This is a primary reason why eyepieces such as the Tele Vue 31 mm Nagler, Explore Scientific 30 mm 82 Series, William Optics 28 mm UWAN, Explore Scientific 25 mm 100 Series, and the

Tele Vue 21 mm Ethos have become more popular choices, producing smaller exit pupils in telescopes than the longer focal length eyepieces such as the 56 mm Plössl.

5. Optimizing the Search for Deep Sky Objects

Finally, to determine the low magnification eyepiece that will be most effective for hunting faint deep sky objects, many observers recommend using an eyepiece that will produce an exit pupil between 2 and 3 mm. At these smaller exit pupils the background sky appears very dark, providing a higher contrast between the background sky and the faint deep sky object. This higher contrast then allows these faint deep sky objects to be more easily seen. In addition to the higher contrast of the view, if the observer also plans to use this eyepiece to "hunt" for deep sky objects, it is best for him or her to use one with as wide of an AFOV as possible so a larger portion of the sky is visible (i.e., a larger TFOV).

To illustrate the importance of maximizing the eyepiece AFOV when hunting for deep sky objects using smaller 2–3 mm exit pupils, let's calculate the numbers using an operative example. Let's assume the observer is using a popular variety of the 203 mm (8 in.) f/6 Dobsonian telescopes available today. With that type of telescope, an eyepiece that would produce a 3 mm exit pupil would need to have a focal length of 18 mm (exit pupil = eyepiece focal length ÷ telescope focal ratio). Performing the calculations for magnification, an 18 mm eyepiece in this telescope would produce 67×. At that magnification, the TFOV of an 18 mm Plössl with its 50° AFOV would only be about 0.75°, which is quite small. If an 18 mm 82° wide-field eyepiece was used instead, the TFOV visible would be 60 % larger (e.g., 1.2°), allowing coverage of a larger region of the sky and improving chances of finding elusive targets.

Exit pupil calculations add more complexity to the observing experience, and as a result many observers dislike taking advantage of what they can provide. However, if an observer possesses a full range of eyepiece focal lengths from 40 mm to 4 mm, then it is certainly easy enough to simply find the eyepiece that provides the best exit pupil for observing through trial and error without doing any calculations. The disadvantage of trial and error is that it can be time consuming, and many observers find knowledge of exit pupils and skill in applying that knowledge when planning an observation can actually allow them to devote more time to observing. Regardless of whether an observer chooses to use an exit pupil methodology or not to guide their observing, exit pupils remain a valuable tool for the amateur astronomer.

High Magnification

Rarely is magnification discussed in the context of an observing strategy. More often, magnification, especially high magnification, is discussed in the context of what may be maximum magnification limits for a particular telescope. There are also many excellent books and online communities where the many aspects of what affects the magnification capabilities of a telescope are detailed (e.g., Dawes limit, the Rayleigh criterion, the mean transfer function, the seeing and transparency of the atmosphere, and the thermal acclimation of the optics to the outside environment). In the context of the eyepiece, however, the considerations are not what magnifications can be achieved but more about how eyepiece features can be advantageous or disadvantageous when conducting high magnification observing.

The four primary attributes of an eyepiece that impact its usability for high-magnification observing are:

- Eye relief
- Apparent field of view (AFOV)
- Off-axis performance (e.g., lack of aberrations/distortions that reduce sharpness of the image)
- Transmission and contrast

Each of these attributes are more or less important when using the eyepiece, depending on the particular observer, the target they are observing at high magnification, and the stability and tracking capability of the telescope's mount.

Eye relief becomes important from the standpoint of comfort during long observing sessions. Typically, as the eyepiece's focal length becomes shorter, so does the eye relief. This is especially true for the classical designs that use a minimum number of elements such as the Plössl, Abbe Ortho, König, Brandon and similar simple designs. As an example, a typical modern 25 mm Abbe Ortho eyepiece has an eye relief of about 20 mm. However, the same eyepiece design in a 4 mm focal length will only have an eye relief of only about 3.5 mm. These designs, with their minimum eye relief, are considered by many observers to be best for short duration observing tasks only. When a high performing eyepiece that contains the least amount of glass to maximize perceived brightness and contrast is the advantage to leverage, then these classic designs excel. However, if the task requires long observation time, such as when studying an object in detail or when sketching, the short eye relief can be counterproductive to the objectives of the observation. Optimizing eye relief for high magnification observing is therefore dependent on the exact type of observing task planned, and can vary with long eye relief being an advantage, or short eye relief being acceptable.

The second attribute to consider when using a high magnification eyepiece is apparent field of view. This attribute is important depending on the target being observed. If the target is a planet or double star, as an example, then there is little to be gained by using an eyepiece with a wide AFOV, since there is rarely any context around these objects to observe. In fact, many observers comment that they prefer the smaller AFOV eyepieces for these type targets, as they provide fewer distractions to observing the primary target. However, if the plan is to conduct high magnification observations of other targets, such as globular clusters, then it may be an advantage to use a high power eyepiece that also has a wider AFOV. An example of this situation is when observing globular clusters. These objects often have stars extending a significant distance from their core, so a wider AFOV is a distinct advantage when observing them at high magnification. The wider AFOV eyepiece affords a distinct advantage in this circumstance, as the greater TFOV around the object is preferable to have in the field of view of the eyepiece.

In addition to the context of the surrounding field of view of the target being observed, another important reason to use an eyepiece with a wider AFOV for high magnification observing is when the telescope's mount does not track the object automatically, or if it is not stable. When using very high magnifications, the TFOV visible in the eyepiece naturally becomes smaller. Consequently, if the telescope's mount does not automatically track, the target will drift through the field of view in just a view seconds, potentially resulting in only a very brief observation. Use of a larger AFOV eyepiece means that the object takes longer to drift through the field of view and therefore the observation can be longer. In addition, if the mount is not stable, then higher magnification observing can accentuate this instability, and with a narrow AFOV eyepiece even the simple act of focusing may cause the target to severely jitter or even move completely out of the field of view. Therefore, if the telescope's mount is not stable or if it does not have automatic tracking, then a wider AFOV eyepiece is generally a better choice if high magnification observing is planned.

The third consideration for high magnification observing is the performance of the off-axis of the eyepiece with the telescope. Like one of the AFOV considerations, off-axis performance is generally more important if the telescope's mount does not have automatic tracking. In these circumstances, since the target will drift rapidly through the field of view at high magnifications, it is customary to place the target at one end of the field of view so observing time can be maximized as it drifts across the entire field of view. If the eyepiece chosen has an off-axis with aberrations and distortions (i.e., field curvature, astigmatism, lateral color, etc.), these will reduce the sharpness of the image, further limiting the time during the drift that the object will appear sharp for a productive observation. Therefore, for high magnification observing it is important to understand how the off-axis of the eyepiece chosen will perform in a telescope that does not have a tracking mount.

As an example, the TMB Monocentric eyepiece, one of the very best performers for high magnification planetary observing, has a poor performing off-axis in short focal ratio telescopes. In a short f/5 Dobsonian, as much as 50 % of TMB Monocentric's field of view may be out of focus. Although this eyepiece is an excellent choice for a short focal ratio telescope with tracking, where the planet is maintained in the center of the field of view, this eyepiece becomes a poor choice for short focal ratio non-tracking telescopes.

The last characteristic we will highlight that can be an important consideration when conducting high magnification observations is the perceived transmission and contrast performance of the eyepiece. It is important to note that these characteristics of an eyepiece are very difficult to assess; therefore these performance characteristics are usually only found in the field observing reports sometimes provided by other amateur astronomers. Where the transmission and contrast characteristics of an eyepiece come into play is for the most part in two very specialized areas of observing: planetary observing and faint galaxy/nebula observing. For those observers where planetary or detailed observations of faint deep sky galaxies/nebulae are a passionate pursuit, finding and using eyepieces that perform best on these targets is a never-ending quest. For these observers it is a constant challenge to attain the best possible eyepieces for the task, scrutinizing all new entrants to the field that could possibly render these targets a little bit better when viewed through their favorite telescopes.

For the general observer, however, this nuance of eyepiece performance is not always that important of a consideration because most modern eyepieces perform at exceptional levels of transmission and contrast, so having the best-of-the-best is not a passionate pursuit. Therefore, only for those observers who depend on very small gains in brightness and contrast to bring out the most challenging aspects of a planet, galaxy, or nebula will these difficult to assess characteristics of an eyepiece become important.

When considering an eyepiece's transmission, we know from a theoretical standpoint that each air-to-glass interface in the optical design loses a small portion of light. Therefore, those eyepieces with more glass groupings in the design will theoretically have a lower transmission than those with fewer lens groups. As an example, a Plössl design has four elements of glass in two groups. These two groups then have four air-to-glass interfaces. When compared to another eyepiece that has more than two groups in its design, the Plössl will transmit a little more light (as long as the anti-reflection coatings and glass types in each eyepiece are the same). Therefore, the general assertion one often hears is that eyepieces with simpler designs, having less glass, will be brighter, transmitting more light.

This is actually very true from a theoretical standpoint. As long as all other aspects of the eyepieces are the same (e.g., level of polish of the lenses, coatings, internal baffling, etc.) eyepieces with fewer lens groups will transmit slightly more light and also have slightly higher contrast. However, although this may be true from a theoretical standpoint, it cannot be reliably extended to production eyepieces because each manufacturer builds its eyepieces using coatings of different efficiencies, spending more or less cost on internal baffling and polishing of the glass, etc. Even a same line and focal length of eyepiece can have different levels of anti-reflection coating efficiency if the manufacturer upgraded the coatings over time. It is therefore impossible to generalize that an eyepiece with less glass will actually be better in terms of transmission and contrast, or that even the same brand eyepiece of different production vintages will have the same transmission and contrast.

Given these difficulties, how can the transmission and contrast characteristics of an eyepiece be determined? Unfortunately, manufactures do not indicate the test results for transmission and contrast from their production eyepieces, so there are no quantitative reference sources to determine a particular eyepiece's transmission or contrast characteristics. However, there is a wealth of qualitative information on these attributes for eyepieces—the multitude of user reports and reviews on how select eyepieces performed when used in specific telescopes observing specified targets. Over the years, based on a predominance of observer reports and reviews, several eyepiece lines have gained a strong reputation as being top performers, providing views that are perceived by the observers as having slightly improved levels of both brightness and contrast. These reports, as with any observation report, do not tell if the actual transmission or contrast of an eyepiece is different, but they can convey an observer's subjective perception of image brightness and perceived contrast. And these reports also cannot tell if it was solely the eyepiece responsible, as it is nearly always impossible to separate characteristics of the telescope from the eyepiece.

So when reading observation reports understand that they are, in reality, optical "system" reports instead of "eyepiece" reports, since the entire optical chain is involved in the observations being noted by the observer and not just any one element. However, with all these caveats in place, over many years there were developed a handful of eyepieces that have consistently obtained accolades from observers related to how sharply they are perceived to perform, their lack of perceived scatter, their perceived brightness, and their perceived contrast. Examples of these are the Zeiss Abbe Orthos (ZAO) versions I and II, the TMB Supermonocentrics, the vintage Carl Zeiss Jena (CZJ) Orthos (.965 in. barrels only), the Pentax SMC Orthos (.965 in. barrels only) and Pentax XOs, and the Astro-Physics Super Planetary (AP-SPL). This elite group of eyepieces have the reputation of providing the brightest, highest contrast, and most detailed views for both planets and many faint deep sky objects. The optical designs for all these eyepieces, coincidentally, fall into the "less glass" variety, with all of them being two or three group designs with only five or fewer glass elements. These eyepieces also, coincidentally, fall into the class of having extremely high quality builds, with premium optics and premium prices.

Although the elite planetary group of eyepieces are considered the best-in-class, this does not mean that there are not more complex optical designs that do not come very close to the performance of the elite planetaries. It is worth it to note that many observers today report that a select few eyepiece lines with many more optical elements than the elite planetaries can come very close in performance. These eyepiece lines include the Docter UWA, the Leica VARIO 25 50× Aspheric Zoom, the Pentax XW, the Tele Vue Delos, and the Tele Vue Ethos, all of which are optical designs incorporating as much as 10–12 glass elements. Many observers report superb planetary and faint object deep sky performance from these very complex optical designs, demonstrating that when a manufacturer pays special attention to every aspect of an eyepiece's design and build, even the most complicated designs using many elements can rival the best-in-class minimum glass specialty eyepieces.

Fig. 2.19 Premium classic planetary eyepieces (Image by the author)

Since there is no single eyepiece design that fully maximizes all the eyepiece attributes of eye relief, apparent field of view, off-axis aberration and distortion control, and brightness and contrast performance for high magnification observing, many observers choose a path of maintaining multiple short focal length eyepieces. This approach is centered on treating the eyepiece as what it is, a tool for a highly specific task. Taking this approach means it can be advantageous to maintain several eyepieces in overlapping focal lengths to best match the eyepiece's unique characteristics to the task at hand. An observing situation that illustrates this is when the evening's observing is to include some lengthy observation of a planet such as Jupiter. For this target, given that it is a planet, the AFOV of the eyepiece to choose would normally not be important. However, during this evening Jupiter's moons happen to be positioned widely apart from the planet, so a wider AFOV high magnification eyepiece will be more appropriate. Finally, since the plan is to conduct some lengthy observing and sketching of this single target, an eyepiece with comfortable eye relief is required as well.

During the course of the observation, it is noted that there is a hint of a structure called a barge on the edge of one of Jupiter's main equatorial cloud belts. Unfortunately this barge is faint and of low contrast, so the comfortable, wide-field eyepiece being used is just not showing the barge decisively or distinctly. In this situation, if the observer maintains specialized eyepieces for specialized tasks, he or she would then turn to a specialized high quality, narrow field, tight eye relief, minimum lens count eyepiece to determine if this highly optimized design better shows the suspected barge. In this scenario a TMB Supermonocentric eyepiece could be an appropriate choice.

Although the observer can no longer see all of Jupiter's moons because of the eyepiece's narrow AFOV, and its very short eye relief means the viewing is uncomfortable and not suited for extensive sketching, in our scenario (and as reported in real observation reports), this specialized eyepiece showed the barge very distinctly! After enjoying the "catch" of this elusive barge, the observer can then continue using the high magnification eyepiece with more comfortable eye relief to observe other aspects of this wonderful planet and continue sketching. This is an example of a real-life scenario that observers have encountered and demonstrates that when observing extremely challenging aspects of a target, the slight edge a highly optimized eyepiece can provide can be worth the extra expense when an observer is passionate in particular observing tasks. Of course, this approach is not appropriate for all, and for other observers who do not need to go to these extremes to satisfy their observing goals, a better solution could certainly be the more comfortable and larger AFOV design of a quality wide field.

High magnification observing is a specialized niche in astronomical observing. Whether the goal is tracking down challenging features on planets, going as deeply as possible into the cores of globular clusters, trying to tease out as much definition as possible in the spirals arms of distant galaxies, or trying to split a difficult double star, sometimes these tasks can be done best when using multiple eyepieces of the same focal length that have demonstrated themselves to perform best in certain very narrow observing circumstances. The more care an observer takes to fully understand and exploit the unique aspects of an eyepiece, its design, and its operational characteristics, the greater rewards are there to be found. Although there is no "best" eyepiece or eyepiece design for high magnification observing, what is "best" is when the particular strengths of a given eyepiece or eyepiece design can best match the needs of the observer for the specific observing task he or she wishes to pursue.

Chapter 3

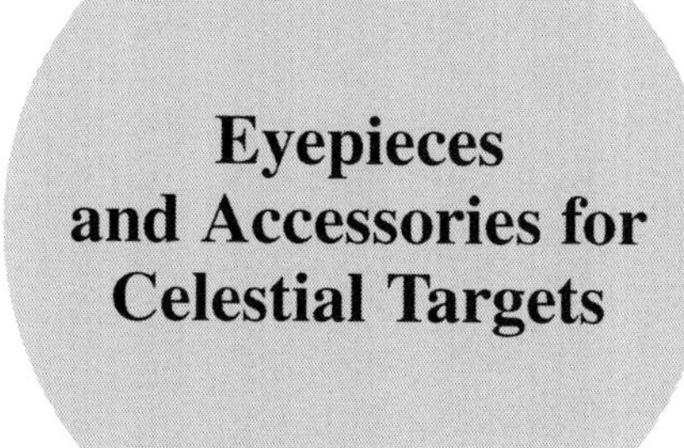

Eyepieces and Accessories for Celestial Targets

Although one often hears about eyepieces for planetary observing, it is not common to hear about eyepieces specialized for any other celestial object category. However, just because it has never become popular to think of a deep sky object (DSO) eyepiece or a globular cluster eyepiece, etc., this does not mean that there might not be certain eyepiece characteristics that can best exploit the observation of these targets.

Since celestial targets vary greatly in size, shape, brightness, and color, it is only reasonable to assume that their unique properties will show differently depending on the unique properties of a particular eyepiece. As an example, when observing the subtle color shades in the polar regions of Jupiter, even though two eyepieces may be of the same focal length and same AFOV size, if they each add a slightly different tone to the image (sometimes also referred to as hue or tint), then they may each show the colorations in Jupiter's polar regions differently. For many observers, they perceive distinct tone differences between eyepiece lines, some being characterized as showing the view with more warm tones, more cool tones, or being tone-neutral. Given that some observers note tone differences, this has also led them to develop likes and dislikes when using eyepieces for different celestial objects based on the tonal quality they perceive from the eyepiece. Therefore, if an observer perceives tonal differences, then this quality can be used to their advantage, as many observers report that for some planets, like Jupiter, warmer toned eyepieces show the bands in its atmosphere more distinctly.

To illustrate another scenario, such as observing the cores of globular clusters, eyepieces with a wider AFOV will show an aesthetically improved view many times, since at the high magnifications necessary to resolve their cores if the

W. Paolini, *Choosing and Using Astronomical Eyepieces*, The Patrick Moore Practical Astronomy Series, DOI 10.1007/978-1-4614-7723-5_3,
© Springer Science+Business Media New York 2013

eyepiece's AFOV is narrow then the context of the outer regions of the globular cluster can be lost. There are actually many characteristics of an eyepiece that, when together, can greatly impact its overall capability for a selected target type. Therefore, although there may be no popular characterization of eyepieces for deep sky observing like there is for planetary observing, with a little attention to the characteristics that are important to each celestial target type, an observer can easily assemble a number of eyepieces that for them can become "lunar" eyepieces, or "globular" eyepieces, etc. The following presents those eyepiece characteristics that for many observers prove to be advantageous for the Moon, stars and clusters, nebula and galaxies, and planets.

Celestial Targets

The Moon

When using eyepieces specifically for the Moon, there are several considerations. First, the Moon is unlike any other evening target, having extreme brightness and a wide array of very high contrast features. It can therefore take much more magnification showing a clear and detailed view than many other targets.

Many find that the Moon may be able to successfully use 1.5× or more magnification than a typical maximum magnification used for planets. Therefore, the first order of business in using an eyepiece for the Moon is to make sure the right focal length is on hand for extreme magnifications. As an example, if a 100 mm telescope with a focal ratio of f/8 and focal length of 800 mm provides good views up to 160× on planets (e.g., about a .63 mm exit pupil), for the Moon this same telescope should be able to use magnifications as high as 1.5×, or 240× when the seeing and transparency of the sky are good. Therefore, for lunar observing it is generally a good idea to have an eyepiece on hand that will produce very high magnifications for some extreme lunar exploring.

Fig. 3.1 Full Moon (Astrograph courtesy of Mike Hankey, Freeland, MD, USA—www.mikesastrophotos.com)

Next, many observers enjoy viewing the Moon in its entirety as it provides a wonderful perspective being able to see the entire Moon in the field of view of the eyepiece. The angular size of the Moon is approximately 0.5°, so to have the capability to view the Moon from this unique perspective requires an eyepiece-telescope combination that provides at least a full 0.5° of TFOV. However, having additional context is often more aesthetically pleasing, so an eyepiece-telescope combination producing 0.75° of TFOV may prove more advantageous.

To round out an observer's lunar exploring plans, in addition to full Moon capability and extreme magnification capability, a few intermediate focal lengths between these two extremes are also recommended. A wide variety of magnifications is often not necessary for enjoyable lunar observing, therefore only two or three select eyepieces and a Barlow are all that is generally needed for an enjoyable and productive evening of lunar observing.

Finally, many observers find a wide-field eyepiece as the preferred lunar eyepiece, especially at high magnifications. Since the Moon is so very rich in details, it is simply easier to fix the eyepiece on one area of the Moon and spend a significant amount of time viewing all the details throughout the field of view. When the AFOV of the eyepiece is narrow, more repositioning of the telescope would be required versus when a wide-field eyepiece is used.

Stars and Clusters

Stars and clusters, while all being similar in that they are composed of stars, are still somewhat varied, and these variations drive the use of different eyepiece qualities to maximize observation of these celestial objects. Double stars are probably the most specialized target from this class of objects. The usual paradigm of many observers is to approach double stars as challenge targets, trying to split the closest double possible with their eyepiece and telescope. This being the case, eyepieces that produce higher magnification in your telescope are a natural choice to use. For double star observing, considerations should be the comfort of the eye relief of the eyepiece if long observation times are anticipated, and AFOV considerations depending on whether your telescope's mount does or does not have motorized tracking capability (more AFOV is advantageous if the telescope does not have automated GOTO and tracking).

Fig. 3.2 M10 (Astrograph courtesy of Mike Hankey, Freeland, MD, USA—www.mikesastrophotos.com)

A secondary consideration can also be the tonal quality of the eyepiece (e.g., the tone of the image it produces: warm, neutral, or cool). Tone can become important mainly when observing colorful double stars as the neutral to cooler toned eyepieces tend to show double stars more vividly to the perceptions of many observers.

Since human color perception varies greatly from individual to individual, it is strongly recommended for observers to determine for themselves the impact of eyepiece tone on their perception of color rather than depend on the observations of others. Regardless of any eyepiece tone, however, a good trick to make star colors a little more obvious is to very slightly defocus the star point. The additional size of the star point that results from a slight defocus is sometimes all that is needed to make the star's color appear more strongly.

Moving to open clusters, the TFOV of the eyepiece-telescope combination can be very important, as many open clusters are large in size, requiring a significant TFOV to adequately see the entire cluster. This being the case, if open clusters are on the evening's observing plan, then the use of lower magnification, larger AFOV, and larger TFOV eyepieces becomes an advantageous choice. To illustrate, one of the several showcase open clusters that are very large, such as M45, the Pleiades Cluster, requires well over two full degrees or more TFOV to adequately frame. With TFOVs this wide, many telescopes require an eyepiece that produces very low magnification with as wide of an AFOV as possible. Therefore, for these very large open clusters a better choice is to use eyepieces with 2″ barrels, as these will maximize the TFOV potential of the telescope.

Like double stars, the tonal quality of the eyepiece used for open cluster observing can also be a consideration, particularly if the cluster contains one or more carbon stars that typically have a rich orange to deep red hue. Again, for some observers eyepieces with reputations for being more neutral toned sometimes show carbon stars more vividly than warmer toned eyepieces.

The final category of stars are those that are in globular clusters. Globular clusters, or simply "globs" as many observers call them, can be spectacular objects with a unique beauty all their own. Since many globs are distant, faint, and small when viewed, eyepieces that produce higher magnifications are better suited for these targets. Generally, eyepieces that produce an exit pupil of between 0.5 and 1.0 mm are good choices for globs, especially if the observer's intent is to observe deeply into the core of the glob to resolve as many faint stars as possible.

In addition to higher magnifications and smaller exit pupils, globs are one of the few targets that seem to always show better in wider AFOV eyepieces than narrower AFOV eyepieces. This is true because outside the densely packed core of the glob there are a multitude of more loosely scattered stars that provide a richer context for the core. Therefore, higher magnification eyepieces with wider AFOVs to show as much TFOV as possible are typically the recommended choice for enjoyable viewing of globular clusters.

Nebulae and Galaxies

Eyepiece choices for nebula and galaxies will vary greatly, as these targets can range in size from several degrees, like M31, to a very small fraction of a degree, like M57. To effectively observe these greatly varied celestial objects, the observer will therefore need to use eyepieces that produce very low magnifications, showing as much TFOV as possible, to higher magnification eyepieces that produce exit pupils from 1 to 3 mm. As many deep sky observers note, higher magnification is often a benefit for some of these targets since the smaller exit pupils darken the background sky and allow the internal structures of some of these faint objects to be seen. Examples where this is beneficial is when observing the dust lanes and spiral arms of galaxies or the structure within nebulae.

An easy and popular target to see this at work is with the Great Orion Nebula, Messier 42. This target is a breathtaking wonder to behold using an eyepiece that will produce a TFOV of about 1°. This size TFOV will frame many of the stars in the Sword of Orion where M42 is positioned in the center. At this large TFOV, broad patches of this milky-white nebula will be visible sweeping across the black background sky. However, as magnification is increased these broad milky patches will be transformed into a mottled structure of dark and light regions weaving throughout. At these higher magnifications, in addition to observing the intricate mottled structure of M42, they are useful to observe the beautifully tight grouping of four stars in the nebula referred to as the Trapezium. To the keen eye on steady evenings using higher magnifications, two fainter companions to the four stars of the Trapezium can be revealed, transforming the Trapezium into a striking asterism of six stars. These two faintest stars, referred to as Trap-E and Trap-F, are a challenge object that has thrilled observers throughout history.

Fig. 3.3 M51 (Astrograph courtesy of Mike Hankey, Freeland, MD, USA—www.mikesastrophotos.com)

Eyepiece choices for nebulae and galaxies can therefore run the gambit from needing low to high magnifications, with the two ends of this range being perhaps more appropriate than the middle capabilities. And in addition to the eyepiece, observing these objects can often be significantly accentuated through the use of the specialized accessories such as high contrast filters, light pollution filters, hydrogen beta filters, and oxygen-III filters.

The Planets

Eyepieces for planetary observing are probably one of the hottest topic areas in amateur astronomy. Classically, a planetary eyepiece is considered one that uses the least number of glass elements in its design (often only 3–5) and has a small and well corrected field of view that maintains orthoscopic qualities (e.g., things appear as they are without distortion across the field of view, linear features do not bend or bow, and angles do not change). The reason for the importance of the lack of distortion characteristic is that the round shape of a planet or the Moon, or linear features on its surface, will distort so that linear features appear bent, or its round shape will turn into an oval or egg shape when rectilinear and angular magnification distortions are present. Since these distortions are common to wide-field designs, some observers prefer eyepieces with narrower AFOVs for planetary observing.

Fig. 3.4 Jupiter (Astrograph courtesy of Mike Hankey, Freeland, MD, USA—www.mikesastrophotos.com)

Another important consideration for a planetary eyepiece has to do with the fact that most planetary observing uses higher magnifications than is typical for other celestial targets. Magnifications of between 150× and 300× are considered normal for productive planetary observing. Since the magnifications need to be high, this also impacts eyepiece choices, such as, is the eye relief comfortable enough for extended observations, or is the AFOV of the eyepiece wide enough to allow for adequate drift time across the field of view if the telescope is using a manual non-driven mount. Understanding that planetary observing requires higher magnifications necessitates that the observer plan on using an eyepiece with eye relief appropriate to his or her long observing session preferences and an AFOV size appropriate to the telescope's tracking ability. If the observer does not account for these, then planetary observing can become a frustrating experience.

An area of much controversy in the observing community has always been the contention that a best-in-class planetary eyepiece must have the least number of lenses in its design as possible. This is sometimes called a "minimum glass" eyepiece. The rationale for minimizing the lens count is often not obvious but can be quite valid. Planets are unique celestial objects having a mix of both high and low contrast areas, features that vary in size from very small to very large, features that possess a wide variety of colors and hues, and finally features that do not remain constant but change—and sometimes change very rapidly.

Because planets have such a dynamic and broad range of features, an eyepiece that reduces scatter to a minimum has the maximum light throughput possible and can provide the highest contrast image possible without unwanted light artifacts such as flare and ghosting. Such eyepieces are considered optimum to these types of observing tasks. Given that each additional element of glass in an eyepiece slightly degrades each of these desired optical strengths needed for critical planetary observation (e.g., its transmission, scatter, and contrast), this validates the logic of why eyepieces with less glass and highly corrected fields of view, such as the classic Abbe Orthoscopic design (which has only four glass elements in two groups) or the Monocentric design (which has only three elements in a single group) have been highly sought out by planetary enthusiasts. For those observers who have a special passion for planetary observing, even the smallest gains in contrast and brightness from an eyepiece are often well worth the usual investment these specialized eyepieces command.

Since an eyepiece that minimizes the number of optical elements is classically considered best for planetary observing, does this mean that those eyepieces with many glass elements are not suited for planetary observing? The answer to this question is a resounding "No!" The truth of the matter is that any quality eyepiece serves very well in the role of planetary observing, and many highly complex designs such as the Tele Vue Ethos and Leica Vario Aspheric Zoom are examples of extremely capable eyepieces for planetary observing. The minimum glass argument simply means that if all other things are equal in the production of an eyepiece (e.g., the polish of the lenses, the precision of the prescription of the lenses, the exactness of their placement, the quality of the application of the coatings, and the internal baffling and stray light control inside of the eyepiece), then an eyepiece with fewer elements will very slightly outperform an eyepiece with more elements. The key phrase to take note of in this last statement is "if all other things are equal," as this is one reason why a minimum glass approach is not automatically a better choice.

Each eyepiece manufacturer builds its eyepieces to varying levels of precision and quality, and all manufacturers use different coating technologies from different suppliers, some highly proprietary in nature. Given the variability that exists between manufacturers on how finely they figure and polish the lenses for their eyepieces, and how exactingly they build and baffle their eyepieces, an eyepiece with a more complex multi-element design can easily outperform a minimum glass design that is not built and executed to a similar high standard. Selection and use of a planetary eyepiece should therefore rest more on the reputation that the eyepiece has attained for planetary observing over the years, and after that, how that reputation has proven itself to be true in the eyes of each individual observer!

Since specialized planetary eyepieces are typically very expensive, and the observational gains will always be small and subtle, only the unique passion that each observer has for planetary observing can determine if these gains are worth the investment. For some observers, the gains to be realized by these specialized eyepieces are either not noticed or are considered too small to be important. However, for others these small gains fulfill a critical need in their observational capabilities. In the end, whether the gains that can be realized from an elite-class planetary eyepiece are significant and worth the extra expense is entirely a subjective assessment that is unique to each observer.

A final, and often overlooked, attribute of a planetary eyepiece is the tonal quality of the image the eyepiece produces. For example, do the optics of the eyepiece impart any tint, hue, or coloration to the image? Some eyepieces have a reputation as being either warmer toned, neutral toned, or cooler toned. As reported by some observers, these tonal differences can also impact the lenses' performance on some aspects of particular planets. As an example, a warmer toned eyepiece will accentuate the cloud bands of Jupiter very nicely, making them more easily seen than when using a cool toned eyepiece. However, the same warm-toned eyepiece will show the subtle gradations of shadings and hues in Jupiter's polar region less well, whereas a cool toned eyepiece will show these color changes more distinctly.

Fig. 3.5 A collection of high-end planetary eyepieces from Astro-Physics, Brandon, Nikon, Pentax, TMB, and Zeiss (Image courtesy of Bill Rose, Larkspur, CO, USA)

Whether an elite-class planetary eyepiece will be beneficial to an observer depends largely on how he or she approaches planetary observing. For some planetary observers planets are simply a casual interest, and for others they are a lifelong passionate pursuit. The best "planetary eyepiece" is therefore greatly impacted by how an observer approaches planetary observing. For some observers it makes no sense to use an eyepiece with a small AFOV and tight eye relief to observe a planet, as it appears just as wonderful in a comfortable wide-field eyepiece. However, for other observers, only best-in-class planetary eyepieces are considered adequate for their very demanding approach to planetary observing (e.g., Zeiss Abbe Orthos, Carl Zeiss Jena Orthos, Pentax SMC Orthos, Pentax XOs, TMB Supermonocentrics, Astro-Physics Super Planetary, or VERNONscope Brandons).

Comparing One Eyepiece to Another

Finally, an activity that many amateur astronomers enjoy is to compare one eyepiece with another to help them make a choice on which eyepiece to buy or sell. Observers will often borrow eyepieces from their fellow hobbyists or bring their own eyepieces to star parties to compare against other interesting eyepieces that are shared at the telescopes.

Fig. 3.6 Two 5 mm eyepieces ready for a shootout competition (Image by the author)

Whenever an observer chooses to compare two eyepieces, it is critical that he or she level the field as much as possible, so the observations can be as valid as possible. As a first rule, observers should always strive to compare two eyepieces in the same telescope, on the same evening. This ensures that all the telescope optics being used are exactly the same, that the sky conditions are the same, and that their own health and visual performance will be the same. In this way, the only variable in the test is narrowed to the eyepiece.

It is also important to ensure the eyepieces compared have freshly cleaned optics. Too often an observer will not realize that an older eyepiece has accumulated a layer of fine dirt or haze that is not readily apparent, whereas the newer or less used eyepiece being compared is thoroughly clean. Any difference in cleanliness of the eye lens or field lens of one eyepiece can therefore make an observer incorrectly conclude that the eyepiece is not as bright as the other eyepiece and even wrongly conjecture that the eyepiece must have better coatings. In addition, any small speck of dust on the eye lens of one eyepiece could severely compromise how sharply it presents the view compared to another eyepiece that has a clean eye lens. Cleanliness of the optics is therefore critical.

In addition to cleanliness, it is important that the focal lengths of the eyepieces being compared are exactly the same. If the focal lengths are off even by a small amount, it can mean that one eyepiece is producing an exit pupil that might be 10 % brighter, even though the focal lengths are almost equal. As an example, comparing an 8.0 mm eyepiece to an 8.5 mm eyepiece will not provide valid results because this 0.5 mm difference in focal length means that the exit pupil of the 8.5 mm eyepiece will show an image 13 % brighter than the 8.0 mm eyepiece. The observer may then wrongly conclude that one eyepiece is brighter and also appears higher in contrast because of some difference in the coatings or the glass used, when in fact the difference is from nothing more than this very small difference in focal length. They may even wrongly conclude that the 8.5 mm eyepiece seems warmer in tone compared to the 8 mm eyepiece, all due to the small brightness differences in the exit pupils these two eyepieces produce in the telescope.

Although comparison of eyepieces may be an enjoyable and rewarding activity for many, caution should always be taken to use observing strategies that minimize or eliminate any differences in the entire optical chain, from the sky conditions all the way to the eye of the person making the assessment, as even the smallest of differences can have significant and unrealized impacts. And it is also important to realize that once the field has been leveled so that the eyepiece is the only changing component, these comparisons are really a comparison of differences of the optical "system" and not of just the eyepiece. Since it is very difficult to know what the telescope versus the eyepiece is contributing to the image, the results of comparison tests can only be extended to other optical chains that are very similar. As an example, if a comparison of two 5 mm eyepieces was conducted using a single 8 in. SCT telescope, the observer's reports on the quality of the views through each of the eyepieces, as well as their differences, can more reliably be extended to what other observers who have 8 in. SCT telescopes may find, but not what other observers who use Newtonian, refractor, or even larger or smaller aperture SCTs may find. Therefore, when reviewing the comparison reports of other observers, it is important to note the telescopes they used, as well as all other conditions during their comparison to make an assessment of how closely their observed differences may or may not be true for your unique optical chain of equipment.

Eyepiece Accessory Considerations

Just as the eyepiece is considered an accessory to a telescope, there are also devices that are considered accessories for the eyepiece. The Barlow lens and the amplifier lens (sometimes referred to as a telecentric lens), along with filters are the most common types of eyepiece accessories. A lesser known accessory, the Zoomset, is a new development created by Günther Mootz in Germany, which is an accessory that brings a new and exciting level of flexibility to the eyepiece.

Barlows and Amplifiers

The Barlow and amplifier lens are devices used to increase the magnification produced by the eyepiece with the telescope. Barlows are generally a cemented negative doublet (e.g., two lenses cemented together that diverge light rays), which in addition to providing more magnification can also slightly alter the AFOV and eye relief characteristics of an eyepiece.

An amplifier, on the other hand, is generally made of a pair of cemented doublets (e.g., two pairs of cemented doublets separated by an air gap). Amplifiers generally preserve the AFOV and eye relief characteristics of the eyepiece. Amplifiers sometimes go by the term of "telecentric lens," and a well known amplifier is the Tele Vue Powermate line.

Fig. 3.7 Examples of various Barlows. Left to right: APM 2.7× ED, Tele Vue 2×, Siebert 2.5×, GSO 2.5×, University Klee 2.8× (Image by the author)

Barlows and amplifiers are marked with a magnification factor that indicates how they will affect the eyepiece. Some common magnification factors available are 1.6×, 2.0×, 2.5×, 3×, and 5×. If an eyepiece produces 100× in a telescope, when that eyepiece is used with a 2× Barlow, the resulting eyepiece and Barlow combination will now produce 200× in that telescope. An alternate way to think about how a Barlow or amplifier will impact the eyepiece is focal length instead of magnification. In the previous example, if the eyepiece that produced 100× in the telescope had a focal length of 15 mm, then when the eyepiece is used with a 2× Barlow, the new focal length of the eyepiece and Barlow combination can be thought of as the focal length of the eyepiece divided by the magnification factor of the Barlow (e.g., 15 mm ÷ 2×=7.5 mm). The following formulas describe how to calculate the impact of using a Barlow or amplifier with your eyepiece:

1. $$\text{Magnification[new]} = \text{Magnification[eyepiece]} \times \text{Magnification[barlow]}$$

Notes: Magnification[new] means the increased magnification when using the eyepiece with the Barlow in the telescope. Magnification[eyepiece] means the magnification the eyepiece produces in the telescope. Magnification[barlow] means the magnification factor that is on the Barlow (e.g., 2×, 3×, etc.).

2. $$\text{FL[eyepiece]} = \frac{\text{FL[eyepiece]}}{\text{Magnification[barlow]}}$$

Notes: FL[eyepiece] means the focal length of the eyepiece in millimeters; Magnification[barlow] means the magnification factor that is on the Barlow (e.g., 2×, 3×, etc.).

3. $$\text{FL[new]} = \text{FL[telescope]} \times \text{Magnification[barlow]}$$

Notes: FL[new] means the new effective focal length of the telescope with the Barlow. FL[telescope] means the focal length of the telescope in millimeters. FL[telescope] means the focal length of the telescope in millimeters.

To illustrate these formulas in action we will use them with our previous example. If an eyepiece with a 15 mm focal length produces 100× in a telescope, what is the new magnification and new focal length of the eyepiece and Barlow combination using a 2× Barlow? Using formula #1 above we calculate the new magnification as 100 × 2 or 200×. Using formula #2 above we calculate the new focal length of the eyepiece and Barlow combination as 15 mm ÷ 2× or 7.5 mm. This is often called the "effective" focal length of the eyepiece and Barlow combination. One can also think of the Barlow as changing the focal length of the telescope instead of the eyepiece and arrive at the same results (formula #3). If a telescope had a focal ratio of 1,500 mm, with a 2× Barlow one could think of the telescope as now having a 3,000 mm focal length and the resulting magnification calculations with various eyepieces would also be correct.

It is important to note that Barlows have been misunderstood over time. Observers will often hear from others that a Barlow will degrade the performance of the eyepiece. Consequently, some amateur astronomers will profess that it is always better to purchase, as an example, a 5 mm eyepiece rather than use a 10 mm eyepiece with a 2× Barlow. How this misunderstanding developed is open to conjecture, but some believe it developed due to the use of poor quality lenses in Barlows during the early days of their use. More plausibly, however, this misunderstanding is probably more due to the improper use of the Barlow by inexperienced observers. Many times, doubling or tripling the magnification of the telescope and eyepiece can be beyond either what the telescope can handle or what the atmospheric seeing will allow. The truth about the use of Barlows is that given the current quality of optical manufacturing, in the hands of experienced observers, rarely will any degradation in their eyepiece's performance be detected when using a Barlow properly. Barlows should actually be considered a valuable and integral part of an observing system. When a quality Barlow is purchased, rest assured that with proper use and attention to magnification limits, it will provide a quality view as good as the eyepiece alone.

Beyond the capabilities of a Barlow or amplifier with a single eyepiece, these add-ons can also be part of an effective strategy when building an entire eyepiece collection. As an example, suppose an observer determines that he or she requires six eyepieces of 30 mm, 20 mm, 15 mm, 12 mm, 8 mm, and 6 mm to cover all their magnification needs. If he or she were to integrate a Barlow into the solution, the number of eyepieces needed can be reduced by as much as one half! So instead of purchasing six eyepieces, the same range of focal lengths can be achieved by simply purchasing the first three eyepieces with a 2.5× Barlow. When the 2.5× Barlow is used with the 30 mm eyepiece, then the new effective focal length of this combination is 12 mm (using formula #2 we calculate this as 30 mm ÷ 2.5× = 12 mm). Similarly, the 20 mm and 15 mm eyepieces when used with a 2.5× Barlow will provide the other focal lengths desired, 8 mm and 6 mm.

Given how effective a Barlow can be, why don't observers always use a Barlow to achieve their eyepiece focal length needs? There can be several reasons for this, with the most prominent probably being related to observing ergonomics. Some observers do not prefer having the eyepiece extending far out from the focuser, which using a Barlow would cause. If the eyepiece is heavy, then moving it further out from the focuser could affect the balance of the telescope, requiring a re-balance or tightening of movement controls to hold the telescope tube steady. Finally, some observers don't like how the use of a Barlow may affect their "flow" while observing. Some find it interrupting or inconvenient to add components to the eyepiece and perform the focal length or magnification calculations when switching from one eyepiece to another with the Barlow. Instead, they find it simpler and more enjoyable to keep the calculating and manipulating down to a minimum and instead purchase the extra eyepieces to avoid all the flipping and manipulations. So the effectiveness of using a Barlow can be a highly personal preference that is entirely unrelated to any optical considerations, and it is sometimes more of a matter of simply what an individual observer likes and dislikes while observing.

For those that do choose to use a Barlow, a common question is why one would choose to use a Barlow over an amplifier, or choose an amplifier over a Barlow. There are several circumstances where the choice of one or the other may be of better advantage. If the observer is planning to use a Barlow or amplifier with an eyepiece that has eye relief that is long enough that it makes eye positioning difficult, then an amplifier may be a better choice than a Barlow. A Barlow will increase the eye relief of an eyepiece slightly, whereas an amplifier's design will in most instances keep the eyepiece's eye relief at or near to the original design of the eyepiece.

This is typically only an issue if the eyepiece in question already has eye relief long enough to make maintaining proper eye position difficult. For these eyepieces, using a Barlow can exacerbate the situation and the amplifier is the better choice.

Fig. 3.8 The Tele Vue 2.5× Powermate, Dakin 2.4× Barlow, and TMB 1.8× Barlow (Image courtesy of William Rose, Larkspur, CO, USA)

Another reason why an amplifier may be a better choice is if the observer is choosing to use it with a longer focal length wide field eyepiece. Many times longer focal length eyepieces with AFOVs larger than 65° vignette when used with a Barlow. A popular example is the Tele Vue Optics Panoptic line. These 68° AFOV eyepieces, particularly the 24 mm Panoptic, when used with a Barlow will not show a sharp field stop when viewing. Instead, the field stop will be non-distinct and fuzzy in appearance, with the light transmission diminishing in the field of view closer to the field stop. However, when an amplifier such as the Tele Vue Powermate is used, or when the Tele Vue Panoptic-Interface adapter is used in conjunction with the Barlow, then the image of the field stop will be sharp and defined with no perceivable light reduction. Luckily, for shorter focal length eyepiece, even for those with very large AFOVs of 82° or more, there is rarely a similar issue, and a Barlow typically works very well, negating the need to use the generally more expensive amplifiers. As an example, an 11 mm Tele Vue Nagler T6 or Explore Scientific 82 eyepiece, both of which have an 82° AFOV, will show a sharply defined field stop with almost any quality 2× Barlow.

As was previously mentioned, the use of a Barlow extends the eye relief of an eyepiece. Unfortunately, this extension of the eyepiece's eye relief is not always consistent between Barlows, with some extending the eyepiece's eye relief by approximately 20 %, while others may extend it by approximately 40 % (e.g., shorter Barlows, like the University Optics 2.8× Klee Barlow, have been measured to extend eye relief by 39–42 %). Therefore, the observer should not depend on this attribute if the eye relief of an eyepiece is extremely short, as a smaller percent increase may not be detectable as being more comfortable. The following table illustrates how increased eye relief varies between Barlows, and the increase is not dependent on the magnification factor of the Barlow.

	Eyepiece eye relief increase with Barlow			
	12.5 mm Plössl		25 mm Plössl	
	eye relief		eye relief	
Eyepiece's native eye relief	**8.0 mm**	**Increase**	**17.0 mm**	**Increase**
With Tele Vue 2× Barlow	9.5 mm	19 %	20.4 mm	20 %
With APM 2.7× Barlow	9.5 mm	19 %	20.4 mm	20 %
With Klee 2.8× shorty Barlow	11.1 mm	39 %	24.1 mm	42 %

If an observer needs to get more eye relief over what a current eyepiece is giving, the best tactic would be to choose an eyepiece whose design has longer and more comfortable eye relief instead of counting on this result from a Barlow. This is especially true for eyepieces with very short eye relief, as adding 20–40 % more eye relief when it is already very short, around 5 mm for a typical 8 mm Plössl, will still keep its eye relief under 10 mm, which many observers find less than comfortable. Instead, for those observers requiring longer eye relief eyepieces, there are several eyepiece brands and lines where the eye relief is fixed regardless of the focal length of the eyepiece, and these would make better choices. The Pentax XW and Tele Vue Delos are two such lines where the eye relief is a generous 20 mm regardless of the focal length of the eyepiece.

A final consideration when using a Barlow or amplifier is understanding how the location of the field stop in the eyepiece may impact the advertised magnification factor of the Barlow or amplifier. Many observers do not realize that the stated magnification factor of a Barlow and amplifier is only applicable when the field stop of the eyepiece is located precisely at the shoulder of the eyepiece which rests on top of the Barlow or amplifier housing (the shoulder of the eyepiece is where the eyepiece rests when placed into a focuser or into a Barlow of amplifier). If the field stop is above the shoulder, and therefore above the top housing of the Barlow or amplifier, then the magnification factor may increase. If it is below the top of the Barlow or amplifier housing, then the magnification factor may decrease. This is true even for some amplifiers, which are not supposed to be sensitive to offset distances like Barlows are. For example, the Tele Vue 4× Powermate amplifier will increase to only about 4.1× when the field stop is 20 mm above its housing, whereas the Tele Vue 5× Powermate will function at close to 5.6× when the field stop is offset 20 mm above the top of its housing.

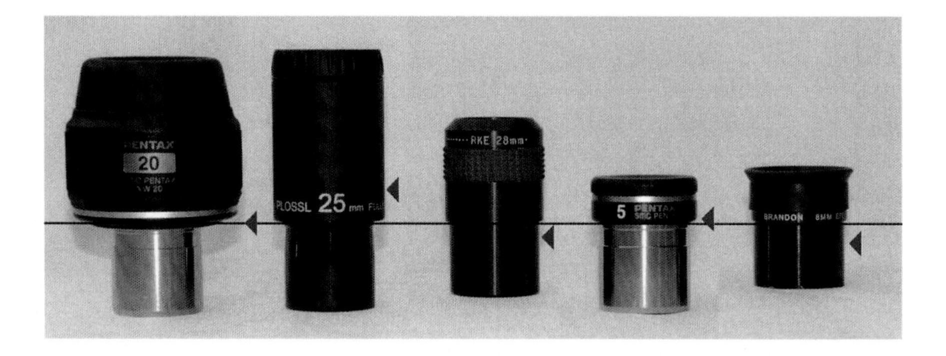

Fig. 3.9 *Red arrow* shows field stop location in various eyepieces. Barlows will produce more magnification when the field stop is above the shoulder, less if below (Image by the author)

If knowing the exact magnification factor is critical for the observing task at hand, then it is also critical to know where the eyepiece's field stop is positioned in the eyepiece's housing, and how any particular Barlow or amplifier being used will behave as the field stop's distance from the top of the Barlow or amplifier's housing is extended or reduced. Unfortunately, the majority of eyepiece manufacturers do not list the position of the field stop relative to the shoulder of the eyepiece's housing. To compound the issue, field stops are sometimes offset by 25 mm or more, depending on the focal length of the eyepiece. Longer focal length eyepieces generally have larger offsets. Tele Vue is an exception, and their website lists not only the field stop offsets for all their eyepieces but also shows the magnification factor behavior of their Barlows and Powermate amplifiers when the field stop is offset by given distances.

The following table shows how several Barlows by different manufacturers operate at different magnification factors with different eyepieces. These magnification differences are all due to the different positioning of the field stop, above or below the eyepiece's shoulder, for each of the indicated eyepieces. Take note of how sensitive the University Optics Klee 2.8× Barlow is to these positional differences. Depending on the eyepiece listed, the magnification factor could be as little as 2.6× or as much as 3.1× when the stated magnification of this Barlow is 2.8×.

	Actual magnification factor					
Barlow	18 mm Baader Genuine Ortho	17 mm Sterling Plössl	14 mm Pentax XW	10 mm Pentax XW	10.5 mm Meade Research Grade Ortho	8 mm Astro-Physics Super Planetary
2.7× APM	2.7×	2.8×	2.7×	2.6×	2.6×	2.6×
2.8× Klee	2.9×	3.1×	2.9×	2.7×	2.6×	2.6×
2.5× GSO	2.2×	2.2×	2.1×	2.1×	2.1×	2.1×
2.5× Siebert	2.6×	2.7×	2.6×	2.5×	2.5×	2.5×
2.0× TV	2.2×	2.2×	2.1×	2.1×	2.1×	2.1×

If the Barlow's focal length is known (usually expressed as a negative number, and the negative is removed when using the formula), then it is easy to determine how the Barlow's magnification will change using the standard formula for calculating a Barlow's magnification:

$$M = 1 + (D \div FL)$$

where FL is the focal length of the Barlow and D is the distance from the top housing of the Barlow to the center of the Barlow lens.

As example, the 2.7× APM Barlow in the table above has a focal length of FL = −62 mm. The distance from the center of the Barlow's elements to the top of the Barlow's housing is approximately D = 105 mm. Using this standard formula the magnification, M, is $1 + (D \div FL)$ or $M = 1 + 105/62$ or $M = 1 + 1.7$ or 2.7×. Unfortunately, most manufacturers do not supply the focal length of their Barlow's, and most manufacturers also do not supply the offset of the field stop of the eyepiece above or below the shoulder of the eyepiece, so without this information this formula cannot be used to calculate the exact magnification a Barlow will produce with a given eyepiece. Fortunately, there is a simple technique that an observer can use to determine the magnification produced with any eyepiece in any Barlow or amplifier.

If an observer wishes to determine the exact magnification that a Barlow or amplifier will produce with any eyepiece he or she owns, he or she should first set up a telescope during the daytime and point it at a tape measure that is placed far enough away so the telescope can come to focus on the tape measure. For this process to work properly the tape measure must completely fill the field of view of the eyepiece, and the entire field stop must be sharply defined in the field of view. With the image of the tape measure positioned across the center of the field of view, count the number of tape measure divisions that are visible from left to right.

To illustrate, assume the observer counted exactly 12 in. on the ruler as it was viewed in the eyepiece's field of view, so the zero inch line was exactly at the leftmost edge of the field of view and the 12 in. line was exactly at the rightmost edge. When the Barlow is added to the eyepiece, the now magnified image of the tape measure shows only 6 in. marks across the field of view, again with the zero inch line exactly at the leftmost edge of the field of view and the 6 in. line at the rightmost edge.

To calculate the magnification factor of the Barlow, simply take the first count, 12, and divide it by the second count with the Barlow, 6, to determine the exact magnification factor of the Barlow.

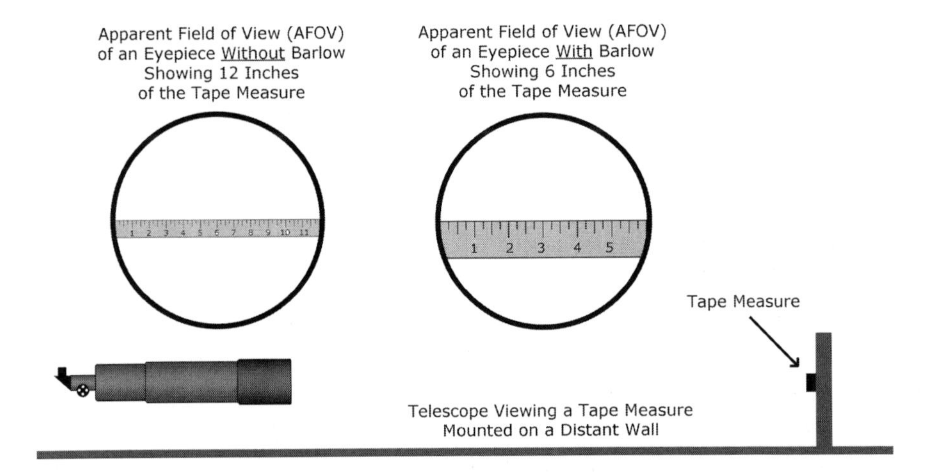

Fig. 3.10 Determining a Barlow's exact magnification factor (Illustration by the author)

In this example, the Barlow's magnification would be:

$$\text{Inches Counted with no Barlow} \div \text{Inches with Barlow} = \text{Magnification Factor}$$

or

$$12 \div 6 = 2.0\times$$

Using the exact same process as above, an observer can also determine how much more or less magnification a Barlow or amplifier will produce for any reasonable offset. To do this he or she would simply repeat the above process, but instead of seating the eyepiece fully into the Barlow, it would be seated 10 mm high to make the observation and calculation. Then he or she would repeat the process again seating the eyepiece 20 mm high. Doing this would then produce three data points to graph to obtain the magnification slope of the Barlow, one with the magnification when the eyepiece was fully seated, then 10 mm high, then 20 mm high.

For those observers who use Tele Vue Barlows and amplifiers, the data is already available on the Tele Vue website. Reviewing the Tele Vue Barlow and Powermate charts, it is seen that 3× Barlow produces 4× magnification when the field stop of the eyepiece is offset approximately 45 mm above the top housing of that Barlow. Therefore, rather than purchasing a completely new Barlow, an observer could instead purchase a much less expensive extension tube that is at or near 45 mm in length. Inserting this extension tube into the 3× Barlow would then convert it to function at 4× whenever needed. Exploiting this common operating parameter of the Barlow, observers can implement many cost effective alternatives rather than purchasing new Barlows or new eyepieces. Note that there is a limit to this capability, as the distance between the Barlow and the eyepiece is extended, off-axis aberrations can increase or the edge of field can vignette due to light loss at the edge. Below is a graph of the magnification changes for several popular Barlows.

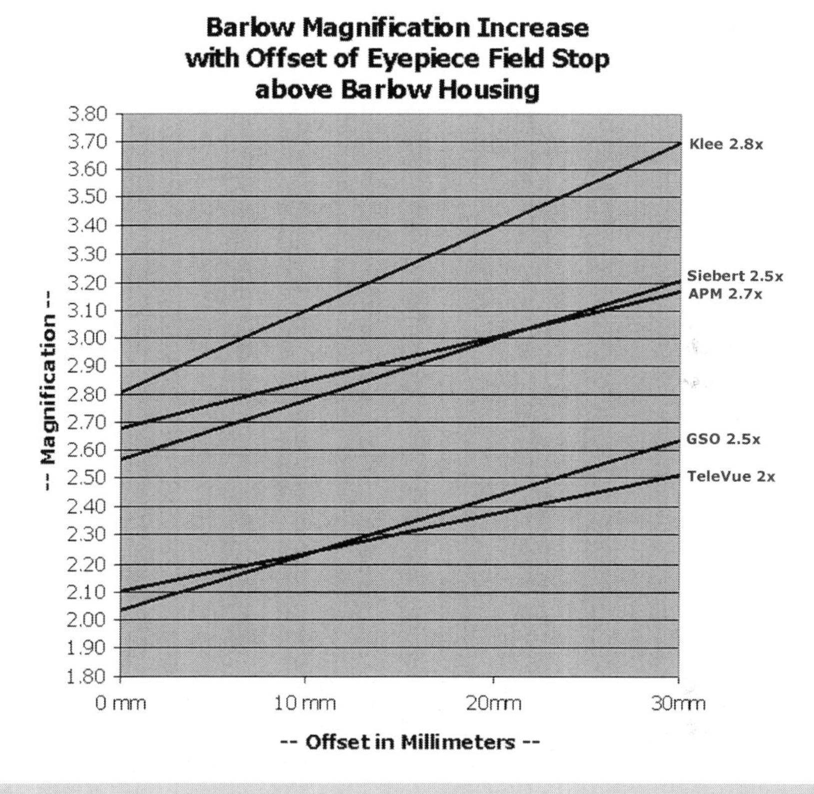

Fig. 3.11 Barlow magnification variation by eyepiece field stop offset (Illustration by the author)

Overall, the Barlow or amplifier is a valuable accessory and can be used to extend the magnification capability of an observer's current eyepieces without having to purchase new eyepieces or worry about any loss of quality in the image produced. As long as the resulting magnification produced with the eyepiece and Barlow combination is within the performance limits of the telescope and within the limits of what the evening's atmospheric seeing will reasonably permit, a Barlow or amplifier becomes a valuable tool, providing the observer with many flexible observing options.

Filters

The use of visual observing filters in conjunction with eyepieces can greatly enhance observations. Eyepiece filters are typically made in both 1.25 in. and 2 in. formats for the two predominant barrel sizes of eyepieces. The filter is typically screwed directly onto the eyepiece barrel; alternately it can be screwed onto the end of most diagonals or onto the end of most 2–1.25 in. focuser adapters. When attached to diagonals or focuser adaptors, the tedious chore of removing the filter and reattaching it each time the observer decides to change eyepieces in the focuser is eliminated. The use of filters offers the observer the added flexibility of potentially reducing glare, increasing contrast, increasing definition, and lessening eye fatigue during an observing session. Today there exist a myriad of filters for just about every observing application, from those for very specific celestial target types, such as emission nebulas, to broad general use filters, such as light pollution filters.

Filters work by blocking out selective undesirable wavelengths of light and allow desirable wavelengths to pass through to the eyepiece. Depending on the wavelengths that are blocked or pass through, the filter will have a different effect. For example, if the observer's desire is to view the famous Horsehead Nebula in Orion, this is best achieved by using a filter that will block all visible wavelengths except those that the nebula reflects back the brightest. The reason why filtration is important for this target is that it is very dim, and the surrounding bright background sky can easily mask its presence. A specialized observing filter called a hydrogen-beta filter was developed that only allows a narrow range of visible light wavelengths—470 to 500 nm—to pass through.

The isolation of this limited spectrum can prove quite useful in assisting the observer to detect faint nebula such as the Horsehead Nebula, the Cocoon Nebula, and the California Nebula. Specialized filters such as these, which only allow narrow parts of the visible spectrum to pass, are generally called narrowband filters. When the filtration task requires a broad spectrum to pass with only narrow parts blocked, these are generally referred to as broadband filters. An example of a broad band filter would be a light pollution filter. These filters typically are designed to block only the strongest part of the spectrum emitted by streetlights, which are often a major contributor to light pollution and sky glow.

Regardless of which filter is used by the observer, it is important that they be made of optical glass and are polished flat to a quarter of a wave accuracy; otherwise the sharpness of the image may be degraded. In addition, some higher quality filters are also multi-coated to suppress any back reflections and ghost images when used. Multi-coating is important when stacking filters to suppress as much as possible the reflections that can bounce between the multiple flat surfaces of stacked filters. Multi-coating is also important because uncoated glass loses approximately 4 % of the light on each surface. Since filters are designed to reduce brightness in specific wavelengths, losing another 8 % of the light across all wavelengths from two uncoated glass surfaces can be counterproductive. It is generally recommended to look for filters that are either fully coated or multi-coated so the maximum brightness of the wavelengths designed to pass is achieved.

Finally, understanding how to properly use each filter is critical to having success with it. The following descriptions of filter classes serve as a good starting point to determine what filters may be most appropriate for specific observing needs.

Calcium II H/K Line Solar Filters: Designed to reveal the Sun's chromosphere by showing the light emitted by ionized calcium (Ca II). The Calcium K-Line filter is generally very narrowly centered around a band pass of 393.3 nm in the deep blue region of the spectrum. The Calcium H-Line is generally centered at a band of 396.9 nm. Since these wavelengths represent a region where the human eye is not that sensitive, if observers are unable to see solar details with the K-Line filter, they should try a H-Line filter instead, as it is closer to where the human eye is sensitive.

Chromatic Aberration Filters: Achromatic refractors, especially short focal ratio designs, suffer from varying amounts of "chromatic aberration," sometimes called false color. This is generally observed as a violet halo around bright objects, particularly planets, bright stars and the Moon. The halo results from all the wavelengths in the visible spectrum not coming to focus at the same point. Chromatic aberration filters block the violet and blue wavelengths of light, thereby reducing the blue halos. These filters do however alter the colors of many celestial targets, which some observers don't like. When used on the Moon, many of these filters give the Moon a yellow cast.

Fig. 3.12 William optics violet reduction VR-1 filter (1.25″) (Image by the author)

Color Filters: Color filters have a long history of use by amateur astronomers. However, use of color filters is for many an "acquired" taste. If the observer is new to astronomy, it is not recommended that he or she immediately acquire color filters. Instead, it is better to develop observing skills and master the proper use of the telescope and eyepiece combinations for observing favorite celestial targets. Once the new astronomer feels adequately skilled using his or her equipment effectively, then he or she can venture into the world of color filters.

Fig. 3.13 Celestron color filters (Courtesy of www.handsonoptics.com. Image by the author)

#8 **Light Yellow** (83 % transmission). This filter is light yellow in color.
Mars: Increases maria contrast.
Jupiter: Improves details of red features in belts and GRS.
Uranus/Neptune: May help bring out atmospheric features when using telescopes with 250 mm (10 in.) of aperture or larger.

#11 **Yellow-Green** (78 % transmission). This filter is yellow-green in color.
Mars: Darkens maria regions.
Jupiter: Helps brings out details in belts.
Saturn: Improves contrast of blue and red colored features, including the polar region.
Uranus/Neptune: May help bring out atmospheric features when using telescopes with 250 mm (10 in.) of aperture or larger.

#12 **Yellow** (74 % transmission). This filter is moderate to deep yellow in color.
Moon: Increases contrast between lunar features in telescopes of 6 in. and larger.
Mars: Increases contrast between red-orange surface features and any blue-green areas such as high clouds by lightening the red-oranges and darkening the blue-greens.
Saturn: Lightens the predominant red-orange features of the atmosphere and darkens the more subtle blue features near the poles.

#15 **Deep Yellow** (67 % transmission). This filter is medium orange-yellow or amber in color.
Venus: Helps to bring out cloud details.
Mars: Further darkens the red-orange surface in a similar way to the #12 and helps in viewing the polar caps and limb haze.
Jupiter/Saturn: Improves visibility of the cloud bands and festoons.

#21 **Orange** (46 % transmission). This filter is medium orange in color.
Mars: Further increases contrast between upper clouds and surface features, like the #15, and help boundaries of maria look more defined.
Jupiter: Helps bring out cloud belt details and GRS details.

#23A **Light Red** (25 % transmission). This filter is medium red in color.
Mercury/Venus: Helps reduce bright blue sky background when observing Mercury during daylight/twilight.
Mars/Jupiter/Saturn: Same results as #15 and #21 filters only more effective than those when used with telescopes with 5 in. and larger apertures due to the reduced transmission of this filter.

#25A **Red** (14 % transmission). This filter is deep red in color and is best for telescopes of 8 in. of aperture or larger due to its reduced transmission.
Mars: Enhances polar caps and dark maria features.
Jupiter: Enhances any lighter hued cloud features that may appear featureless when unfiltered.

#38A **Dark Blue** (17 % transmission). This filter is very deep blue in color and is best for telescopes of 8 in. of aperture or larger due to its reduced transmission.
Venus: Helps bring out cloud details.
Mars: Improves visibility of dust storms.
Jupiter: Strongly darkens any red features in belts and GRS.
Saturn: Strongly darkens reds and oranges in the rings.

#47 **Violet** (3 % transmission). This filter is very deep violet in color. Best for telescopes of 8 in. of aperture or larger due to its reduced transmission, or in smaller aperture 'scopes for very bright objects such as Mercury and Venus.
Venus: Enhances visibility of any transient upper atmosphere phenomenon.
Mars: Improves visibility of the polar caps.
Saturn: Helps make individual rings more distinctive.

#56 **Light Green** (53 % transmission). This filter is medium green in color.
Moon: General enhancing of lunar details.
Mars: Helps bring out details in polar caps and dust storms.
Jupiter: Helps bring out details in polar regions and in cloud belts.

#58 **Green** (24 % transmission). This filter is deep green in color and is best for telescopes of 8 in. of aperture or larger due to its reduced transmission.
Moon: General enhancing of lunar details.
Venus: Improves visibility of any cloud features.
Mars: Helps bring out details in polar caps and of dust storms.
Jupiter/Saturn: Helps bring out details in polar regions and in cloud belts.

#80A **Blue** (30 % transmission). This filter is medium to deep blue in color and is best when used in telescopes of 5 in. of aperture and larger due to its reduced transmission.
Moon: General contrast improvement of all features.
Jupiter: Helps bring out details in cloud belts, festoons, the GRS, and in the polar region.
Saturn: Helps bring out details in the cloud belts and the poles.

#82A **Light** Blue (73 % transmission). This filter is light blue in color and is similar to the #80A except works better for telescopes with smaller apertures due to its improved transmission; can also be a useful filter for star field and star cluster observing from light-polluted locations.

Moon: General contrast improvement of all features.
Jupiter: Helps bring out details in cloud belts, festoons, the GRS, and in the polar region.
Saturn: Helps bring out details in the cloud belts and the poles.

Contrast Filters: There are several versions of contrast filters from boosting filters to ultra-high contrast filters. Their general purpose is to improve the visibility of nebulae, particularly from observing sites that have some level of light pollution. Their transmission bandwidth varies by manufacturer, some having a broader spectrum transmitted, others more restricted to the point that they will perform quite similarly for visual observing, such as the O-III filter. The Lumicon O-III filter has a peak transmission in the 490–505 nm wavelengths, fairly close to the Lumicon Ultra High Contrast filter, which has a peak transmission in the 480–520 nm wavelengths, and more restrictive than the Meade broadband filter peaking in the 460–510 nm wavelengths.

Fig. 3.14 Various contrast filters: Baader contrast booster (*left*), Astronomik H-Beta (*center*), and Meade O-III (*right*) (Image by the author)

For those filters in this category that are more narrow-banded, it is best to choose either one of these or an O-II filter but not both, as they will perform very closely. If the observer already possesses an O-III filter, then it is of little to no advantage to also obtain one of the other contrast filters unless it transmits in significantly more wavelengths. The deep sky objects where observers have reported high contrast filters are moderately to significantly advantageous include:

M1—Crab Nebula
M8—Lagoon Nebula
M16—Eagle or Ghost Nebula
M17—Swan Nebula
M20—Trifid Nebula
M42—Orion nebula
M57—Ring Nebula

M76—Butterfly Nebula
M97—Owl Nebula
NGC281—Pacman Nebula
NGC896—Heart Nebula
NGC1514—Crystal Ball nebula
NGC 2070—Tarantula Nebula
NGC2174—Monkey Head Nebula
NGC2237—Rosette Nebula
NGC2264—Cone Nebula
NGC2359—Thor's Helmet Nebula
NGC2371/2372—Gemini Nebula
NGC2392—Eskimo Nebula
NGC2440—Bow Tie Nebula
NGC3242—Ghost of Jupiter Nebula
NGC6210—Turtle Nebula
NGC6445—Box Nebula
NGC6543—Cat's Eye Nebula
NGC6781—Big Ring Nebula
NGC6888—Crescent Nebula
NGC6905—Blue Flash Nebula
NGC6960–6995—Veil Nebula
NGC7008—Fetus Nebula
NGC7009—Saturn Nebula
NGC7293—Helical Nebula
NGC7635—Bubble Nebula
NGC7000—North American Nebula
IC1848—Soul Nebula
IC5067–5070—Pelican Nebula

Hydrogen Alpha Filters: These are essential if the observer wants to observe and photograph solar prominences. This special filter transmits a very narrow region of the visible spectrum, typically 1/3,000 or less of the visible spectrum's full width. The transmission window of these ultra-narrow filters is centered on the Hydrogen-alpha (H-alpha) line of 656.3 nm that is in the red end of the visible spectrum. These filters are often incorporated directly into the telescope making it dedicated to solar observing. These filters typically involve several components used at multiple locations of the light path on the telescope (e.g., they are more complex than a simple filter placed in front of the telescope).

Fig. 3.15 The Coronado Personal Solar Telescope (PST) (Image by the author)

Hydrogen Beta Filters: These filters enhance the views of nebulae by isolating their sometime predominant hydrogen-beta spectral line centered near 486.5 nm. The result can be a significant contrast between the background space and the light needed to view a few select, extremely faint objects. A steady transparent sky, an 8 in. or larger aperture telescope, and eyepiece-telescope combinations that produce a 7 mm exit pupil are recommended when using this filter. The deep sky objects that observers report respond best to this filter include:

M20—Trifid Nebula
M43—De Mairan's Nebula
NGC1499—California Nebula
NCG2327—Seagull Nebula
IC405—Flaming Star Nebula
IC417—Spider Nebula
IC434—Horsehead Nebula
IC5146—Cocoon Nebula

Fig. 3.16 Astronomik H-Beta filter (1.25″) (Image by the author)

Light Pollution Filters: These filters block the most common wavelengths of light pollution (predominantly produced by mercury vapor lamps) and allow higher transmission of critical hydrogen-alpha and hydrogen-beta lines. Bright, light-polluted skies appear much darker in the field of view and contrast is significantly enhanced, particularly emission nebulae such as the M8 Lagoon and M42 Orion.

Neutral Density Filters (Moon Filters): These simply reduce the brightness and glare that can mask details during lunar and planetary observing. They also do so without affecting the natural colors of the object being observed. This filter is generally used to reduce brightness of objects that are uncomfortable to view at their full intensity, such as the Moon. *Caution:* A neutral density filter is not appropriate for solar observing and will cause blindness if used for that purpose.

Nebula Filters: This is a broad term for the class of filters that can assist in observing various nebulae. For specific nebula filter types see light pollution filters, contrast filters, hydrogen-beta filters, and Oxygen-III filters. For a list of objects where these filters may assist in observing, see the list under contrast filters and hydrogen-beta filters.

An Oxygen-III Filter is a special narrowband pass filter used to enhance the contrast of diffuse nebulae by isolating the passing of light around the O-III wavelengths of 500.7 nm and 495.9 nm. Popular O-III filters typically have a peak transmission in the 485–505 nm wavelengths.

Generally, an O-III filter will accentuate the same deep sky objects (DSO) that a contrast filter will, especially if the contrast filter is marketed as a narrowband nebula/contrast filter or a high/ultra high contrast filter (see list of DSOs under contrast filters). Since the O-III filter is more restrictive, transmitting a narrower bandwidth of wavelengths, for larger aperture telescopes (e.g., apertures of 250 mm and larger) an O-III filter may be a more advantageous choice over a high contrast filter.

Fig. 3.17 Meade O-III filter (1.25 in.) (Image by the author)

Polarizing Filters: These are filters that cut down the glare when observing the Moon, making it easier to see ejecta rays in and around craters. A "variable" polarizing filter will allow the observer to "dial in" a specific amount of neutral density light reduction, often making this a more popular choice than a fixed-reduction neutral density filter.

Fig. 3.18 The Owl polarizing filter (Image by the author)

White Light Solar Filters: Other than the Calcium II H/K line and hydrogen-alpha filters, this is the ONLY other filter that will allow an observer to safely observe the Sun. These filters fit snugly over the front of the telescope's tube and are made out of either a specially coated glass or a Mylar material. These solar filters are not made for the eyepiece and should never used on an eyepiece. Most of these filters show the Sun as blue or blue-white in appearance when used. However, the Thousand Oaks brand white light solar filter tints the Sun in its more natural yellow hue.

Fig. 3.19 White light solar filter for a 102 mm telescope (Image by the author)

Astro Zoom Zoomset (www.astrozoom.de)

The Astro Zoom Zoomset is an intriguing device that was recently invented by Günther Mootz from Germany. This device exploits the characteristics of a Barlow, that when the field stop of the eyepiece is elevated above the top of the Barlow, the Barlow's magnification factor increases. With this device, the eyepiece is attached to the top housing, and the observer's Barlow is attached to the lower housing. The top housing then is twisted, and as it twists it extends or lowers in relation to the lower housing. In effect, this smooth changing of the distance between the eyepiece and Barlow mounted in the Zoomset creates a zoom eyepiece out of the two components. As we have seen in the previous section on Barlows, shorty style Barlows tend to be more sensitive to offsets, creating a greater magnification factor change, therefore these type of Barlows work best in the Zoomset.

Fig. 3.20 The AstroZoom accessory (Image by the author)

Unlike conventional zoom lenses, when the distance is altered between the Barlow and the eyepiece, eye relief can change, focus will change, and depending on the Barlow and eyepiece combination used, the AFOV may even change slightly as the field of view may vignette. These detriments, however, are overcome by the extreme flexibility the Zoomset can offer observers, with an appropriate Barlow allowing them to make virtually every 1.25 in. eyepiece they own into a mini-zoom.

There are several versions of this device available for differing needs. One version takes advantage of the positive–negative designs of many of today's modern wide-field eyepieces (e.g., those eyepiece that have Barlow/Smyth elements built into their barrels). Some versions of the Zoomset allow these eyepieces to function with their housings and barrels mounted separately in the Zoomset. When this is done then no additional Barlow is needed as the eyepiece's own barrel is being used with its Barlow-like elements. This can be advantageous when using one of today's many ultra wide-fields so observers can transform one of their 82° wide fields into a ultra-wide zoom.

As an example, when the large upper housing that encases the Meade 5000 UWA is removed, it reveals a slim optical housing. This slimmed version of the Meade 5000 UWA can then be unscrewed from its barrel, and then both barrel and housing separately mounted in the Zoomset to transform it into an excellent performing 82° ultra wide-field zoom. The 6.7 mm Meade UWA, when slimmed, separated, and mounted in the Zoomset, becomes a 4.7–6.3 mm ultra-wide zoom, or by tuning the positions where the separated housing and barrel of the eyepiece sit in the Zoomset can be adjusted to even be 3.5–4.6 mm, making it perfect for tuning the best high magnification view of globular cluster cores and other celestial objects where the extra context afforded by an 82° field of view is advantageous.

As a cautionary note, not all eyepieces with integrated Barlow/Smyth elements in their barrels can have their barrels safely unscrewed from their housings. Some of these eyepieces use their barrel as a lens retaining device for the lenses in the housing. If this is the case, removal of the barrel will mean the individual lens elements in the housing will fall out. Before using the Astro Zoom Zoomset in a configuration that requires the separation of an eyepiece's barrel from its housing, first confirm with the manufacturer, other knowledgeable observers, or with Astro Zoom if the eyepiece is safe to separate. Also be cautioned that the manufacturer's warranty may be voided if this is done. When the Zoomset is used with eyepiece and separate Barlows, there is of course no similar issue or need for concern.

The Astro Zoom Zoomset is an innovative new eyepiece accessory that, like the Barlow, can be used strategically to expand and enhance the existing collection of eyepieces and Barlows used by the observer.

Care and Cleaning of Eyepieces

Cleaning of the eyepiece, particularly its lens surfaces, is often a source of significant apprehension for many observers. After spending what can be a large sum of money for a valued eyepiece, the last thing any amateur wants to do is to damage the lenses in any way. It is therefore commonplace to hear amateurs express how they have never cleaned their eyepiece to keep it in its best condition. What they ignore, or fail to realize, is the significant amount of light loss that can result because the exposed surfaces of the eyepiece's lenses have built up a layer of grease from accidental contact with fingers or eyelashes and have built up layers of small dust particles or tree sap from the blowing wind. These small areas of grease or dust and particulates on the eyepiece will not only reduce its brightness but will also degrade how sharply it shows lunar or planetary images. Proper cleaning of the eyepiece, as well as proper storage when not in use, is therefore critical to maintaining an eyepiece's top performance capabilities.

Since maintaining an eyepiece's exposed glass surfaces as clean as possible is essential to top performance, how should an observer clean their eyepieces to minimize or prevent damage? As with anything, it is not the act of cleaning that will damage an eyepiece's lenses; instead it is improper cleaning that causes the damage. Current day eyepieces are typically multi-coated, and multi-coatings are extremely hard. Virtually no amount of cotton or common optical cleaning liquid will ever damage the multi-coating on the surface of a lens. However, if the lens is improperly cleaned by pressing and dragging particles of dust and grime into the surface, this is what will scratch and damage the eyepiece's lenses. Typically, proper cleaning should be done with the following steps in this sequence:

1. Examine the surface of the exposed lenses with a magnifying glass or jewelers loupe to assess if there are any particles on the surface.
2. Use compressed air or an air bulb to blow any dirt off the lenses.
3. Use an optical quality brush to gently whisk any particles off the surfaces.
4. Use alcohol that is 90 % or greater in purity, or your favorite optical cleaning liquid like V-Vax product's Residual Oil Remover (ROR), and wet the tip of a clean Q-Tip or other clean and never used cotton product.

Fig. 3.21 Cleaning tools for optics: denatured alcohol, residual oil remover (ror), cotton q-tips, micro fiber lens cloth, air bulb lens brush, jewelers loup (for close-up inspection) (Image by the author)

5. Dab the wetted cotton onto the surface of the lenses, one at a time, and allow it to sit for 30 seconds so that it has adequate time to loosen any grease or grime.
6. Gently swab the lens surface with the wetted Q-Tip to brush and gather the loosened grime and dissolved greases.
7. Repeat steps #5 and #6 two or more times, each time with a new clean and unused Q-Tip, until the surface looks cleared of any and all particulates when examined with your magnifying glass or jewelers loupe.
8. Once all particulates are verified through inspection to be removed, use a new clean and unused Q-tip to make a final wipe of the surface with a very light amount of cleaning fluid, then use the dry end to dry the surface and gently rub it clean of any streaks

As long as the observer is careful in the cleaning of their eyepiece optics, and follows proper procedures, the optics will last a lifetime with no harmful effects from the cleaning. Also, the cleaning method described above is not the only cleaning method, as there are many that recommend different cleaning solutions and different materials to apply and remove the cleaning solutions. What is always critical, however, is to ensure that when any pressure to wipe the surface of the eyepiece's lenses is used, any particles of dust or grime have already been removed and that the material being used to clean the surface is itself clean and free of any particles of dust or grime. It is these small particles of dust and grime that can have a hardness level greater than that of the coatings on the eyepiece's lens, and then cause damage when scraped along the eyepiece's surface.

Once an eyepiece is clean, the next process is to keep it clean. Only when the eyepiece is in the focuser of the telescope should it have its protective top cap removed. All other eyepieces should have their caps on until ready for use or be enclosed in a case. If they are not, then the microscopic dust and particles of tree sap and pollen that are constantly moving through the outdoor environment will inevitably find a home on your eyepiece's lenses. Therefore, the best practice is for eyepieces to remain capped or enclosed in a case until they are ready to be placed in the focuser of the telescope. Doing this will greatly minimize the accumulation of dirt on the eyepiece's lenses over time.

The final process for keeping an eyepiece clean involves how to minimize exposing them to dirt and grease when they are actively being used. Some things to be conscious of when you observe is your breathing and the blinking of your eyes. Two practices that will minimize transferring grease and grime from your own body to the eyepiece is to first make a habit of pulling your head slightly away from the eyepiece just before you need to blink your eye. While this sounds like it may be hard to remember to do, or be a bothersome process, it is actually quite easy and becomes automatic after only a few observing sessions. Eye lashes are coated with a fine layer of oil from the body, and just like the oil on fingers they deposit grease on the lens. In addition to the oils, since eye lashes are constantly exposed to the air, they will naturally collect any pollen or other dust particles that they contact in the air. Therefore, any process which minimizes the possibility of having eye lashes contact the surface of an eyepiece's eye lens when in use will dramatically reduce the amount of dirt and grease that accumulates on the surface of your valuable eyepiece's eye lens.

Finally, observers should also be conscious of their breathing and the position of their nose when observing. Many observers do not realize when they briefly pull away from the eyepiece to take a quick break from observing, that they also inadvertently exhale or breathe onto the surface of the eye lens of the eyepiece. Since our lungs contain all the pollen and dust in the air around us, and add moisture as well to that air, any exhaling or breathing onto the surface of the eye lens is a perfect process to deposit dust and pollen and micro droplets of moisture onto the lens surface, which when they evaporate will leave a small micro-stain.

Therefore take some time to be conscious of your breathing and nose and mouth positions relative to the eye lens of the eyepiece at all times as you observe. A small effort here will greatly reduce the dirt and moisture that can find its way onto your eyepieces. This advice is even more critical if the observer decides to smoke tobacco products while observing, as this just adds orders of magnitude of more particulates in the atmosphere around both the eyepieces and telescopes. It is always the best advice to refrain from smoking near your own optics or those of your fellow observers so that the need for cleaning can be reduced to the minimum necessary, and these very valuable eyepieces can provide lifetimes of enjoyments for observers present and future.

Chapter 4

Popular Eyepieces by AFOV Class

Although there are many aspects that can define the eyepiece, the single most important trait for many observers is the apparent field of view (AFOV) of the eyepiece. This one aspect of the eyepiece, more than any other, influences observers first impressions, and continued impressions as they use the eyepiece. In this chapter, the various eyepiece brands and lines will be categorized by the AFOV class they fall within, 40°, 50°, 60°, 70°, 80–90°, 100°, and 120°. For each class, the pioneers, top performers, and some of those with the most distinctive traits are detailed.

The 40° (and Less) AFOV Class

This first AFOV class of eyepieces is primarily the domain of what is popularly termed as the "planetary" eyepiece, which typically use the Monocentric or Abbe Ortho designs. The planetary eyepiece is classically considered one that uses the least number of glass elements in its design (often only 3 or 4), has a well corrected field of view that maintains orthoscopic qualities (e.g., things appear as they are without distortion across the field of view, linear features do not bend or bow, and angles and magnification do not change), provides as bright and high contrast a view as possible, has acceptable eye relief, and can maintain all these qualities even at the very short focal lengths that planetary magnifications often demand. A contributing factor why this class of eyepieces is the home of the planetary eyepiece is that an optical design that possesses all of these special qualities needs to remain small if off-axis distortions are to be reduced to a level where they are visually difficult or impossible to see.

W. Paolini, *Choosing and Using Astronomical Eyepieces*, The Patrick Moore Practical Astronomy Series, DOI 10.1007/978-1-4614-7723-5_4,

© Springer Science+Business Media New York 2013

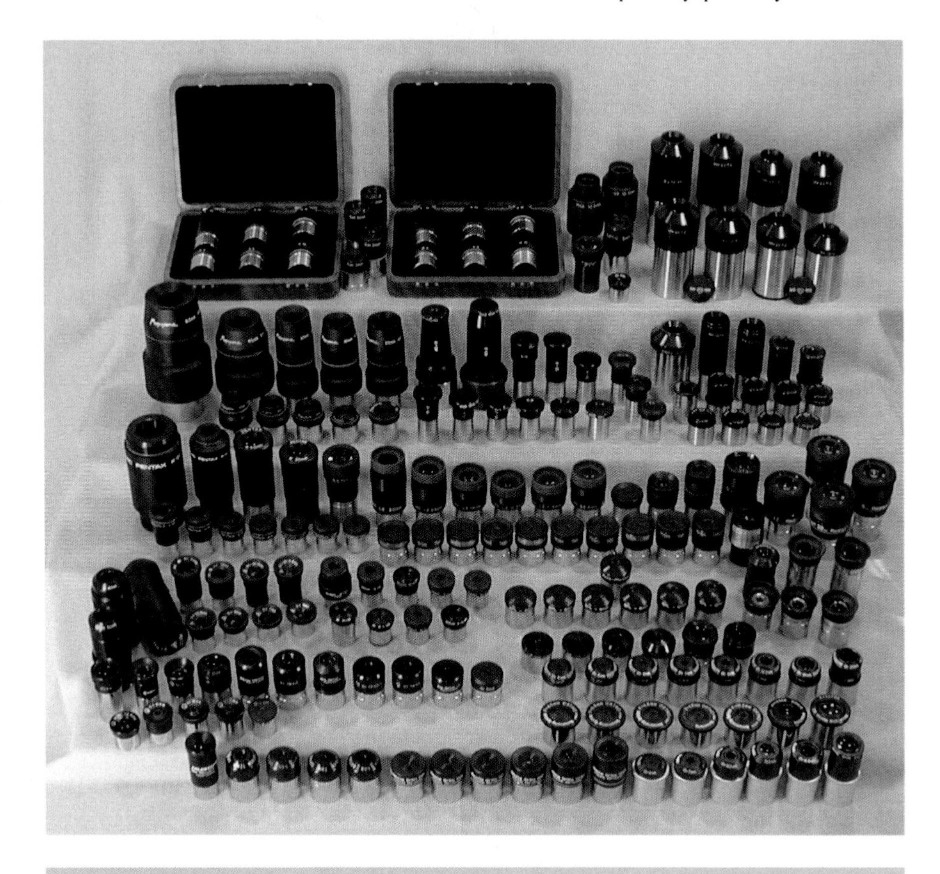

Fig. 4.1 Various high-quality eyepieces for planetary observation by Astro-Physics, Brandon, Celestron, Clavé, Meade, Nikon, Pentax, Takahashi, Tele Vue, TMB, University Optics, Zeiss, and others (Image courtesy of William Rose, Larkspur, CO, USA)

The optical designs that comprise the majority of the most notable eyepieces in this AFOV class are:

Abbe Ortho	(4 elements in 2 groups)
Astro-Physics SPL	(3 elements in 2 groups)
Brandon	(4 elements in 2 groups)
Monocentric	(3 elements in 1 group)
Pentax XO	(5 elements in 3 groups)
RKE	(3 elements in 2 groups)
Sphere Singlet	(1 element in 1 group)

The eyepiece lines in this AFOV class that comprise the above mentioned designs are:

Astro-Physics Super Planetary	Pentax XO
Antares Ortho	Pentax XP
Apogee Super Abbe Orthoscopic	Siebert Optics MonoCentricID

(continued)

(continued)

Baader Classic Ortho	Siebert Optics Planesphere
Baader Genuine Abbe Ortho	Siebert Optics Star Splitter
Cave Orthostar Orthoscopic	Takahashi Ortho
Celestron Ortho	TMB Aspheric Ortho
Couture Ball	TMB Supermonocentric
Edmund Scientific Ortho	Unitron Ortho
Edmund RKE	University Optics HD Ortho
Kokusai Kohki Abbe Ortho	University Optics Abbe (Volcano)
Masuyama Orthoscopic	University Optics O.P.S.
Meade Research Grade Ortho	University Optics Super Abbe
Meade Series II Orthoscopic	VERNONscope Brandon
Nikon Ortho	Zeiss CZJ Ortho
Pentax SMC Ortho	Zeiss ZAO I/ZAO II

By far, the majority of the eyepieces in this category are of the Abbe Ortho design, which is a singlet eye lens group that produces the magnification of the eyepiece followed by a cemented triplet group whose purpose is to correct aberrations. Even though the Abbe Ortho design was not responsible for the visual discovery of any planet (the Huygen design was the eyepiece design in use during most of the discoveries during the classical era of astronomy), it has developed a reputation as "the" planetary eyepiece design. The hallmark of the Abbe design is its lack of distortion. This eyepiece design was invented in 1880 by the German physicist and optical engineer Ernst Karl Abbe. His design goal was to create an optical design that was optimized for taking measurements with a microscope. The design therefore minimizes aberrations that would change the shape, size, or angles of an object anyplace within the field of view. Overall, the Abbe Ortho design has proven to be an excellent choice for the planetary observer. Virtually all of the many Abbe Ortho lines listed in this AFOV class are reported as providing excellent views, attesting to the success of this optical design. Although there are multiple brands listed, examination of many of them will reveal a similar optical mark on many of them. This optical mark is a small circle with the letter "T" in the center that is either stamped into the barrel or applied as a sticker on the eyepiece's housing. Most amateur astronomers believe that this common optical mark indicates that they are from the same original equipment manufacturer in Japan, and observer reports confirm they generally all have consistent performance whether vintage or modern. The Abbe Ortho, regardless of brand name, has earned the reputation as a very good and very consistent performer that rarely receives any negative comments aside from it only having a small 40–45° AFOV.

Fig. 4.2 The circle-t optical mark common to many Abbe orthos (Image by the author)

Within this AFOV class, the most extreme eyepieces are without a doubt the singlet lens eyepieces called the Couture Ball and the Siebert Optics Planesphere. Both of these eyepieces use a single sphere of optical glass as their only lens. This simplest of designs is a planetary extremist's dream as a single lens with only two air-to-glass interfaces and has the theoretical capability of producing the brightest image with the least amount of contrast-robbing scatter. However, as in all optical designs, for every gain there is a loss, so these eyepieces have both very tight eye relief and very small AFOVs.

Fig. 4.3 Couture Ball eyepieces specially built with a whimsical color scheme (Image courtesy of Jim Rosenstock, Fort Washington, MD, USA)

For these eyepieces, although the AFOV can be made to appear larger with an oversized field stop so that targets are easier to acquire for centering in the field of view, in an f/8 telescope only approximately 15° or less of their AFOV will show a sharp image. Because there is only one element, it is impossible to correct the off-axis of these eyepieces. However, while the AFOV is extremely small, some observers report that their on-axis contrast and detail for planetary observing exceeds that of even the most famous line of planetary eyepieces, the Zeiss Abbe Orthos.

Of all the eyepieces in this AFOV class, there are only a small handful that over the years have developed a reputation as being the best-in-class for planetary observing. These are (excepting the sphere singlet, which is in an extreme category due to its very small AFOV):

Astro-Physics Super Planetary (AP-SPL)
Pentax SMC Ortho
Pentax XO

Pentax XP
TMB Supermonocentric
Zeiss CZJ Ortho
Zeiss ZAO I/ZAO II

The first on the list, the Astro-Physics SPL, is another very unique eyepiece in this AFOV class, offered by Astro-Physics from 2004 to 2005. The Astro-Physics Super Planetary is a three-element eyepiece design proprietary to Astro-Physics and/or Aries, the optical company in the Ukraine who produced the SPL lens set for Astro-Physics. This design is purported to keep spherical aberration well within the diffraction limit on-axis for the eyepiece, and off-axis astigmatism is well controlled, being as much as twice as good as other simple two-group designs.

Fig. 4.4 The Astro-Physics Super Planetary (SPL) eyepieces in 12, 10, and 8 mm (Image by the author)

The lenses themselves are very steeply curved, and as a result difficult to manufacture. Although the exact nature of the glasses used or the three-element two-group design itself has never been revealed, design goals for the AP-SPLs were simple—design it to be cleanable with both eye lens and field lens easily accessible to the user, and design it so that internal optics are guaranteed to be "super" clean without any internal debris. In addition to these, the SPL design also had minimal scatter (due to the special attention to have no debris internal in the manufacture), high contrast, and high transmission as design goal characteristics.

The SPL line demonstrates several unique construction elements not typical of most eyepiece lines, including the use of stainless steel instead of chromed brass for the barrel and use of Delrin for the eyepiece housing. The use of stainless steel for the barrel helps reduce scratching from repeated placement in the focuser, as this metal is extremely hard and has no surface coating to wear, and the Delrin housing stays warmer to the touch in very cold weather, reducing the possibility of fogging of the eye lens when used.

The SPLs were produced in 4 mm, 5 mm, 6 mm, 8 mm, 10 mm, and 12 mm focal lengths. The 4 mm focal length was made available in 2005; all other focal lengths were made available the year prior in 2004. The SPLs have an AFOV of approximately 42° so they present a field of view size similar to a standard Abbe Ortho and are broadband multi-coated on all surfaces using electron beam deposition. Eye relief varies between 2.6 mm for the 4 mm focal length eyepiece to 8.9 mm for the 12 mm focal length eyepiece. Due to their short eye relief, these would not be recommended for use while wearing eyeglasses.

The design parameters of these eyepieces were to enable them to be usable in telescopes as least as fast as f/4 focal ratios. The 5 mm SPL's design, as an example, in an f/6 focal ratio light beam yields an on-axis Strehl ratio of .995, 10° off-axis Strehl of .981, and a 15° off-axis Strehl of .945. Although the AP-SPLs were designed to be usable at f/4, their best performance is at f/6 to f/7, which are the focal ratios of the mainstream apochromat refractors.

Transmission efficiency of the SPLs is said to be very high. The coatings on the singlet group have 99.9 % transmission for each surface. The doublet group has 99.84 % transmission on the front surface and 99.75 % on the back surface. Transmission of the internal cemented glass surfaces of the doublet are 99.93 % and 99.98 % for the front and back internal surfaces, respectively. Doing the calculations the overall transmission efficiency of the SPLs is therefore approximately 99.3 %.

Overall, the SPL line had a very short life. Roland Christen stated that they were discontinued because the steep internal curvatures of the lenses did not lend themselves to normal production processes where large batches of lenses could be produced simultaneously. Instead, each lens had to be ground and polished individually, and it was impossible to gain any production advantage. In addition, the final assembly process was overly laborious and required significant assembly-disassembly and re-cleaning to achieve the level of dust-free quality required. Observers generally rank them near the performance of the Pentax SMC Orthos. While generally outstanding in terms of on-axis, some focal lengths exhibit light fall-off near the field stop, as well as some unwanted internal reflections when bright objects are in the field of view. They are an excellent eyepiece in this AFOV class, and the longer focal lengths in particular have a reputation of producing contrast levels as high as what is observed with eyepieces using the single group Monocentric design.

Next, we find that Pentax maintains three eyepiece lines in the best-in-class of planetary performers: the SMC Ortho, XO, and XP. Both the SMC Ortho and XP use the smaller .965 in. barrel standard whereas the XO uses the conventional 1.25 in. barrel. The Pentax SMC Ortho, in addition to its excelling in planetary observations, also has a reputation for showing colorful stars very beautifully. It is therefore an impressive eyepiece to use for viewing colorful double stars or carbon stars. Turning to the Pentax XP, the 3.8 mm focal length is often the showcase eyepiece from this line, as it fills the missing 4 mm focal length spot for the SMC Ortho line. The 3.8 mm XP is reported by observers to have a very low level of scatter and very sharp views, making it perfect for the observer needing a top-tier eyepiece near the 4 mm focal length.

Pentax's last offering in this special class of best-in-class planetary eyepieces, the XO has gained a reputation uniquely its own and is considered by many as among the best planetary eyepieces of all time. These eyepieces, available only in 2.5 and 5 mm focal lengths, were released by Pentax for the 2003 Mars opposition (the "O" in XO stands for visual Observation). Pentax's design goals for the XO series were to create an orthoscopic planetary eyepiece with maximum transmission, maximum contrast, and maximum suppression of optical aberration for telescopes with focal ratios as short as f/4. As a result of these goals, the XO series incorporated high-refraction low-dispersion lanthanum glass elements, Pentax's proprietary SMC full-surface multi-layer lens coatings with the laminated optical elements additionally treated with partial coatings for additional improvements, and the use of computer simulations to determine the best placement for internal baffles and blackening to maximize contrast and fully suppress any stray light.

The 5 mm XO achieves a maximum transmission of 98 %, which means that a 99.6–99.7 % transmission efficiency is required for each air-to-glass interface to achieve the overall 98 % transmission, an amazing achievement for its circa 2003 coating technology. Planetary observers fairly consistently report that the Pentax XO more than met its design goals and that it provides detailed high contrast views on par or exceeding those of even the legendary Zeiss Abbe Orthos. This was reported by the author in a 2009 review of the Pentax XO (available online at http://www.cloudynights.com/item.php?item_id=2065):

> *Where the XO did perform obviously better than all my other eyepieces was in the rendering of white ejecta material in and around craters. As an example, the white wisps of ejecta within the craters around Aristarcus in Schröter's Valley on Luna were both crisper and more detailed using the XO. With my other eyepieces, including the TMB Supermonocentric, the white rays of ejecta within the prominent craters in the valley were observed as a more non-descript lighter and milky-white brighter area. Through the XO, however, a fine latticework of detail was clearly evident within the rays of ejecta extending into the crater, showing significant fine structure and many variations of shading....Overall, I felt the Pentax 5 mm XO was an outstanding world-class performer – and interestingly the only eyepiece I have come across where I can say that its non-planetary purist more than four element design had absolutely no visible degradation to the planetary view to my eye. While not a large AFOV at 44 degrees, it did not feel restrictive like a typical 40° to 42° Abbe Orthoscopic or 30° Monocentric. As a result, I found it very pleasurable to use as a non-eyeglass wearer and I found myself reaching for it more and more to the exclusion of my other eyepieces of similar focal length. As reported, contrast, transmission, scatter, and sharpness were impressive to say the least. As a planetary eyepiece, I felt the Pentax 5mm XO is squarely in the class of the venerable Zeiss Abbe Orthoscopics and could easily stand as the prized part of an eyepiece arsenal for the lunar and planetary enthusiast.*

Next, the TMB Supermonocentric line of eyepieces are considered by most observers to be on par with the very best of anything made for planetary eyepieces, including the Zeiss Abbe Orthos (ZAO). Their brightness and contrast are superb due their minimalist design using only one group of three cemented elements. They are, however, very specialized and not considered as much of a general performer, as they are a specialized planetary performer. This is due to the very limited AFOV of the Monocentric design that is only 30°. In addition, if these eyepieces are used in telescopes with focal ratios shorter than f/6 to f/8, then their off-axis shows moderate field curvature, making the image in that portion of the field of view out of focus. In an f/5 focal ratio telescope, approximately 50 % of the field of view will be out of focus due to field curvature. However, in longer focal ratio telescopes they have a reputation of being elite planetary performers with the unique distinction of having focal lengths only 1 mm apart from 4 mm through to 8 mm.

Fig. 4.5 The complete line of TMB Supermonocentric eyepieces (Image courtesy of William Rose, Larkspur, CO, USA)

Finally, the Zeiss CZJ Ortho and the Zeiss Abbe Ortho (ZAO) are the last two eyepiece lines that are members of this elite best-in-class for the 40° and less AFOV class. Although the CZJ Orthos perform admirably, and are often reported by observers as being on par with the Pentax SMC Orthos, it is the ZAO eyepiece line that is considered the ultimate eyepiece line representing the de facto standard against which all other planetary eyepieces are compared. Since their entrance into the market, the ZAOs have maintained an unparalleled reputation for showing the brightest, scatter-free views with the darkest rich black backgrounds. Besides their planetary excellence, they also provide some of the most striking views of stars. With a complete lack of scatter around stars, their views of open and globular clusters conveys a strong sense of dimensionality, with many more faint stars filling the background and providing the observer with the classic "diamonds on velvet" view that is so often sought after by stellar observers.

The earlier production ZAO-Is (made with high index Lanthanum glass and designed to work well in telescopes with focal ratios of at least f/7) perform on par with the newer ZAO-IIs (designed to work well with telescopes as fast as f/4), with perhaps the slightest edge in brightness going to the ZAO-IIs with their newer anti-reflection coatings. The ZAO-IIs are also distinctive in that their field stops have cut into them small diamond shaped openings at the 90° points around the field of view called micrometric field marks. These openings outside of the field stop are to facilitate locating stars prior to their entry into the field of view for drift timings. Overall, the ZAO-I and ZAO-IIs eyepieces are an example of the pinnacle of perfection in the astronomical eyepiece, and provide a level of precision and clarity in their views that for many observers is without peer.

With the 40° AFOV Class being the domain of the planetary eyepiece, how do some of the members of this class fare against each other? The following excerpt from the author's 2009 competition between eleven different planetary eyepieces provides a good summary of the relative performance among some of the more popular entries in this field (available online: http://www.cloudynights.com/item.php?item_id=1935):

1st Place: The Zeiss Abbe Ortho ZAO-II, TMB Supermono, and ZAO-I;
2nd Place: The Pentax SMC Ortho and Astro-Physics SPL;
3rd Place: Baader Genuine Ortho, Brandon, Radian, TMB Planetary, Sterling Plössl, and University Optics Ortho.

As a set of general conclusions, considering the collective observation results, the following can be confidently asserted:

- Very good atmospheric seeing and transparency are required to bring out many of the rankings and differences noted. On multiple evenings that did not prove good enough to conduct the evaluations, all eyepieces generally showed little difference between them.
- Regardless of an eyepiece's ranking on any test, all the eyepieces produced very sharp, detailed views and every eyepiece served well in the role of lunar/planetary observation.
- Larger aperture instruments or brighter celestial objects are required to show significant differences between eyepieces. Many observing sessions demonstrated this, as reported in results sections of this comparison. This leads to the following sub-conclusions:

  1. Planetary observations can yield significantly improved details using the highest ranked eyepieces in larger aperture instruments.
  2. Lunar observations can yield significantly improved results using the highest ranked eyepieces even in moderate aperture instruments.
  3. The size of an eyepiece's apparent field of view is fairly inconsequential for effective and pleasing lunar and planetary observing. While larger apparent fields of view did make lunar observing more interesting at times, when the focus was on particular lunar features this advantage vanished. This conclusion should be tempered by one's equipment, whether high magnification

tracking is manual or automated, and the observer's skill at manual tracking. If an observer does not enjoy manual tracking and does not have automated tracking, then larger apparent field of views will be an advantage for high magnification lunar/planetary observing.

4. The size of an eyepiece's apparent field of view is an important consideration for viewing extended clusters, nebulae, galaxies, and star fields. As noted in the detailed reporting, when more surrounding context could be viewed for these targets, the observing experience was more dramatic, regardless of other optical considerations.

The ZAO-II ranked highest because it performed every task not requiring a wider apparent field of view the absolute best. It showed the greatest apparent brightness, contrast, and sharpness of all eyepieces. The TMB Supermono was preferred over the ZAO-I because it could pull as much planetary detail as the ZAO-I but would often pull fainter stars. It is also manufactured in 1 mm increments, which is a valued planetary preference. The Pentax and Astro-Physics I considered second tier because they simply would not show as much detail on Saturn as the 1st Place group. Of these two the Pentax was preferred, as it controlled stray light and its scatter was significantly better. In the final grouping, the Baader and Brandon were tied and ranked at the top of their placing because they were subtly sharper on both Saturn and the Moon.

Finally, the Kellner optical design is also a member of this AFOV class for 1.25 in. barreled eyepieces (most 2 in. Kellner eyepieces have AFOVs of 50° or more). Although a very good optical design, it typically requires longer focal ratio telescopes to perform well off-axis, and it is overshadowed by the generally better performing Abbe Ortho. The following eyepiece lines use the Kellner, or Reverse Kellner designs (*Note:* The exception is the Edmund Scientific RKE design, which has radically different curves on the lenses):

Celestron E-Lux (2″ models)	Edmund Scientific RKE
Celestron Kellner	GSO Kellner
Criterion Kellner	Kokusai Kohki Kellner
Orion DeepView	Sky-Watcher Super MA Series
Orion E-Series	Telescope Service RK
Russell Optics (2″ 52 mm and 60 mm)	Unitron Kellner
Sky-Watcher Kellner	

The 50° AFOV Class

The 50° eyepiece class is dominated by the Plössl design, probably the most widely available design on the market. The original Plössl design was invented in 1860 by Georg Simon Plössl and was a pair of asymmetrical plano-convex doublets. In general, the modern Plössl is now two identical symmetrical plano-convex doublets, also sometimes called a Symmetrical or Dialsight design. Typically, these have 50–52° AFOVs, and they work well even in telescopes with fast focal ratios of f/4 to f/5.

The following eyepiece lines fall into the Plössl or Plössl-like category:

Antares Plössl	Meade Series 3000 Plössl
Astro-Professional Plössl	Meade Series 4000 Super Plössl
Astro-Tech High Grade Plössl	Olivon Plössl
Astro-Tech Value Line Plössl	Opt Plössl
Bresser 52° Super Plössl	Orion HighLight Plössl
Carton Plössl	Orion Sirius Plössl
Celestron Omni	Owl Black Night Plössl
Celestron Silvertop Plössl	Parks Silver Series
Clavé Plössl	Sky-Watcher SP-Series Super Plössl
Coronado CeMax	Smart Astronomy Sterling Plössl
Edmund Scientific Plössl	TAL—Symmetrical Super Plössl
Garrett Optical Plössl	Telescope Service Plössl
GSO Plössl	Telescope Service Super Plössl
GTO Plössl	Tele Vue Plössl
Long Perng Plössl	Vixen NPL

As can be seen in the list above, brand names market their Plössl eyepieces as either "Plössl" or as "Super Plössl." In today's market there is no real distinction between the two naming conventions, as they all provide a very similar design using two symmetrical achromatic doublets with an AFOV of 50–55°. The only exception is the Clavé Plössl. This eyepiece line arguably started the consumer fascination with the Plössl design. The Clavé Plössl began manufacture in 1955 from a facility in Paris. Over time it gained a reputation as being one of the best planetary eyepieces made. The design is considered more of a true Plössl as the two doublets are not symmetrical but are asymmetrical. This design is more difficult to manufacture, as it is made of four unique glass lenses cemented together into two different-size achromatic doublets. Therefore, this design requires the manufacture of four different glass elements for each eyepiece, whereas today's more common substitute of the symmetrical design only requires the manufacture of two unique glass elements. Observers today still consider the Clavé Plössl as an outstanding performer, and it is a highly prized find on the used market and often commands extremely high prices.

Fig. 4.6 Multiple generations of the Clavé eyepiece (Image courtesy of William Rose, Larkspur, CO, USA)

As good as the Clavé Plössl's design was, the marketplace eventually settled on the easier to manufacture symmetrical doublet design, and riding on the excellent reputation Clavé created for the Plössl design a multitude of other manufacturers eventually produced eyepieces under the Plössl name. Then, in 1980, a new type of Plössl was introduced by Tele Vue Optics. Using the Clavé Plössl as the standard, Al Nagler, founder of Tele Vue Optics, established a design goal to produce a symmetrical Plössl that would beat the performance of the famed Clavé Plössl. The Tele Vue Plössl's optical design was the result of this goal and is a patented variation of a standard modern Plössl characterized as a Plössl-type eyepiece made with symmetrical achromatic crown and flint doublets, where the external facing flint elements are double concave (U.S. Patent #4,482,217 filed in 1983). This design uses high-index glasses and offers improved off-axis astigmatism and coma control compared to the standard symmetrical Plössl design, which has flat or plano external facing lens surfaces. The Tele Vue design is also purported to be well corrected for spherical aberration and lateral color.

Introduced in 1980 with initial focal lengths of 7.4, 10.5, 17, and 26 mm, the Tele Vue Plössl has come to be considered the "standard" by which all other Plössl offerings are judged. (*Note:* There was a very short number released marked as 10.4 mm, which the original design diagrams indicated, instead of 10.5 mm as was indicated in the patent.) This eyepiece line is an excellent example of a best-in-class eyepiece line in the 50° AFOV class. By 1995, the Tele Vue Plössl line was adjusted and extended to the following focal lengths: 8 mm, 11 mm, 15 mm, 20 mm, 25 mm, 32 mm, 40 mm, and the larger 2 in. barreled 55 mm.

Unlike the current-day models, the very earliest models of the Tele Vue Plössls had smooth housings with very strongly beveled volcano tops and no eyeguards. The initial models produced also had the more conventional flat or plano surface externally facing lenses instead of the concave surfaces noted in the Tele Vue patent filing. Although not yet using the concave surfaced lenses, they still were a lens prescription different from the normal Plössl of the day using special high index glasses to achieve their improved performance.

Fig. 4.7 Rare 10.4 mm Tele Vue Plössl with strong volcano housing and flat/plano lens surfaces (Image courtesy of Tony Miller, Stoney Creek, Ontario, Canada)

The plano surfaced variants were quickly replaced by the patented concave surface lens design (further improving astigmatism correction), and the housing changed to a volcano design that was not so strongly beveled. Sometime during this period, an optical mark also appeared on the housings that amateurs popularly refer to as the Circle-NJ mark.

Fig. 4.8 "Circle-NJ" optical mark that appeared on early Tele Vue eyepieces (Image by the author)

Many of Tele Vue's other eyepiece models in the early 1980s also had this optical marking on the housings. These now vintage Tele Vue eyepieces with the Circle-NJ optical mark have become a much sought after variety by amateur astronomers, and many purport that their performance is slightly superior to previous and subsequent incarnations of the Plössl line. The optical house in Japan that produced this optical mark is also a frequent topic of discussion by amateur astronomers. However, Tele Vue considers the particular optical houses where it sources its optics as proprietary and has never released any information related to this optical mark.

Interestingly, there has been one other non-Tele Vue eyepiece that also carried this Circle-NJ optical mark. For an extremely limited time a small number of Celestron Silvertop Plössls were distributed which had this optical mark instead of their more common Circle-V (Vixen) optical mark. These are a much sought after vintage eyepiece by amateur astronomers and are considered by many to be some of the very best performing vintage Plössls.

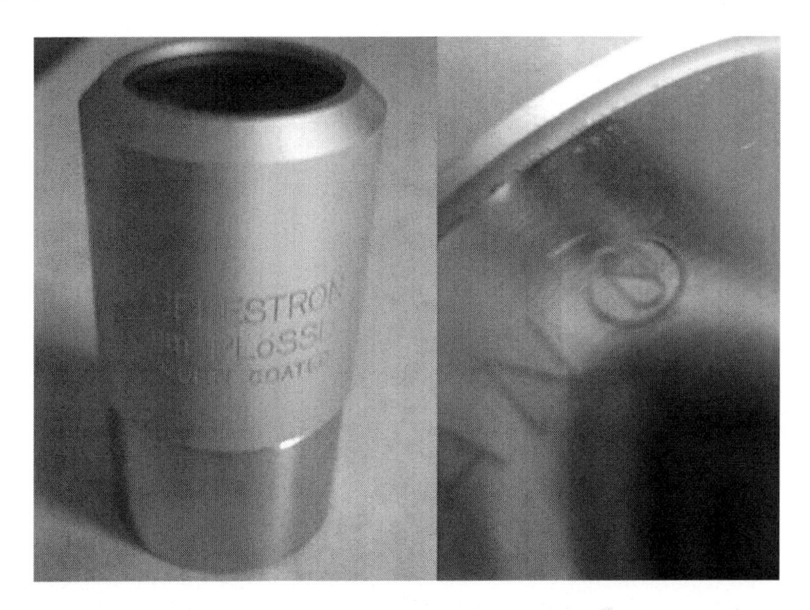

Fig. 4.9 Celestron Silvertop Plössl with extremely rare Circle-NJ stamp (Image courtesy of Mike Bacanin)

To date, the mystery remains as to the source of the Circle-NJ optical mark. Subsequently, these optical marks disappeared from Tele Vue eyepiece lines and only the country of origin stamp remains, which is Japan for the Tele Vue Plössls. (Note that the 55 mm is the exception, as it was originally a Japan sourced optic and later was switched to a Taiwan sourced optic.) The later versions of this line moved from the smooth-sided volcano design to the current day's more ergonomic housing incorporating diamond patterned rubberized grip panels around the housing, foldable rubber eyeguards (adjustable height on the 55 mm), and safety recessed barrels.

Fig. 4.10 Vintage Tele Vue smooth-sided Plössls (Image courtesy of Neville Edward, Norwich, Norfolk, U.K.)

Observers typically report across the board that the Tele Vue Plössl line is an outstanding performer overall, with an excellent high quality build. This eyepiece line provides views that are very sharp both on-axis and off-axis and provides better off-axis performance in shorter focal ratio telescopes than the standard Plössl design offered by most other vendors. High contrast, very high transmission, and low scatter are typically reported as characteristics of their performance. For many observers the Tele Vue Plössls also serve as a very capable classic planetary eyepiece, although limited in some respects as its shortest focal length is 8 mm.

With all of its many strengths, the Tele Vue Plössl line does however have one small weakness. When used with many Barlows, including Tele Vue's own 2× Barlow, the field of view vignettes so the field stop becomes less distinct and the view appears slightly dimmer near the field stop. This issue increases in severity with shorter Barlows such as the University Optics 2.8× Klee Barlow, and appears less severe with longer Barlows. Even though these Plössls vignette when Barlowed, this is not the case if an amplifier instead of a Barlow is chosen to use with this eyepiece line. Therefore, if an observer plans to regularly incorporate a Barlow accessory with their Tele Vue Plössls, use of the Tele Vue Powermate instead will effectively alleviate the vignette issue. The Tele Vue Powermates use a positive doublet lens after its negative doublet lens to redirect the diverging rays towards the normal and thus negates the vignette of the AFOV caused by a typical Barlow with the Tele Vue Plössls. In all other respects, many observers consider the Tele Vue line of Plössls as an example of one of the best, if not the very best, Plössls currently on the market.

In today's market the use of external-facing concave lens surfaces for a Plössl, which is one of the hallmarks of the Tele Vue Plössl's improved design, is shared by a small number of other eyepiece lines as well. Some of the other Plössl lines that use these include: Astro-Tech High Grade Plössl, Smart Astronomy Sterling Plössl, Long Perng Plössl, Meade 5000 Super Plössl, and the Vixen NPL. These particular lines also share reputations for similarly excellent optical performance. The Astro-Tech High Grade Plössl, Smart Astronomy Sterling Plössl, and Long Perng Plössl lines also have a unique distinction over other four-element Plössl lines, having an extended AFOV. All other Plössl lines have an AFOV of 50–52°, whereas the Astro-Tech High Grade Plössl, Smart Astronomy Sterling Plössl, and Long Perng Plössl have extended this to an advertised 55° (closer to 57° when measured). These larger AFOVs, although only a few degrees more than a standard Plössl, are visually more impressive and makes these few Plössl lines perform more like an eyepiece in the next larger AFOV class.

Departing from the traditional symmetrical doublet of the modern Plössl design, there is another design in the 50° AFOV class that is both distinctive and considered an excellent performer. This design is composed of five elements and adds the extra glass element between the two doublets typical of the Plössl.

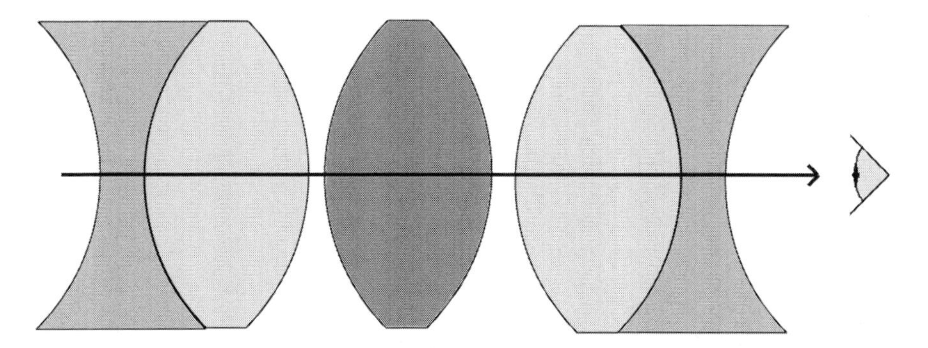

Fig. 4.11 The Basic Masuyama optical design. (lens sizes and curves are illustrative only) (Illustration by the author)

This five-element design was first marketed with the Masuyama eyepieces in the mid-1980s. As the design became successful, many other companies offered designs that were very similar. These have come to be called pseudo-Masuyamas by many amateur astronomers. Most of them have a very solid reputation of providing clean, high contrast views, with a 52° AFOV. The 30 and 35 mm of most of these lines typically has a special reputation for providing very engaging views that have a quality all their own. The eyepiece lines that share this Masuyama or pseudo-Masuyama design include:

- Antares Elite
- Baader Eudiascopic
- Bresser 60° Plössl
- Celestron Ultima
- Meade Series 4000 Super Plössl (pre-1994, smooth sided, 5-elements)
- Masuyama
- Meade Series 5000 Super Plössl
- Omcon Ultima
- Orion Ultrascopic
- Parks Gold Series Plössl
- Takahashi LE
- Tuthill Super Plössl

(*Note:* The Omcon Ultima and Tuthill Super Plössl lines are not detailed in the Desk Reference following this chapter)

The one downside of the Plössl and Masuyama/pseudo-Masuyama designs however are their tight eye relief for the shorter focal length eyepieces. If an observer desires an eyepiece in the 50° AFOV class with longer eye relief, then the following eyepieces lines should be investigated:

- Agena ED
- Astro-Tech Long Eye Relief
- Orion Deep View
- Orion Edge-On PlanetarySky-Watcher LET/Long Eye Relief (LER)
- Smart Astronomy SA's Solar System Long Eye Relief
- Stellarvue Planetary
- Astro-Professional Long Eye Relief Planetary
- Vixen Lanthanum (LV)
- Vixen NLV
- Williams Optics SPL (Super Planetary Long Eye Relief)

Overall, the 50° AFOV class of eyepieces are considered by many observers to be the "workhorse" eyepiece class, providing excellent views at relatively low cost that work well in most all telescope designs and focal ratios. With many of the eyepieces in this class being workhorse, working class eyepieces for many observers, this does not mean they will all provide views that are exactly the same. Instead, each line brings with it a set of unique characteristics making the "feel" of their views slightly individualistic. As the 2012 comparison by the author of a large number of 24–26 mm focal length eyepieces in this class as well as a few from the neighboring AFOV classes revealed, sometimes an eyepiece's individualism proves more important than any cold and clinical assessment of its optical performance (available online at http://www.cloudynights.com/item.php?item_id=2729):

> *At the beginning of this exercise, my thinking was that AFOV would end up being too seductive to ignore and therefore make one of the three 68° wide fields the inevitable obvious choice. But this is not at all the impression I have at the end of this comparison. Instead, I am left with fond impressions provided by a number of the eyepieces because of the particular strengths they each had.*
>
> *The wide fields were of course all wonderful experiences. I particularly liked the precision of the star points in the TV Panoptic, yet often found myself reaching to the 24 mm ES-68 (Explore Scientific 68 Series) as a preferred viewing choice because of the brighter image it provided, impression of a darker background, and the more engaging character it gave the target being viewed. The 24 mm Meade SWA however gave some stunningly wonderful views on particular targets and at times showed better than the wide fields. As an owner of the 24 mm Tele Vue Panoptic used in this comparison, and as much as I hate to say it, I feel of the three wide fields the 24 mm Panoptic would come in last since I am not a stickler for having a perfect star point to the very edge at all times and am more interested in the center field characteristics. But still, the 24 mm Panoptic is an old observing companion so doubt we will easily part, should that happen.*
>
> *The 25 mm ZAO (Zeiss Abbe Ortho) was also a favorite because of the sheer magnificence of the imagery it provided regardless of the target, and how some targets, like Messier 42, were so intense when viewed through the 25 mm ZAO that it made me feel like I was literally within the nebula rather than just observing it from afar. This eyepiece, more than any other, comes closest to being a perfect performer. The 24mm Brandon was also a memorable view, sharing some of the unique attributes of the 25 mm ZAO with so little scatter and providing a "cleanliness" to the view where others fell short. The 26 mm Celestron Silvertop Plössl also ranks high on my list as while not the best of them all, it never failed to satisfy or made me feel lacking. Then the 24 mm UO (University Optics) König, with its reputation for a difficult off-axis, surprised me with the excellence of its image and how marvelous it was when mated with a long focus achromat, showing such a sharp and wide field performance capably coming from such a tiny little eyepiece. Finally, the field of view of the 25 mm UO Ortho, while small and unassuming in comparison to the others, still provided a uniquely special character to its view that was intensely pleasing to behold, not once letting me feel like its field of view, although lacking in size, was ever lacking in heart.*

So my personal winner of this comparison turned out not to be one but many, because each served up something special about their views that others could just not replicate. In the end, the best for my tastes (e.g., 25 mm ZAO, 24 mm ES-68, 25 mm Sterling Plössl, 25 mm UO Ortho) were not determined because of off-axis precision or optical distortion and aberration control but was about where each eyepiece excelled, and how those special characteristics allowed me to understand the celestial target I was observing with both more intimacy and more insight. After my experience with this field of long focal length eyepieces, I can definitely say that more rewards come from the observing journey with the attributes that each eyepiece uniquely possesses, rather than expecting the rewards will come from some destination related to an eyepiece's aberration control or apparent field of view size. As Ralph Waldo Emerson said: "Life is a journey, not a destination." And so it is also true with observing and observing equipment, and each eyepiece was therefore a winner in its own right!

The 60° AFOV Class

The 60° AFOV class of eyepieces, although prolific, has the dubious distinction of being considered by many observers as having an AFOV that is not narrow, yet not what would be classified as a wide field either. As a result, this class of eyepieces often gets overshadowed by either the specialty nature of the 40° and 50° classes or the improved expansive views offered by the 70° and larger AFOV classes. In addition, there is also not any single line within this class that stands out as a line with a strongly loyal following of observers. The eyepiece lines that fall into this AFOV class are:

Agena Enhanced Wide Angle
Agena Wide Angle
APM UWA Planetary
Astro-Professional EF Flatfield
Astro-Professional EF Flatfield
Astro-Tech Paradigm Dual ED
Astro-Tech Series 6 Economy Wide Field
Bresser 60° Plössl
BST Explorer ED
BST—Flat Field
Burgess Planetary
Burgess Wide Angle
Celestron X-Cel LX
GSO (Guan Sheng Optical) Kellner
Kokusai Kohki Erfle
Meade Research Grade Wide Field
Meade Series 5000 HD-60
Meade Series 5000 Plössl
Olivon 60° ED Wide Angle
Olivon Wide Angled Plössl
Orion Edge-On Flat-Field
Orion Epic ED II
Orion Optiluxe (discontinued)
Owl Astronomy High Resolution Planetary
Pentax XF
Rini
Siebert Optics Performance Series
Siebert Optics Star Splitter/Super Star Splitter
Sky-Watcher Extra Flat
Smart Astronomy Extra Flat Field
Surplus Shed Erfles
Surplus Shed Wollensak
Telescope Service Edge-On Flat Field
Telescope Service NED "ED" Flat Field
Telescope Service Planetary HR
Tele Vue Radian
TMB Planetary II
University Optics König/König II/MK-70/MK-80
University Optics Super Erfle

Within this AFOV class, there are several lines that typically receive very favorable comments from observers, and therefore are worth further investigation by any observer considering eyepieces from this class. The Tele Vue Radian is one such eyepiece. This line was introduced in 1998, and in 1999 was actually listed as a Hot Product by *Sky & Telescope* magazine. To-date, this is the only line of eyepieces produced by Tele Vue where the entire line of offered focal lengths was introduced within a single year. The Radian line was also the first Tele Vue line to offer not only long eye relief to better accommodate eyeglass wearers but also a fixed eye relief across all focal lengths, even for the very short focal length 3 mm Radian.

Early observer comments on the Radians agreed that it was a very sharp eyepiece, had one of the better distortion-free orthoscopic fields of view available, and even provided sharp star points right to the field stop even in short focal ratio Dobsonian telescopes. Nevertheless, these reviews also pointed out a moderate level of lateral color, an apparent brightness that was not as good as classic planetary eyepieces like many Abbe Orthos, and a few observers even reported internal reflections.

The Radian's Instadjust adjustable eyeguard system also developed a love-hate relationship among observers, with some liking the mechanism and others finding it too sensitive to accidental repositioning. Finally, like the Nagler T4s, the long eye relief of the eyepiece contributed to its propensity for blackouts when used by many observers. Although these issues presented problems for some observers, particularly the blackouts, many did not encounter these difficulties when using the Radians, as proper use of the adjustable eyeguard could prevent the blackout issues. For the other issues observers were noting, Tele Vue addressed them all—a Pupil Guide device was provided to assist observers with finding the proper exit pupil placement, they published instructions on how to remove the Instadjust housing so observers could tighten the mechanism for themselves so it would not reposition so easily, and newer coating groups were eventually added.

Interestingly, now that many of the Radian focal lengths are discontinued, there seems to be a mini-resurgence of interest in the line by amateur astronomers who frequently purchase their eyepieces on the used market. As time progresses, the Radian line may end up being considered a highly prized discontinued eyepiece as excepting for the lateral color and eye positioning sensitivity issues, it offers a rather uncommon distortion-free 60° AFOV that provides sharp-to-the-field-stop performance in virtually any focal ratio telescope; and all this is accomplished in a relatively compact form factor eyepiece.

Fig. 4.12 Tele Vue Radian eyepieces (Image by the author)

After the Tele Vue Radian was firmly established in the marketplace, a new competitor was introduced, the Burgess/TMB Planetary eyepiece. This line soared in popularity as it provided an AFOV close to the size of the Radian but was offered at a fraction of the cost. Although the Burgess/TMB line was eventually discontinued, it still lived under the following brands:

APM UWA Planetary	Telescope Service Planetary HR
Olivon Wide Angled Plössl	TMB Planetary II
Owl Astronomy High Resolution Planetary	

Today, although similar eyepiece lines are still available under these brands, the TMB Planetary has not maintained its initial popularity. It remains, though, a good value showing a nicely wide AFOV with comfortable eye relief, even in its shortest focal lengths.

Another line that has received consistently good reports from observers in this AFOV class is the Pentax XF series. This line offers only two focal lengths, 8.5 and 12 mm, but it is consistently rated as providing beautiful high contrast views and bright images. Early reports of the Radians compared to the XFs favored the XFs, particularly for deep sky objects such as galaxies and nebulae where the XF showed these faint targets both brighter and in greater extent. However, a group of more recent entries into this class receives observer reports even more favorable than the Pentax XF. These new lines have an optical design that appears very similar to the Pentax XF and are offered under various brandings. (Although it cannot be certain, they are probably all re-brandings of an eyepiece line made by a single original equipment manufacturer.) These newest entries are:

Astro-Tech Paradigm Dual ED	Orion Epic II ED
BST Explorer ED	Telescope Service NED "ED" FlatField
Olivon 60° ED Wide Angle	

These eyepiece lines generally receive slightly better reports than the Pentax XF, stating that the Pentax XF has perceivable field curvature and some slight ghosting on very bright objects, whereas these new offerings do not. These lines are solidly built eyepieces, small in size, with mechanical features not often seen in an eyepiece with such a small form factor. Their housing is completely rubberized, which gives them very secure handling characteristics and also protects them better than full metal housings if accidentally dropped from reasonable distances. All focal lengths sport distinctive colored bands for easy association with their focal length and each has an adjustable twist-up eyeguard.

Considering that their size is only slightly larger than some Plössls, they have additional mechanical features, as well as extended AFOVs and longer eye relief; these eyepiece lines offer a better choice than a standard Plössl for many observers. Their field of view is reported as being well controlled, with very flat fields and little to no field curvature. Lateral color is similarly well controlled and often reported as being non-existent or only very slightly present in the 25 mm focal length. The only distortion of note that is reported is rectilinear distortion, which is a common distortion seen in most eyepieces that are adept at handling shorter focal ratio telescopes. The shorter focal lengths tend to show more rectilinear distortion than do the longer focal lengths of these lines.

Although the Astro-Tech Paradigm Dual ED, BST Explorer ED, Olivon 60° ED Wide Angle, Orion Epic II ED, and Telescope Service NED "ED" FlatField handle shorter focal ratios rather well, they are still not perfect. A focal ratio of f/5 is probably a practical limit for these lines, as some observers report that at that focal ratio, only approximately 85 % of the field of view remains sharp. Compared to other eyepiece lines, there are several observers who have reported that a quality Plössl may show a very slightly cleaner on-axis view, but of course it will have a smaller AFOV and much tighter eye relief, particularly in the shorter focal lengths. There are mixed reports whether these lines show better views than the TMB Planetary line; however for those observers who already have the TMB Planetary or similar eyepieces, the Astro-Tech Paradigm, BST Explorer ED, Olivon 60° ED, Orion Epic II ED, and Telescope Service NED "ED" FlatField offer a nice complement to the TMB Planetary and similar lines offering longer 12 mm, 18 mm, and 25 mm focal lengths.

On various celestial targets, observers report the Astro-Tech Paradigm, BST Explorer ED, Olivon 60° ED, Orion Epic II ED, and Telescope Service NED lines to be excellent generalists, performing well on double stars, star clusters, nebula, and even in a planetary role. They provide vivid color performance for colorful double stars and have excellent transmission for dim nebulae and bringing in dim stars in globular clusters. Given their very good performance across many target types, medium wide 60° AFOV, and low cost, many observers find these excellent candidates for binoviewers and a nice step up in performance and ease of use over a typical Plössl eyepiece.

Finally, the Celestron X-Cel LX and Meade HD-60s are other well regarded lines in this AFOV class. For the X-Cel LX, user reports generally comment that they control off-axis distortions and aberrations well, and that they maintain pleasingly sharp star points off-axis, even in shorter focal ratio telescopes.

Overall, although the 60° AFOV class of eyepieces shows no single line that is clearly distinctive or superior, it does show a contingent of very solid performing offerings. And although this class of AFOV does not provide a "wide-field" experience, it does offer a much larger than Plössl AFOV with an off-axis that has less distortion than a typical wide field and with eye relief more comfortable than a typical eyepiece in the 40° and 50° AFOV classes. If an observer is not looking specifically for optimum planetary performance but is instead looking for a solid performing generalist eyepiece that works exceedingly well in a broad range of focal lengths with medium-wide views, then the 60° AFOV class offers numerous excellent choices.

The 70° AFOV Class

Like the 60° AFOV class, this AFOV class is also very prolific, containing a broad range of entries and capabilities:

Agena Super Wide Angle	GSO Superview
Antares Classic Erfle	GTO Proxima
Antares W70	Long Perng 68° Wide Angle
Astro-Professional SWA	Meade 4000 SWA
Astro-Tech AF Series 70	Meade 5000 SWA
Astro-Tech Titan II ED	Meade Series 4000 QX
Astro-Tech Wide Field	Nikon NAV-SW
Baader Hyperion	Olivon 70° Ultra Wide Angle
Celestron Axiom	Olivon 70° Wide Angle
Celestron Ultima LX	Opt Super View
Denkmeier D21/D14	Orion Expanse Wide-Field
Explore Scientific 68 Series	Orion Premium 68° Long Eye Relief
Garrett SuperWide Angle	Orion Q70 Super Wide-Field
Orion Stratus Wide-Field	Tele Vue Wide Field
Owl Astronomy Enhanced Superwide	Telescope Service Expanse ED
Pentax XL	Telescope Service RK
Pentax XW	Telescope Service SWM Wide Angle Eyepieces
Russell Optics 2″ Series	TAL Super Wide Angle
Siebert Optics Observatory	Telescope-Service Wide Angle
Siebert Optics Ultra	Tele Vue Wide Field
Sky-Watcher AERO	TMB Paragon
Sky-Watcher PanaView	University Optics 70°
Sky-Watcher Ultra Wide Angle	Vixen Lanthanum Superwide (LVW)
Surplus Shed Wollensak	William Optics SWAN
Tele Vue Delos	William Optics WA 66°
Tele Vue Panoptic	

Although there are many eyepieces in this AFOV class, the one eyepiece line that arguably set the standard for this class was the Tele Vue Wide Field. The various focal lengths of the Tele Vue Wide Field were introduced from 1982 through 1984: 15 mm, 19 mm, 24 mm, 32 mm, and 40 mm (see U.S. Patent #4,525,035). These were an innovative design for the period that was a very successful line with observers due to its significantly sharper field performance over the popular "wide field" Erfle and König eyepieces of the time. Its off-axis, however, had moderately large amounts of uncorrected rectilinear distortion, angular magnification, as well as some astigmatism. With the 1989 introduction of the Paracorr coma corrector, Al Nagler became increasingly intolerant of the off-axis distortions and aberrations in the Wide Field line. Consequently, from 1992 through to 2003 Tele Vue replaced the aging Wide Field line with an improved version that had better off-axis performance, including reduced off-axis astigmatism and a slightly larger

AFOV of 68°. This new line was called Panoptic. The full Panoptic line was comprised of the following focal lengths: 15 mm, 19 mm, 24 mm, 22 mm, 27 mm, 35 mm, and 41 mm.

Over the years, several of the Panoptics have gained a significant reputation among observers. The 24 mm Panoptic probably holds a most notable position as being one of the best performing, maximum TFOV, 1.25 format wide-field eyepieces available. Even though the 24 mm Panoptic was brought to market as far back as 2002, there is still no 1.25 in. eyepiece in this focal length/AFOV class today that produces a sharper view across the entire field of view. It has become the de facto wide-field standard for a maximum TFOV1.25 in. barreled eyepiece (e.g., its field stop, like that of a typical 32 mm Plössl, is the maximum size feasible for a 1.25 in. barrel eyepiece). In addition to this distinction, it also has a nicely compact form factor and fairly comfortable eye relief as well, although it and a few others in the Panoptic line may have eye relief too tight for some eyeglass wearers.

In addition to the 24 mm Panoptic, the 27 and 35 mm Panoptics have similarly developed very loyal followings, with many observers considering these also as highly engaging best-in-class eyepieces that they have no hesitancy in recommending to others. Finally, the 41 mm Panoptic shares the distinction, along with the 55 mm Plössl, as having the largest field stop in a 2 in. barrel. The 41 mm Panoptic's field stop measures a full 46 mm in diameter, allowing it and the 55 mm Plössl to generate more true field of view than any other Tele Vue offered.

Fig. 4.13 24 mm Tele Vue Panoptic, a favorite for many with binoviewers (Image courtesy of William Rose, Larkspur, CO, USA)

Finally, the now discontinued Panoptic is distinctive from others of the line in that it had a dual 1.25 in./2 in. skirt where the 2 in. portion could be removed converting the eyepiece into a 1.25 in. only barrel. Tele Vue no longer uses a skirt like this, and all of their eyepieces with dual skirts are fixed.

Fig. 4.14 Tele Vue 22 mm Panoptic with its removable 2 in. skirt (Image courtesy of Phil Piburn, Excelsior Springs, MO, USA)

Overall, the Panoptic line represents a "safe bet" when choosing an wide-field eyepiece. Rarely are observers disappointed with their performance regardless of the telescope they are used in, even with a focal ratio of f/4.5. The 15 mm Panoptic, now discontinued, is probably the only exception, as its eye relief was much tighter than any others of the line. As a result, it never developed a following like the rest of the line. The Tele Vue Panoptics, after more than 20 years since their initial introduction with the 35 and 22 mm focal lengths, continue to be a well-loved and well-performing line among amateur astronomers, providing impressively pleasing views despite the emergence of the many newer technology 100+ degree ultra-wides available today—a long-standing stature few other eyepiece lines have maintained.

Although the Tele Vue Wide Field and Panoptic established the standard for high performance characteristics of this AFOV class, these are by no means the only top performers in this AFOV class and today are both challenged and surpassed by others. The top-tier of performers for this AFOV class today includes the following eyepiece lines:

Denkmeier D21/D14	Pentax XL
Explore Scientific 68 Series	Pentax XW
Meade 4000 SWA	Tele Vue Delos
Meade 5000 SWA	Tele Vue Panoptic
Nikon NAV-SW	

Of these lines, the Explore Scientific 68 Series, Meade 4000 SWA, and Meade 5000 SWA are the most direct competitors to the Tele Vue Panoptic, all providing outstanding performance in telescopes of all focal ratios. The Nikon NAV-SW is the newest entry and with initial observer reports shows itself as a very refined performer. The Denkmeier maintains an outstanding reputation, particularly the 14 mm, which shows an AFOV larger than its advertised 65°, but it is a limited line having only two available focal lengths. However, the longest standing competitor in this AFOV class that has earned a reputation as the second standard, besides the Panoptic, to be compared against for 70° eyepieces is the Pentax XW line.

Pentax introduced the predecessor to the XW line in 1996. This line had a 65° AFOV and was called the XLs. They would eventually set the stage for Pentax's improved design that Pentax introduced in 2003, the 70° AFOV the Pentax XW. Today, the Pentax XWs are considered by many as the new de facto standard of performance in the 70° AFOV class for the 10 mm and shorter focal lengths. The reputation for these focal lengths is a comfortable 20 mm eye relief, low scatter, high contrast, color neutral images, and performance that is sharp to the edge of the field of view. In the 14 mm and longer focal lengths the Pentax XWs show varying degrees of field curvature so they cannot claim as good off-axis performance as the Tele Vue Panoptics. However, many observers still prefer the Pentax XWs due to their comfortable viewing ergonomics and superbly engaging views. Overall the Pentax XWs are arguably the best-in-class for 70° AFOV eyepieces, commanding a very large and very loyal following of amateur astronomers and being able to handle all observing situations excellently, from wide-field vistas to the very demanding role of planetary observing.

Today, the newest top-tier entry into this AFOV class that is challenging all the other top competitors is the 72° Tele Vue Delos. This newest line has positioned itself to challenge the Pentax XW's current position as the arguable leader of the short focal length 70° eyepiece class. The inspiration behind the Tele Vue Delos line was to create an eyepiece with Tele Vue Ethos-like performance but with a constant 20 mm of eye relief. Tele Vue optical designer Paul Dellechiaie developed the design, and Al Nagler named the line to recognize Paul Dellechiaie as the principal designer of both the Delos and Ethos lines (e.g., "Del"os and "Del"lechiaie).

Observer reports confirm that many of the marketing claims about the Delos are indeed true, with performance being sharp to the very edge of the field, no visible lateral color, and no discernible off-axis aberration. Light transmission is also reported as being very high, with experienced observers placing it between the performance of the Pentax XW or Tele Vue Ethos and that of the famed Zeiss Abbe Orthoscopic. Some field reports indicate that with critical observation, a greater portion of extended objects such as galaxies and nebulae are visible in the Delos versus the Ethos or Pentax XW. Some also indicate that difficult "averted vision only" stars in the field of view are more easily maintained in the Delos than in the Ethos or Pentax XW.

For observers who are familiar with the Pentax XW line, which has a reputation for being extremely comfortable to view through, some observers that have focal lengths from both eyepiece lines have commented that the Delos is close to the comfort of the Pentax XW but not quite the equal. However, it needs to be noted that eye positioning sensitivity (e.g., eyepiece exit pupil behavior) can many times be highly impacted by physiological differences in the human eye between individual observers. This may explain why there are conflicting reports, with some observers finding the Pentax XWs less sensitive to eye position, while other observers find the Delos to be less sensitive.

For eyeglass wearers, the Delos is perhaps the most accommodating eyepiece of the many Tele Vue lines. The Delos has a constant 20 mm eye relief across all available focal lengths. Generally, observers who wear eyeglasses report that eye relief in the 15–17 mm range is the minimum needed to still see the entire AFOV of a wide-field eyepiece while wearing eyeglasses. The Delos's longer 20 mm eye relief therefore accommodates eyeglass wearers very well. For those observers who have astigmatism and do not wish to wear their eyeglasses while observing, the Tele Vue Dioptrx is also compatible with the entire Delos line.

Overall, a substantial majority of observer reports agree that the Tele Vue Delos is a more refined Ethos, having improved ergonomics, handling, and mechanics. The critical observer may also see slight improvement in transmission and contrast over a same focal length Ethos—all of this coming from an eyepiece that when compared to the Tele Vue Ethos is smaller, weighs less, and has the added benefit of an adjustable eyeguard. And for those observers who have some of the newest extreme ultra-short focal ratio telescopes, such as the Webster/Lockwood 28 in. f/2.75 Dobsonian, the Delos is reported as being the best performing wide-field in this new class of ultra-short focal ratio telescopes.

Besides the very top performers, many of the other eyepieces in this AFOV class will only perform as well in their off-axis when used with telescopes that do not have very short focal ratios (typically f/6 and longer). These include, but are not limited to:

Astro-Tech Titan II ED	Sky-Watcher AERO
Baader Planetarium Hyperion	Sky-Watcher PanaView
Celestron Axiom	Telescope Service Expanse ED
Celestron Ultima LX	TMB Paragon
Orion Stratus	Vixen LVW

Overall, the 70° AFOV class offers the observer with varied choices, running a broad range of features and focal lengths comprehensive enough to satisfy most observers. This AFOV also serves as an excellent example of wide-field observing that can be accomplished at better price points than is available from some of the larger AFOV classes of 80° and more. For many observers the 70° AFOV is considered optimum, providing a better balance of both wide-field immersion and ease of use, than those eyepieces that offer larger AFOVs.

The 80°–90° AFOV Class

In the market today, the 80–90° AFOV eyepieces come in two distinct varieties, those that provide well corrected off-axis performance in both short and long focal ratio telescopes (referred to as Class-1 for the purposes of this discussion), and those better suited for only the longer focal ratio telescopes of f/8 or greater (referred to as Class-2 for the purposes of this discussion).

Typical Class-1 Competitors

Antares Speers-WALER Series
Astro-Professional UWA
Celestron Axiom LX
Celestron Luminos
Docter UWA
Explore Scientific 82°
Meade Series 4000 Ultra Wide Angle
Meade Series 5000 Ultra Wide Angle
Orion MegaView Ultra-Wide
Sky-Watcher Nirvana UWA
Sky-Watcher Sky Panorama
Takahashi UW 90°
Tele Vue Nagler
Williams Optics UWAN

Typical Class-2 Competitors

Agena Ultra Wide Angle
Apogee Widescan
Astrobuffet 1RPD
BW Optik Ultrawide
Kokusai Kohki Widescan
Moonfish Ultrawide
Olivion 80° Ultra Wide Angle
Owl Astronomy Knight Owl Ultrawide
Sky-Watcher Ultra Wide Angle
Surplus Shed Wollensak
TAL—Ultra Wide Angle
University Optics 80°
University Optics Widescan

In the late 1970s, Al Nagler, founder of Tele Vue Optics, designed a sufficiently radical new optical design for an astronomical eyepiece in this AFOV class that he then patented (U. S. Patent #4,747,675 and #4,286,844). This design was given his name, following the proud tradition within the industry to name significant design milestones after the inventor (e.g., Erfle, Kellner, König, Plössl, and others). Based on his design, the first Nagler eyepieces, introduced in 1980, were a scaled series in the focal lengths of 4.8, 9 and 13 mm. These first Naglers are often referred to by amateur astronomers as the Type-1 Naglers; however, Tele Vue refers to them as "the original series." The Nagler line of eyepieces then set the standard in 1980 for the Class-1 category of competitors within this AFOV class, and it remains the standard to this day.

Fig. 4.15 The Nagler Original Series in 4.8, 9, 11, and 13 mm (Image courtesy of Steven Cotton, FL, USA)

The original 13 mm Nagler was actually patented as a 90° eyepiece. The initial 13 mm production model released had an 84° AFOV. Later models were then stopped down to the advertised 82° AFOV for sharpness, eye relief, and exit pupil reasons. The original production series of Naglers also only had the eye lens multi-coated, with all other glass surfaces single-coated. Tele Vue continued to evolve and improve the Naglers with both more focal lengths, newly designated design types, and fully multi-coated optics, as this became a standard in the marketplace. From 1982 through to 2003 additional focal lengths of the original series were introduced, as well as the improved Type-2, Type-4, Type-5, and Type-6 Nagler designs. Although never released, Tele Vue also developed a prototype of a 25 mm Type-2 Nagler, but it required too much focuser in-travel to be practical and therefore never made it to production.

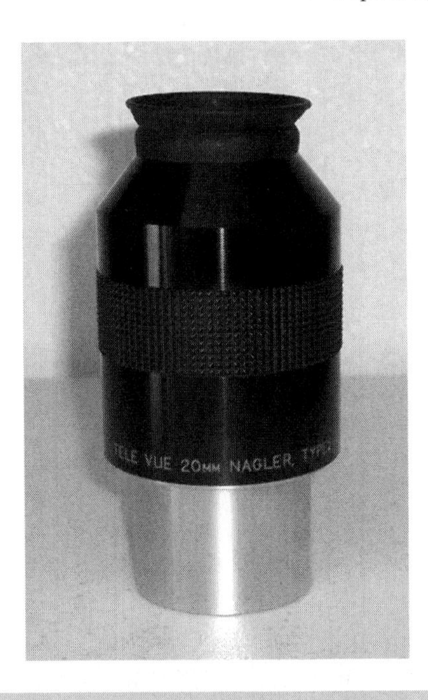

Fig. 4.16 Discontinued Nagler 20 mm Type-2 (Image courtesy of Steven Cotton, FL, USA)

Fig. 4.17 Prototype 25 mm Nagler Type-2s (Image © 2013 Tele Vue Optics—www.televue.com)

The original design goal for the Nagler eyepieces was to achieve a wide-angle perspective similar to that of naked-eye vision with high contrast and a sharp image across the entire field of view that was also comfortable when viewing. The desire was to try to make the telescope virtually "disappear" for the observer, and provide the sensation of a "spacewalk" among the stars, instead of a porthole view of the stars.

The Nagler line finally stabilized with a full range of 82° AFOV eyepieces being available in focal lengths of from 2.5 to 31 mm, and this line is currently comprised of the Type-4, Type-5, and Type-6 design designations, and two zooms. (Note: The two Nagler zooms only have 50° AFOVs.) The Type-5 (long focal length) and the Type-6 (short focal length) designs provide the full range of focal lengths required for productive observing. The Type-4 design provides a set of mid-range focal lengths, which have longer eye relief and variable height eyeguards for observers who require those special attributes.

Fig. 4.18 The long eye relief Type-4 Nagler (Image courtesy of Konstantinos Kokkolis, Piraeus, Athens, Greece)

Overall, observers report very positively on the Nagler eyepieces. This line very much established the Tele Vue reputation for providing eyepieces with off-axis field of views that are sharp to the field stop. Today, the Naglers remain a benchmark eyepiece for off-axis performance, against which all other ultra wide-field designs are ultimately compared.

As good as the off-axis correction of the Nagler line is, this does not mean that there are still not minor issues to the critical eye. Observers report that for some focal lengths, there is sometimes a little more lateral color than preferred (e.g., the 5 mm Type-6 and 12 mm Type-4). Lateral color tolerance is, however, a very personal matter, so while some observers report it as sometimes bothersome in these eyepieces, others report that it is minor and goes unnoticed during routine observing.

Eye positioning sensitivity is also a common report by observers for some of the focal lengths of the Nagler line (e.g., mostly for the long eye relief Type-4 designs). The most common report is related to kidney beaning and blackout occurring when the eye is not in the correct position. Also reported is that eye positioning affects lateral color visible in the eyepiece, so lateral color may only occur if the eye is not in the correct position for the eyepiece.

When an eyepiece has long eye relief, such as the Nagler T4s, it is common for some observers to experience these issues. Many times, these issues are caused not by any design fault with the eyepiece but simply by the observer having a more difficult time keeping his or her eye positioned over the eyepiece at that long eye relief point. To aid observers with acquiring and maintaining proper eye position with long eye relief eyepieces, Tele Vue will generally have adjustable height eyeguards on these eyepieces and also supply an accessory called the pupil guide. This device is simply a plastic disk with a hole cut in the center that is inserted onto the top of the eyepiece. The smaller aperture hole of the pupil guide then becomes a natural guide to help observers position their eyes correctly over the precise center of the eye lens. Observers do not have to worry that this device will block any of the image formed by the eyepiece when the pupil guide is positioned correctly, and the 15 mm diameter hole in the pupil guide is therefore more than sufficient.

To properly use adjustable eyeguards and pupil guides, the observer should have the pupil guide in place and then start with the eyeguard fully extended. Starting this way prevents the observer from initially getting too close or "inside" the exit pupil. While looking on-axis, the observer should then lower the height of the eyeguard until he or she can just detect the field stop using peripheral vision. At this point the observer has attained the proper eye position for the eyepiece.

Although not the original intent of the pupil guide, the device does have other positive effects besides aiding the observer in obtaining the proper eye position over the eyepiece. With the pupil guide in place, contrast is also improved by blocking stray light from entering the eyepiece through the eye lens. The device also helps prevent dewing of the eye lens in humid conditions, and it helps to keep the eye lens clean by providing a measure of protection from accidental contact by fingers or eyelashes. Given that observers who choose to use the pupil guide generally report that it does successfully assist in reducing eye positioning issues, if purchasing the Type-4 Naglers on the used market it is therefore advisable for the purchaser to ensure the supplied pupil guide is included with the eyepiece. Otherwise, the purchaser can contact Tele Vue customer support to purchase a replacement pupil guide if so desired.

For many observers, the Type-4 design is also reported as the Nagler design that provides the most engaging viewing experience and the strongest "spacewalk" experience. Many observers attribute this to the longer eye relief and, as a result, the more comfortable viewing. For the Type-6 (12 mm eye relief) and for the shorter focal length Type-5's (10–12 mm eye relief), it is easy to find reports from observers that there is sometimes difficulty viewing the entire 82° field of view unless they position their eye uncomfortably close to the eyepiece. If an observer needs more generous eye relief, then a careful selection from the Nagler line with the integration of a Barlow would be all that is needed to build an effective range of Nagler focal lengths with longer eye relief. To illustrate, an observer in need of more generous eye relief could choose the 31 mm Type-5 (19 mm eye relief), 22 mm Type-4 (19 mm eye relief), 17 mm Type-4 (17 mm eye relief), and 12 mm Type-4 (17 mm eye relief), then add a 2× Barlow for use with the 17 and 12 mm Naglers to achieve effective focal lengths of 8.5 and 6 mm. This scenario, as well as others utilizing a Barlow, can provide the entire Nagler experience with satisfying long eye relief.

In addition to the special reputation the Type-4 Naglers have among observers, there are two other Naglers that also enjoy a special reputation. These are the 16 mm Type-5 and the 31 mm Type-5. More than anything else, observers generally marvel at the power the 16 mm Type-5 packs into such a compact package. It is by far the smallest of the Naglers, yet given its small size it gives up nothing related to the impact of its expansive view.

Fig. 4.19 The smallest Nagler—the Tele Vue 16 mm Type-5 (Image courtesy of Tony Miller, Stoney Creek, Ontario, Canada)

Second, there is the 31 mm Nagler Type-5. This Nagler has the distinction of providing the widest TFOV of any Tele Vue eyepiece excepting the 55 mm Plössl and the 41 mm Panoptic. Even though those eyepieces show a slightly larger TFOV, the expansive field of view of the 31 mm Nagler means an observer can view almost the same TFOV and do it with higher magnification, a smaller exit pupil, and as a result obtain a richer dark background due to its increased magnification over the 41 mm Panoptic or the 55 mm Plössl. Observers generally accept the 31 mm Nagler Type-5 as "the" benchmark eyepiece for maximum TFOV ultra wide-field eyepieces. Many observers also consider it a must-have eyepiece. This is particularly true for observers with large aperture short focal ratio Dobsonian telescopes that require the largest TFOV possible using focal lengths shorter than 40 mm so exit pupils can stay at reasonably small diameters to avoid potential shadowing from the secondary mirror.

Turning from astronomical observing to daytime terrestrial observing, there is one anomaly observers report for the 31 mm Nagler Type-5—the field stop is reported to show what observers have termed as a "ring of fire." This ring of fire is typically a bright blue glow right at the field stop of the eyepiece. Because of this phenomenon, observers who do significant daytime terrestrial observing often recommend the 35 mm Tele Vue Panoptic in lieu of the 31 mm Nagler for this type of viewing. The ring of fire has no other optical impact other than it being a distraction for some observers during daytime observing. However, a few observers have reported the phenomenon under certain circumstances during astronomical observing as well, such as when observing the full Moon. For the most part, though, this phenomenon is not considered to take anything away from the 31 mm Nagler's otherwise spectacular performance, especially in the astronomical observing setting for which it was designed.

Overall the Nagler line of eyepieces is an excellent performing ultra wide-field and has earned the reputation for being the benchmark eyepiece line for sharp off-axis wide-field performance. Any new eyepiece entering the marketplace to compete in the 82° AFOV class will inevitably be compared against the venerable Nagler. Observers new to the Nagler can expect the famed "spacewalk" experience, especially if they are only used to eyepieces showing smaller AFOVs. Although the eye relief of the Nagler line is not generally characterized as tight, observers should expect a tighter than comfortable experience on some models. However, with practice, most quickly overcome any reluctance to view so closely while taking in an expansive 82° field of view. Although the field of view of the Nagler has been surpassed with newer generation 100° and 120° eyepieces, the Nagler's sharp-to-the-edge spacewalk field of view continues to receive accolades from observers.

Besides the trend-setting Nagler, there are many other noteworthy members of this AFOV class providing extremely well corrected fields of view. All 15 eyepiece lines listed in the Class-1 category can be counted on to provide very good views even on shorter focal ratio telescopes, making any an excellent choice. Noteworthy is the Explore Scientific 82° Series. More than any other eyepiece line, the Explore Scientific 82's tend to often compete head-to-head with the Nagler line. Observer reports frequently attest to the Explore Scientific 82's as having 95 % of the performance of the corresponding Nagler focal length at much less cost. Overall the Explore Scientific 82° Series maintain an excellent reputation among observers.

The Astro-Professional UWA, Orion MegaView Ultra-Wide, Sky-Watcher Nirvana UWA, and Williams Optics UWAN are an interesting sub-class in the 80–90° AFOV class. They all share a very similar if not identical form factor, the exact same focal lengths, and consistent performance reports from observers. The 28 mm focal length of these lines is often compared to the Nagler 31 mm Type 5 as a candidate for a 2 in.eyepiece to provide the maximum TFOV possible. The 16 mm of these lines often receives reports placing it on par with the Nagler 16 mm T5 in terms of optical performance.

Fig. 4.20 The Nagler 31 mm Type-5 compared to the competing William Optics 28 mm UWAN (Image courtesy of Tamiji Homma, Newbury Park, CA, USA)

Finally, the two newest members of this class are the Docter UWA and the Takahashi UW. Both of these receive extremely high praise as modern standard-setters having the most up to date optical technologies incorporated and besting some of the older favorites. The Docter eyepiece, however, has received a number of accolades from observers stating that its pinpoint star images, ability to bring in structure of galaxies, and contrast is simply breathtaking, even when compared against other best-in-class lines in this AFOV class. Unfortunately there is only a single focal length made in the Docter series, but many observers indicate that it is a most desirable member of the 80–90° AFOV eyepieces available.

The 100° AFOV Class

In 2007 the amateur astronomy community was introduced to a new kind of production eyepiece, with a larger AFOV than even that of the then largest, 82°, AFOV eyepiece lines. Today, although the 100° AFOV class is not as prolific as the smaller AFOV classes, it is still represented by two sub-classes. (Note that the Meade XWA is not represented in either of the lists because it has only recently been released.)

Explore Scientific 100°	Tele Vue Ethos
Nikon NAV-HW	

Those 100° eyepieces more suited for longer focal ratio telescopes for the best performance off-axis are:

Agena Mega Wide Angle	TMB 100
Orion GiantView 100°	Zhumell Z100

Tele Vue Optics, with their introduction of the Ethos line, pioneered for the amateur astronomy community the 100° production eyepiece that was also well corrected in the off-axis. Early in 2006, this concept and definition for a line of fully corrected 100° eyepieces was first proposed by Tele Vue president David Nagler to Paul Dellechiaie, Tele Vue's then protégé optical designer. Paul then designed the basic eyepiece to the point that the original design goals were fulfilled. The design was then presented to Al Nagler, who offered additional guidance to fine tune the design. In an interesting side note, Al Nagler's evaluation of Paul Dellechiaie's initial design spurred him to look at his own alternative design for a fully corrected 100° eyepiece. Ultimately, Al Nagler chose Dellechiaie's design as the one that best fulfilled all of David Nagler's design criteria. The Ethos was then introduced to the amateur astronomy community at the 2007 Northeast Astronomy Forum (NEAF) as the first production eyepiece for amateur astronomy with a 100° AFOV. The 3.7 and 4.7 mm focal lengths of the Ethos line were then later added and are specially designated as Ethos-SX, since their AFOV is set at a larger 110°. The "SX" designation stands for "Simulator Experience," as Al Nagler had developed a 110° field of view for the U. S. Apollo program's Lunar Landing Simulator.

Fig. 4.21 Example of Ethos AFOV Advantage—100° (*left*) 82° (*center*) 70° (*right*) (Underlying Messier 31 astrograph courtesy of Mike Hankey, Freeland, MD, USA—www.mikesastrophotos.com. Illustration by the author)

In the field, amateur astronomers give high praise to the Ethos. Many report amazement at the transmission and contrast of the line, and that counterintuitively its performance seems to rival eyepieces with significantly fewer glass elements. Some observers also report that compared to the Nagler line, the Ethos line shows tighter star points, blackout or kidney beaning is much better controlled than in some of the early Nagler models, and viewing comfort is much improved. Finally, a few observers report that the color rendition of the Ethos line appears superior to that of the Nagler line. The difference in color rendition between the 13 mm Ethos and 13 mm Nagler, as an example, is described by some as an "eye opener" with the Ethos presenting carbon stars, red giants in globular clusters, the pinks in Messier 42, and the colors in Jupiter's Great Red Spot as all being better rendered with the Ethos.

Although the eye relief of the Ethos line is fixed at 15 mm, a number of observers report that it is either barely sufficient, or not sufficient, to view the entire field of view when wearing eyeglasses. However, Tele Vue makes a device called the Dioptrx that fits to the top of most of their eyepieces to correct for astigmatism in the human eye, so eyeglass wearers with astigmatism can view without their eyeglasses. The Dioptrx is not compatible with all Tele Vue eyepieces, but it is compatible with the entire Ethos line. If an observer chooses not to use a Dioptrx and wear eyeglasses instead, then some of the longer eye relief Naglers may be a more effective choice for long focal length, ultra wide-field viewing (e.g., 12 mm Nagler T4, 17 mm Nagler T4, 22 Nagler T5, and 31 Nagler T5 all have eye relief greater than 15 mm).

The Ethos was, and still is, a groundbreaking eyepiece development. It provides a stunning 100° AFOV that is well corrected from center to edge, even in very short focal ratio telescopes. Given the performance and AFOV milestones the Ethos design has achieved, it has developed a unique following among amateur observers. Many actually have based their entire eyepiece strategy around this line, since its massive 100° AFOV allows some interesting advantages.

Although most of the discussion among amateur astronomers is typically around its ultra-wide AFOV, the Ethos line has several other attributes not often found together in a single eyepiece line. These advantages are: a large and consistent-sized eye lens, the ability to physically convert the entire line to a uniform 2 in.-only barrel size (using a supplied accessory in the case of the Ethos SX models or the optional 2 in. extender for the dual skirted Ethos), a uniform weight distribution between the various focal lengths that reduces telescope balancing issues, uniform optical performance and eye relief, and finally, focal length spacing that provides fairly uniform TFOV increments. The Tele Vue Ethos line is, therefore, not only a very capable performer but is also an extremely well thought out design for use at the telescope.

Following suit with the introduction of the Tele Vue Ethos, the Explore Scientific 100° and Nikon NAV-HW lines were subsequently released. Like the Ethos, these lines also provide a very refined level of performance that includes excellent off-axis performance in shorter focal ratio telescopes. The Explore Scientific 100° Series, like the company's other lines, are often reported by observers as providing 95 % of the performance of the standard-setting Ethos at a reduced price point. Like the Ethos, the Explore Scientific 100° Series also has a loyal following of observers, and this line has a unique distinction among the 100° AFOV eyepieces of also being completely waterproof (it can be safely submerged in water) and being nitrogen purged to prevent internal fogging of the elements in humid conditions.

The Nikon NAV-HW line has the distinction that when purchased it operates at two different focal lengths, so it is like getting two eyepieces. Each eyepiece includes an "EiC" tele-extender lens that is specifically designed for each Nikon NAV-HW focal length. This EiC shortens the focal length of the 12.5 mm Nikon NAV-HW to 10 mm, and shortens the focal length of the 17 mm Nikon NAV-HW to 14 mm.

The remaining eyepieces in this 100° AFOV class, while offering cost-effective alternatives for consumers, require longer focal ratio telescopes to show a well corrected off-axis. In the case of the Zhumell Z100, observer reports indicate that the 16 mm unit shows approximately 25 % of its field of view with moderate off-axis aberration, which then becomes severe in the last 10 % of the field of view when used in telescopes with shorter focal ratios. However, at focal ratios of f/10, the off-axis is reported to clear considerably, showing good performance over 90 % of the field of view. Reports for the 9 mm generally indicate significantly better performance with the off-axis being sharp to the field stop even in shorter focal ratio telescopes.

Given that the Agena Mega Wide Angle, Orion GiantView 100°, and TMB 100 eyepieces all come in the same focal lengths and have similar if not the exact same appearance, many contend that the performance reported for the Zhumell Z100 is equal to these lines as well. Therefore, observers can probably expect similar experiences with softer off-axis performance in short focal ratio telescopes but very good off-axis performance in slower focal ratio telescopes. Some observers report the eye relief feels tighter than expected on the TMB and Zhumell lines and that the eyeguard needs to be folded down to see the entire field of view. These line have an advantage, however, in that they are lighter weight than the premium 100° lines, which gives them an advantage where telescope balance is concerned. Overall, these lightweight, low cost 100° eyepieces, while not providing a perfect off-axis performance, are still reported to provide an engaging spacewalk experience. The 9 mm reports rather consistently as being the best of the line even in f/6 focal ratio telescopes; however, the 16 mm needs an f/10 long focal ratio telescope for an acceptable off-axis that is sharp over 90 % of the field of view.

The 120° AFOV Class

This is a brand-new AFOV class for production astronomical eyepieces. There is only one current entry in this class, the Explore Scientific 120° series, which is only available in the 9 mm focal length as of late 2012. Although only a very few initial observer reports are available, those that have reported say that it takes the "spacewalk" to an entirely new level. One observer reports that the AFOV is so wide that their peripheral vision is insufficient and that they actually need to turn their head to observe the edge of the field of view. Another observer comments that this large of an AFOV conveys an impression that objects nearer the edge of the field of view seem to be physically further away in space, a perception that sometimes accompanies naked-eye viewing when there is a lack of reference objects close by, such as in the desert.

Fig. 4.22 The 9 mm Explore Scientific 120° Series next to a 7 mm TMB Supermonocentric for size comparison (Image courtesy of Tamiji Homma, Newbury Park, CA, USA)

For the one eyepiece in this newest AFOV class it is exciting that observers are reporting it as having superb performance, showing an excellent level of detail, contrast, and aberration correction. The only negative comments are with the weight, as this 9 mm focal length eyepiece weighs in at more than even a Nagler 31 mm T5. Given what the potential size and weight of focal length 120° AFOV eyepieces may be, future focal lengths may be restricted to shorter, rather than longer. Only time will tell as manufacturers continue their bold innovative ventures into these exciting new realms of astronomical eyepieces.

Chapter 5

Advice from the Amateur Astronomer Community

Astronomical observation is a proud tradition that dates back much further than the development of the telescope. Ancient Babylonian texts dating back more than 3,500 years record 21 years of the planet Venus' rising and setting. More than 3,000 years ago astronomical observers in China during the Shang Dynasty made observing reports of solar eclipses. A diary of a 12-year-old child in Japan during the 1600s records 4 months of observations of Comet C/1664 W1. Later, large societies organized in the early 1800s, such as the Astronomical Society of London, conducted astronomical research using "gentleman astronomers" instead of professional astronomers.

Then, in 1890, the still active British Astronomical Association was founded specifically for amateur astronomers. Following the British lead, Chicago teenager Frederick Leonard organized the Society for Practical Astronomy in the United States in 1909. Today, the amateur remains a vibrant and active participant in astronomy whether assisting with scientific research with organizations such as the American Association of Variable Star Observers (AAVSO) or assisting the public to become more interested in recreational astronomy through such organizations as the Society for Popular Astronomy. In the series of essays to follow, amateur astronomers from around the world share their many valuable insights, conveyed to you in their own voices. These amateurs range in experience from the beginner to the seasoned, with more than half a century of observing experience, each sharing their own unique perspectives into the world of choosing and using an astronomical eyepiece.

W. Paolini, *Choosing and Using Astronomical Eyepieces*, The Patrick Moore Practical Astronomy Series, DOI 10.1007/978-1-4614-7723-5_5,
© Springer Science+Business Media New York 2013

20 Years of Personal Eyepiece History (1992–2012)

By Christoph Bosshard, Zürich, Switzerland (temporarily in Tokyo, Japan)

My first serious telescope in 1992 was a used 1980s 4 in. Pentax Achromat that came in a nice wooden box including 0.96″ eyepieces. These simple Ortho and Kellner designs seemed good as long as I had no comparison. After several observations with decent 1.25 in. multi-coated Plössl eyepieces, this changed and I bought myself a set to replace the original eyepieces that came with the telescope.

The next step came with a 10 in. Dobsonian that was mostly used observing from remote dark sites. This non-driven mount made observation with the 50° FOV of the Plössl's a bit cumbersome when compared to my friend's all new Nagler eyepieces. So out went the Plössls and in came a whole set of ultra wide-field Nagler's. Fast forward to 2012 where a big refractor on a motorized mount from light polluted skies focuses my observations on planets, double stars and small planetary nebulas. To get the last bit of sharpness and contrast out of my telescopes I am going back to simple and efficient designs such as the Ortho's that I first owned 20 years ago.

My conclusion is that there is not one perfect eyepiece design but a set of eyepieces that suit your telescopes, preferred objects and viewing habits, all of which will change over time.

Achieving Focus with Long Focal Length-Rated Eyepieces

By James Spriesterbach, Los Angeles, CA, USA

If you find yourself using an eyepiece rated for long focal ratios (such as a VERNONscope Brandon), in a 'scope that is of low focal length (such as a short Newtonian or Dobsonian), then you might notice that you have problems achieving focus. A quick fix you can try is pulling the eyepiece up the focuser or diagonal shaft about a quarter to half way (don't pull the eyepiece all the way out of the focuser or diagonal of course!), and readjust the focuser until the eyepiece comes into focus. Once you have found the appropriate spacing to achieve focus, you can then use rubber bands or women's hair ties placed around the eyepiece to hold the spacing.

Astronomical Eyepieces: Objects of Pleasure

By Konstantinos Kokkolis, Piraeus, Athens, Greece

Visual amateur astronomy, as a hobby, is an activity undertaken mainly for pleasure, during our leisure time. One can extract pleasure from many aspects of this multidimensional activity, including learning, meditating, collecting, enjoying the outdoors, constructing and communicating with fellow amateur astronomers. However, since we are interested in visual astronomy, the main pleasurable component is observing itself, and, therefore, most positive reinforcement is elicited from the sense of vision. This might explain the fact that, although all the elements of the optical tract play their role and most people agree that the most significant part is the telescope with its design, aperture, and geometry parameters, many visual observers think a lot about astronomical eyepieces. The eyepiece plays the role of the visual interface; it is what the eye looks at and what the eye looks through. Besides being more manageable, generally cheaper and easier to obtain than the rest of the telescope, it is also the most reachable fitment for the eye: the eye reaches the glass, limits are repealed and the telescope feels like a bionic expansion of one's self. In order to enjoy this often ritualistic experience one should take into account some factors other than the absence of aberrations and realism of image. When the exit pupil diminishes the presence of annoying floaters, the eye relief prevents eye strain, the apparent field of view approximates the eye's perimeter and the design and materials used ensure a natural visual feeling, free from major perceivable distortions, then the sense of visual pleasure is maximized. Between the many available eyepieces that are capable of providing such experiences, one of my personal choices, reached through a process of trial and error, is the famed Tele Vue Nagler T4 line. But only your eye is entitled to tell you which might be your "feel good" eyepiece.

Benefits of Wider AFOV Eyepieces in Undriven 'Scopes

By Chris Mohr, Raleigh, NC, USA

In 'scopes mounted without tracking drives, using eyepieces with wider apparent fields of view has practical benefits beyond just providing a more impressively panoramic view. In particular, for any given focal length of an eyepiece, the wider the apparent field of view (AFOV) the eyepiece is designed to have, the wider the true field of view (TFOV) it can be designed to have, at least to a roughly proportionate extent. The wider the true field of view an eyepiece shows, the less often it will be necessary to be distracted from observing an object by having to nudge the 'scope's aim to keep it within the field of view and then wait for any vibration thereby induced to subside before steady observing is again possible. For a given eyepiece in a given undriven telescope, the maximum interval during which an object can remain within in the edge-to-edge field of view before Earth's rotation moves it beyond the field of view is called its "drift time." So how much proportionate gain in "drift time" comes with using an eyepiece with a larger AFOV versus one of the same focal length and a narrower AFOV?

There is a natural yardstick and a useful optical formula that are in combination helpful toward answering the above question about gain in drift time with increase in AFOV. First, the time it takes for 1° TFOV of sky to rotate past a given point is readily determinable: Earth rotates 360° per day, 15° per hour, or 0.25° per minute (or 1° every 4 min). Second, the true field of view produced by any given eyepiece depends on the focal length of the particular telescope it's used in:

TFOV = arcTangent (Field stop diameter in mm ÷ Telescope focal length in mm)

Note: The acrTangent portion of the calculation above can be approximated by multiplying the result of the division by 57.3.

So where's the link in this between drift time, TFOV and AFOV since there's no explicit mention of AFOV in these formulas? The link is that widening the AFOV of an eyepiece design permits widening the field stop (light intake aperture at the bottom of the eyepiece).

However, it is not quite so simple to state a generically applicable formula linking gain in drift time from gain in AFOV, for at least three reasons. First, the extent to which the expansion in field stop corresponds with expansion of AFOV is not reliably uniform across different designs and makes of eyepieces. Second, many eyepiece makers fail to provide accurate (or indeed any) specification of field stop diameter for their eyepieces, and third, there is variance among different designs and makes of eyepieces in the extent to which the field of view outside the central area is sufficiently free of aberrations to provide useful viewing of detail, especially when used in shorter focal-ratio telescopes.

Fortunately, an illustrative, accurately reliable real-world example is available. Tele Vue not only publishes accurate specifications for its eyepieces but conveniently produces 17 mm focal-length eyepieces in 100 deg AFOV (Ethos), 82 deg AFOV (T4 Nagler) and 70 deg AFOV (Delos) designs. Note that although technically the Delos is a 17.3 mm eyepiece, that's close enough for our illustrative purposes.

TELESCOPE: Tele Vue NP-101, focal length = 540 mm, aperture = 101.7 mm

Eyepiece design	TFOV	TFOV difference	AFOV difference
17 mm ETHOS (100 deg)	3.14 deg	–	–
17mmT4 Nagler (82 deg)	2.58 deg	18 % Less	22 % Less
17.3 mm Delos (70 deg)	2.25 deg	28 % Less	30 % Less

Binoviewing on a Budget

By Brendan Cuddihee, Liberty Township, OH, USA

The majority of my experience in eyepieces for binoviewers has come over the last 7 years or so where I have become an avid user of a very high quality recent vintage C8. Most of my viewing at the present is done using Denkmeier binoviewers with the Power Switch as I find that using two eyes not only gives me a sense of three dimensionality but in addition, and perhaps most importantly, a more relaxed and natural viewing experience. The Power Switch gives me three magnifications with each eyepiece I use, a .65 reducer, 2× magnification, and straight through modes. With this type of set up I require sets of two when it comes to eyepieces, thus beginning my quest for eyepieces that were top performers but were easy on the bank account. None of the EP's I use are exotic in any way, the most complex being a five-element design, but the C8 is, by nature, very easy in terms of its demands on an eyepiece concerning off-axis aberrations. The eyepieces I have ultimately kept or in my case bought, sold and re-bought, may not perform as well when used with faster focal ratio 'scopes such as short focus refractors or newts. I also prefer eyepieces with longer eye relief. With this being said, here is a shortlist list of eyepieces that I keep coming back to.

28 mm Edmund Scientific RKE

The RKE's are a three-element modified Kellner design with excellent throughput. The 28 mm has a unique way of framing the star field as one that floats in front of your eyes. In essence, the eyepiece seems to disappear, and all you see is a pool of stars. It is the only eyepiece that I have ever used that gives this effect. It is fantastic for open clusters and nebulae. M42 is a must see in this eyepiece. The 21 mm focal lengths are also extremely sharp but lack the floating effect of the 28s.

24 mm Brandons

These are more expensive than the RKE's but are hand-crafted by Don Yeier of VERNONscope and feature a higher degree of polish than most of their contemporaries. Their attributes include a low level of light scatter, essential for viewing brighter objects such as Jupiter, Saturn and the Moon. They are my eyepieces of choice when it comes to these objects. They also excel in fainter deep sky objects.

Edmund Plössls

I own the 28 and 15 mm pairs of these eyepieces and find that although light throughput is a bit lower than some fully multi-coated eyepieces, their slightly warmer tones and excellent contrast characteristics render Jupiter, Saturn and Mars most beautifully. The color saturation is so rich that they bring out subtler details not seen in other more neutral eyepieces. They give a more "toned" view than the Brandons, but are gorgeous in their own right.

19 mm Smart Astronomy EF's

This is one of the EP's I have bought and sold several times, always hoping to find something a bit better. A couple of years ago I also had purchased a pair of 19 mm Tele Vue Panoptics to test against these. The difference in my C8 was so insignificant as to be inconsequential to me, especially in light of the increased cost of the Tele Vue's. These are my only wide-field EP's with a 65° FOV. Light throughput is very good, and I find that with the Denkmeier power switch set on the reducer mode, I can comfortably fit the entire double cluster in the FOV.

These are a few of the eyepieces that I consistently use and will always have a spot in my eyepiece kit. They are definitely not the most cutting edge, nor are they the most expensive, but I have found that keeping the choice of eyepieces simple is gratifying in itself. Another thing to keep in mind is that these days it is really very difficult to find a poor eyepiece. Most will perform well, with performance variations being more subtle than that of changing the design or size of the telescope being used.

Choosing a Low Magnification Wide-Field Eyepiece

By Erika Rix, Liberty Hill, TX, USA

For incredible, high-contrast wide field views, I use the Astro-Tech 38 mm Titan 70° wide field 2″ eyepiece. It has five elements/three groups for a good field edge and a great 20 mm eye relief. The lenses have high transmission, anti-reflection multi-coatings and the barrels have a black matte finish resulting in low internal reflections and high contrast. At night, when your fingers are cold with minimal light to see by, there are features that can make your observing session stress free such as parfocal eyepieces to reduce time spent refocusing, rubber grips on the eyepieces, and safety grooves in the barrel to prevent it from falling out of the eyepiece holder if the thumbscrews loosen. The Titan sports all of those. The drawback is its weight at 24.8 ounces, resulting in minor adjustments to balancing. I also need to move my head side to side slightly to take in the entire field of view, but feel this is a small price to pay for being able to fit multiple galaxies, large nebulae, or large clusters in the field of view.

Choosing a Zoom Eyepiece

By Erika Rix, Liberty Hill, TX, USA

As a sketcher when time is of the essence to render the views, no eyepiece case should be without a good quality zoom so that an observer may quickly adjust the magnifications for seeing conditions, not to mention having a range of focal length all in one eyepiece. The Baader Hyperion Mark II 8–24 mm zoom is one of my most used eyepieces for ease of use. It gives sharp, high contrast views and has a

50–68° field of view depending on the focal length used, sporting click stops for each adjustment. Clickable stops are a bonus for keeping track of magnification without the need to refer to the markings on your eyepiece, breaking your concentration and hindering your dark adaptation. For observers that have multiple 'scopes, the Hyperion zoom comes with both 1.25- and 2 in. barrels to fit either eyepiece holder. A parfocal eyepiece is desirable to reduce refocusing between settings.

Couture Ball Eyepieces (The Do-It-Yourself Planetary)

By Steve Couture, Belle Mead, NJ, USA

I consider myself to be primarily a planetary observer, though I spend a considerable amount of time also observing globular clusters and the Moon. No expense has been spared over the years to collect the best planetary eyepieces our hobby has to offer. Many sets have come and gone, been found lacking and eventually sold to fund other astro needs, but one set will always remain—the Couture Balls. Locker-room humor aside, I laugh every time I drop eyepieces printed with my name into the tray alongside the venerable Zeiss and Nagler brands.

Fig. 5.1 The Couture eyepiece set (Image courtesy of Steve Couture, Belle Mead, NJ, USA)

After reading William (Bill) Paolini's ideas concerning the most minimal glass lens, the sphere, I decided to build a ball eyepiece for my own use. I didn't own an inexpensive eyepiece that I felt comfortable disassembling for the cause, so I decided to go the cheap route, minimal material and effort for a minimal lens cheap. Centering a ball lens into a rubber stopper inserted into a 1.25 eyepiece barrel seemed like minimal material and effort for sure. Throw in an O-ring around the barrel to prevent the eyepiece from plunging into the diagonal mirror and I'd be ready to conclude this experiment.

Fig. 5.2 A minimally disassembled eyepiece (Image courtesy of Steve Couture, Belle Mead, NJ, USA)

The decision to go with Edmund Scientific as a supplier was based on their low minimum order requirements, they operated in New Jersey where I lived, and they could provide a high tolerance version of BK7 spheres. I wasn't sure what other suppliers could provide based on their website advertisements. I decided to use BK7 based solely on the fact that I own binoculars made with the same material and was pleased with the result.

Eye relief was a concern from the beginning, so I decided to have the ball lens protrude from the top. I'd adjust the ball height accordingly, based on observed comfort. Sure, it would be difficult to keep the lens clean, but the rubber stopper design provided easy removal of the ball for cleaning. As it would turn out, all future versions would have ball lenses protruding from the top, providing increased eye relief comfort.

Fig. 5.3 A minimally assembled eyepiece (Image courtesy of Steve Couture, Belle Mead, NJ, USA)

The night of reckoning finally arrived with Mars as my primary target in my TEC140 mounted on a Losmandy G-11. My eyepiece tray for that night contained the following eyepiece f/l's—6 mm Zeiss Abbe Ortho II, 6 mm TMB Supermonocentric, and a 5.7 mm ball. I didn't have high expectations for the ball eyepiece when I dropped it into the diagonal before the others. Its narrow usable AFOV was apparent from the beginning, about 10–15° I estimated. Not a good start, I thought. Imagine my surprise when both the Zeiss Ortho and Supermonocentric eyepieces failed to show the same level of contrast and sharpness as the lowly ball eyepiece. While Mars' northern polar region was observed in all three eyepieces, the southern polar region was barely discernible, if at all, in the former two. It just popped bright white in the ball eyepiece. There was only one serious flaw in the design that would limit the eyepiece's use from my perspective. The eyepiece's rubber stopper stunk to high heaven; hold your breath while viewing type of stink. Nevertheless, the design evolved to include a rubber grip on top that would accommodate a lens cap to protect the ball lens in storage.

Fig. 5.4 The ball eyepiece Version 2 (Image courtesy of Steve Couture, Belle Mead, NJ, USA)

I was so smitten by the ball lens that I decided to build four focal lengths for personal use, 2.9 mm, 4.4 mm, 5.9 mm, and 7.3 mm. The winged grip, purchased from Orion telescopes, was altered to accommodate spare #365 caps that I owned. The black barrels were purchased from Surplus Shed. Research into a suitably comfortable nose clip continued. Meanwhile, I reported my viewing observations and posted pictures of the ball eyepiece on Cloudy Nights. I realized that someone would inevitably request that I build and sell some ball eyepieces. I turned a few people away, opting instead to describe how easy it was to build ball eyepieces with a 1.25 in. barrels and rubber stoppers, all the while warning about the smell. Eventually, guilt feelings crept in, and I searched for a way to accommodate people, create a decent product without the smell, and account for materials, time, and effort. Labor charge was a big concern. There were two issues; I really didn't desire to make money on the project, and I'm not a nice enough person to give my time away without compensation. Mr. Sensitivity, I'm not!

Finally the decision was made to donate my time to my favorite charity, a school for children with learning disabilities. For the price of materials and a $50 donation/

eyepiece to the school, I'd build any focal length desired. I'd like to thank Cloudy Nights for allowing me to get the word out without being labeled a vendor. Well over $1,200 was raised for The Bridge Academy in Lawrenceville, NJ.

The motivation to do some good was there, but so was the eyepiece smell. Something had to be done about the smell. I located a bulk supplier of a name brand eyepiece with sufficient lens opening that could accommodate all ball focal lengths with minor modifications. The eyepiece could accept a much smaller and less odorous rubber insert to hold the ball lens.

Fig. 5.5 The ball eyepiece Version 3 disassembled (Image courtesy of Steve Couture, Belle Mead, NJ, USA)

Holes in the rubber insert were drilled slightly smaller than the lens diameter using a drill press. The original rubber stoppers were used, only now less care was devoted to centering the drill in the stopper, since so much rubber was filed away in the final process. Also, the drill bit acted as a mandrel, holding securely onto the stopper after going completely through. While still spinning on the drill bit, using a utility knife I would cut horizontal sections in the stopper of about ¼ inch to ½ inch depending on ball diameter needs. One stopper could produce 4–5 inserts. I removed the sections, cut the diameters down using HD cutting shears, and then proceeded to reattach the rubber stopper to the drill bit for final filing and sanding. The result was the far less stinky eyepiece seen here.

Fig. 5.6 The ball eyepiece Version 3 assembled (Image courtesy of Steve Couture, Belle Mead, NJ, USA)

Four or five painted examples of this rubber stopper adapted eyepiece were sold. Painting became an issue. I had to sand the old name and focal length off the housing and cover up the result. A layer or two of primer and at least two layers of red paint requires at least 1 or 2 weeks, if not more, to properly harden. I didn't factor that into the delivery schedule. That's where the Brother Label Maker came in handy. The white label wrapped around the red painted lens barrel helped resist finger print formation while handling the soft, newly painted surfaces to a certain extent.

Shortly after producing the initial batch of ball eyepieces, I rediscovered a bar of Delrin that I had purchased years earlier. It occurred to me that I could cut, drill, and shape the material to hold the ball lens in its eyepiece housing. I also considered that the material would be more durable than the rubber stopper material I had been using. Though more labor intensive, Delrin spacers made more sense from a quality perspective, so the decision to upgrade to better building materials was easy. I cut the Delrin into roughly 1/8 in. thick pieces, from which I could produce four spacers.

Fig. 5.7 The raw delrin spacers (Image courtesy of Steve Couture, Belle Mead, NJ, USA)

The difficult part of the process was creating the beveled inside portion of the top spacer, which allowed the ball lens to protrude from the top of the eyepiece housing. I used the same drill bit/mandrel procedure in my drill press for rounding the corners of the Delrin spacers to allow the spacers to fit inside the eyepiece housing. I also notched the spacer on top with an Exacta blade while it was spinning so that the spacer would fit level with the eyepiece housing. After using a slightly smaller drill bit to make the initial hole appropriate to hold the lens, I used a larger drill bit to bevel the inside so that the ball protruded slightly from the top.

Fig. 5.8 The ball inside the delrin/spacers (Image courtesy of Steve Couture, Belle Mead, NJ, USA)

Sanding the Delrin spacers with 2000 grit sandpaper complete the process. Spacers and ball lenses were secured with the retaining rings that originally came with the eyepieces.

Fig. 5.9 The final product (Image courtesy of Steve Couture, Belle Mead, NJ, USA)

Eyepiece Basics I

By Neil English, Stirlingshire, Scotland, U. K.

Eyepieces have traditionally generated enormous interest among contemporary amateur astronomers. Some are obsessed by them, while others seem to get by with very modest offerings. The truth of the matter, in my opinion, is that eyepieces are to telescopes as amplifiers are to hi-fi systems. There is little point in acquiring a good telescope and outfitting it with a sub-standard ocular. The discerning amateur ought to invest in the best set of eyepieces for his or her intended purpose. But therein lies the conundrum. Which eyepieces should you choose?

Well, for starters, most amateurs can actually get by with just three carefully chosen oculars. Throw in a good Barlow lens and you'll have enough range in magnification to meet any application. The advances in anti-reflection coating technology, lens edge blackening and proper housing have enabled modern wide angle eyepieces, with as many as eight or nine elements to rival the performance of more simple, traditional designs. In addition, the marketing of models with greater eye relief has empowered many eyeglass wearers, who can finally enjoy crisp, full-field views at the telescope.

As a double star observer who does not wear eyeglasses, I have found using traditional designs more than adequate for my particular needs. Other amateurs, using telescopes with shorter f-ratios, are best advised to seek out high quality modern designs with better edge of field correction. The current line of Tele Vue eyepieces cover every available niche, but they are quite pricey in relative terms. That said, there are welcome signs that other companies can produce similar quality models for substantially less financial outlay. As ever, it pays to do your research and, if at all possible, try before you buy.

All of my instruments have fairly high ratios (f/9 or greater), allowing me to make use of high quality orthoscopics. I own a complete set of Baader Genuine Orthoscopics and have been more than pleased with their sharp, high contrast views. For wide-field sweeps, I also employ 2 in. models. For example, my lowest power ocular is embodied in the corpus of a 40 mm Erfle, which works very well at f/9, and, most recently of all, I have been mightily impressed with the new suite of 100° eyepieces available from Tele Vue, Explore Scientific, and Meade. These new eyepieces deliver an altogether unique experience of the deep sky, enabling very large swathes of the starry heavens to be scrutinized comfortably and at fairly high magnifications, rendering vistas quite beyond the kind of earlier generations of amateurs.

What will be the shape of things in the future? That's hard to say, but it is not unreasonable to expect a greater marriage between electronics and optics, particularly in the development of a new generation of image-intensified eyepieces with ultra-high dynamic range that can deliver compelling views of even the fainter deep sky objects from the comfort of a small backyard telescope. It is my fondest hope that such technology will 1 day be made available at a modest price. Of course, as with all technology, it's best to watch this space!

Eyepiece Basics II

By Tony Miller, Stoney Creek, Ontario, Canada

Eyepieces are key components in the optical train in a telescope. Many choices are out there to sift thru and decide what your likes and dislikes are. Key features I look at for in my eyepiece selections are:

- AFOV—Apparent field of view, or how much sky is visible; for planetary I really don't consider AFOV, but for a push to non tracking 'scope the more corrected AFOV the better. I find the 70–82° AFOV well corrected enough for me.
- Elements in the Eyepiece—More elements seem to cut down transmission in an eyepiece; when selecting a planetary EP try and get a quality minimal element eyepiece.
- Color Saturation—Some eyepieces tend to be dull without life. The colors of stars don't really pop. Some excellent color saturation eyepieces are the Tele Vue Plössl smoothies, which are absolutely wonderful bringing out the banding on Jupiter and my favorite EPs while viewing the gas giant.

- Eye Relief (ER)—Tricky here, as many manufacturers indicate lots of ER but usually not the case. I found close to 20 mm of ER is what I need while viewing with glasses to see the field of view nicely.
- Edge Correction—For faster 'scopes (say F6 and below), this is where some serious glass is needed to get a clear sharp view to the field stop.
- Eye Placement/Comfort—Some eyepieces tend to be very finicky and hard to keep the view. Either blackouts or distorted images appear if your eye position is not centered over the EP.

Bottom line when choosing an EP, what do you think about it and how does it suit your eye because you're the one looking through it, and if it's pleasing then it's the right one for you.

I have been through many eyepieces, and here are some recommendations in different classes:

- 35–50 AFOV—ZAO's, TMB SMC, CZJ's, Pentax SMCs, XOs, UO HD's, BGO, TV Smoothie Plössls, Meade 4000 smoothies
- 65–70 AFOV—Pentax XW's, Vixen LVWs, Meade SWA 5000s
- 80–84 AFOV—Nagler T1, T2, T4, and T5 and some of the older Naglers are my favorites, Meade UWA 4000s, William Optics UWANs
- 100 AFOV—ES 100s, TV Ethos
- Zooms -Antares SW 5-8, Baader Zoom

My personal favorites:

- Planetary: XO and CZJs
- Wide-Field General Use: 14 UWA 4000 smoothie and the 20T2 Nagler
- Finder Eyepiece: 28 WO UWAN really well corrected down to F4.5. A sleeper in my book.

Eyepiece Designs for the Critical Lunar and Planetary Observer

By Carlos E. Hernandez, Pembroke Pines, FL, USA

I have been observing the Moon and planets for over 40 years using a variety of instruments at different observing sites. The critical planetary observer, and deep sky observer as well, must attempt to prepare the instrument in order to take advantage of that moment of steady atmospheric seeing that allows a faint and/or low contrast feature to be detected over the surface of the Moon, ruddy surface of Mars, or atmosphere of Jupiter, among many other objects. The observer must be using a diffraction-limited instrument that is collimated and acclimated to the surrounding temperature in order to take advantage of the steady moments of seeing (atmospheric).

The eyepiece is an integral part of the equation, and if not properly matched for the object being observed or atmospheric conditions (e.g., too much magnification) then detail will be missed by the observer. I have used a variety of eyepiece designs over the years, such as the Abbe Orthoscopic, VERNONscope Brandon, Tele Vue

Plössl, Edmund RKE and other novel designs as well. The critical lunar/planetary observer is attempting to pull out that last photon of information from the image produced by the instrument during an observing session. In general, the lower number of elements that an observer uses (e.g., Abbe Orthoscopic eyepiece uses four elements in its design) will typically allow the detection of low-contrast features over the surface or atmosphere of a planet.

While any eyepiece design may produce a bright, sharp, and apparently high contrast image on-axis the designs employing a smaller number of elements (e.g., Abbe Orthoscopic, VERNONscope Brandon, Tele Vue Plössl) will allow the detection of that faint and elusive feature that the momentary steadiness of the atmosphere may provide. The lunar and planetary observer does not require a large apparent field (e.g., 82° field provided by the Tele Vue Nagler) in the study of the Moon or planets, but the larger field eyepiece designs available, such as the Tele Vue Nagler, Explore Scientific 82° eyepiece, and Baader Hyperion (among other excellent designs) provide excellent views on-axis, although employing several elements (up to eight elements) in their design. The critical observer will enjoy many excellent observing sessions using the classic designs mentioned above but should not avoid other newer designs employing more elements. Last but not least the observer must train his or her eye in the detection of faint detail over the Moon and planets, as experience will allow the use of non-classic designs under less than ideal atmospheric conditions.

Eyepieces: Less Is More

By Erik Bakker, Haren, The Netherlands

In these days of abundance, every amateur astronomer has to choose how to best put that into his or her observing fulfillment. At first, most of us start expanding our eyepiece collection and sooner or later end up with at least one case full of different eyepieces. I know I did. The nights are then spent endlessly changing eyepieces in order to find the "perfect" match between object, 'scope, sky conditions and the observer.

Over the decades, my most memorable evenings under the stars were spent observing. Not changing eyepieces. Eventually, I spiraled towards observing with two eyepieces and one Barlow. The 'scope that showed me the power of that route was my Questar 7. Fewer eyepieces means more observing.

Another important thing to consider with eyepieces is contrast, gaining importance as we increase magnification. Where at low power overall correction of the field is key, generally achieved by more glass elements, the high power planetary image needs maximum on-axis contrast. Here, too, less glass is more. Especially superbly if polished.

I have now applied these principles to all of my observing, deep sky and planetary. Both mono and bino. And my small and big 'scopes. I have come to appreciate longish focal length oculars with comfortable eye relief and an apparent field of view of between 52° and 70°.

Do match eyepieces to your particular needs and preferences. They perform differently in different 'scopes and magically interact with the observer's eye and soul.

So here is what less is more looks like for me:

4″ f/8 APO:
Mono: 16.7 and 12.8 mm Zeiss Victory Diascope WW eyepieces with 68° AFOV and Tele Vue 2″ 2× BIG Barlow
Bino: Tele Vue Bino Vue and 2× corrector with a pair of 25 mm Zeiss orthoscopic microscope eyepieces with 52° AFOV and the Tele Vue 2″ 2× BIG Barlow

16″ f/5 Dob:
Mono: Tele Vue 22 mm Nagler T4 with 82° AFOV and Tele Vue 2″ 2× BIG Barlow
Bino: Tele Vue Bino Vue and 2× corrector with a pair of 25 mm Zeiss orthoscopic microscope; eyepieces with 52° AFOV and the Tele Vue 2″ 2× BIG Barlow

So I invite you to muster your courage to stop changing your eyepieces. Dare to pick only two for the night and enjoy them to the fullest. You may be surprised at how much more you see…with less.

General Comments on Eyepieces

By Ed Ting, Amherst, NH, USA

My first telescope was an old copier lens taped to the end of a cardboard tube. For the eyepiece, I took the eyepiece off an old toy microscope. This rig served me for years as a kid, and I never wanted for anything else. So the last thing I am is an eyepiece snob. Having said that, I will never forget my first look through Tele Vue's original 13 mm Nagler. Whoa! I found the claims for the "spacewalk" experience to be accurate. The view was immersive, like watching an IMAX movie. What's more, unlike other, previous wide-field eyepiece designs, the view sharp and contrasty right out to the edge. I was immediately hooked. Over time, I gradually saved up money and wound up getting the entire Nagler series. Eyepieces are an investment. Good eyepieces tend to stay with you, even as you buy and sell your 'scopes. It's like your mother used to say—buy it once. I've owned dozens of telescopes through the years, but my 7 and 13 mm Naglers, and 19 and 27 mm Panoptics have never left me.

Jolly Small Eyepieces

By Alexander Kupco, Ricany, Czech Republic

Since being a kid I have always been attracted by the beauty of the night sky. Only recently, for about last 3 years, have I been doing more systematic visual observation again. On the way to finding equipment that would best serve my observational habits, I encountered the small format 0.965 in. eyepieces. By chance, a telescope I purchased last year, a Zeiss AS80/1200, came with a nice eyepiece turret and two Zeiss 0.965 in. orthoscopic eyepieces. I then poked through the astro equipment I still have from my youth and found another two excellent 0.965″ eyepieces: a Zeiss Huygens 25 mm and an ATC 8 mm Erfle. Fortunately, the East German optics are far more common in my country and it took me only about a year to collect the remaining Zeiss orthoscopic eyepieces from 40 to 4 mm (except the 8 mm, which is rare).

I'm now finding myself observing with these small 0.965 in. eyepieces more and more. Not only with the AS80 but all my other telescopes as well. I do most of my observations from a backyard in a small town just at the border of a city with a population of 1.5 million. Low power wide angle views are not that inspirational under such conditions. But I like to do all other things—double stars, the Moon and planets—and I also like the challenge of pulling out the faint fuzzies as well. I find the small 0.965 in. eyepieces are suited to all those needs in a rather excellent way. Being of the Abbe orthoscopic design, they also provide sharp and high contrast views of planets and the Moon. The views are to my eyes basically on par with the views through some of the most highly regarded planetary eyepieces such as the TMB Supermonocentric (7 and 16 mm) and the Pentax XO5.

To my surprise, these orthoscopic eyepieces also excel in observing faint fuzzies. For example, one night I quickly compared the appearance of two faint galaxies (NGC4666 and NGC4845) in my 100 mm f/9 ED telescope. They were visibly better defined in the Zeiss 16 mm orthoscopic eyepiece than they were in the Tele Vue 15 mm Plössl. Also, the nearby faint stars were more easily spotted in the orthoscopic eyepiece. Then, during one exceptional night, I was even able to spot from my backyard in the same 100 mm telescope and Zeiss 16 mm Ortho a few very faint galaxies in the hearts of Abell 2197 and Abell 2199 clusters. Since it is always difficult when observing small planetaries to focus at high powers under low light conditions, and even more difficult when using UHC and OIII filters (which are often needed for these objects), it was very helpful that the Zeiss 0.965 in. eyepieces were parfocal.

Last but not least, the feature that I like on these 0.965 in. eyepieces is their handiness. They are small and light, so I can fit six of them just in the box for a single 27 mm Panoptic eyepiece. They are therefore very convenient in reducing the number of trips needed to the backyard when setting up the telescope. The jolly small 0.965 in. eyepieces have quickly become my favorite everyday workhorses.

Learning to Use Hyper-Wide AFOV Eyepieces

By Don Pensack, Los Angeles, CA, USA (www.EyepiecesEtc.com)

Though the hyper-wide apparent fields of view in many contemporary eyepieces are enticing to observers, many other observers, new to the use of such exceptionally wide fields of view, have trouble using them. Our normal vision, per eye, has a peripheral range of up to 140°, and the use of a 100–120° field eyepiece does not appear to be intuitive, so some explanation is called for.

Our instinct is to move the eye back and forth to look farther to one side or the other, but this is not the technique one uses with 100° field eyepieces. Being normal eyepieces, they form exit pupils behind the eyepiece's top lens, and it's still necessary to hold the pupils of our eyes in position in order to see the fields of view. Hence, when changing where the eye is focused in such a large field, simply moving the eye to look at the edge doesn't work because it moves the pupil of the eye away from the exit pupil of the eyepiece. It's necessary to hold the pupil of the eye at the exit pupil's position and roll the head around that point so the eye is still in the same position relative to the eyepiece when looking directly at the edge of the field. This technique is also used for narrower fields of view (such as the 82° eyepieces), so if moving to 100° from 82°, this issue might not arise. But many observers move from 50° fields, where one simply glances back and forth to see the edges, to 100°, where the technique of use is slightly different.

This technique is an easily learned skill, but it is a skill. The reward of learning it, however (and it's not hard), pays off with spacewalk views of space through the telescope. That's what makes learning to use the hyper-wide eyepieces so rewarding!

Learning What Eyepieces Satisfy Your Needs

By Judson Mitchell, NC, USA

It takes a while to learn what eyepieces (EPs) are required to satisfy your observing needs. You must be cognizant of your telescope's limitations and aware of observing conditions. Also it is important to consider your viewing choices. My choices happen to focus primarily on deep sky objects (DSOs), namely galaxies and planetaries. After owning several dozen EPs I have found most situations require a very limited range. Using a 12 in. Dobsonian (Dob) reflector my functional EPs have been a 40 mm Paragon, 20 mm T5 Nagler, 17 mm T4 Nagler, and 13-11-9 mm T6 Naglers. These provide 99 % of my viewing pleasure. Under our upstate South Carolina skies seeing is usually moderate to mediocre. Thus, during average local conditions, images of DSOs commonly break down and provide no additional detail below my 13 Nagler. Occasionally excellent seeing allows me to use my 7 and 5 mm TMB Planetaries on Solar System objects as well as DSOs.

My primary choice of Naglers was dictated by quality, availability (all were purchased used), and field of view (FOV). With a non-tracking Dob, the 82° field of view of the Naglers is imperative to me. This allows viewing of an object for tens of seconds before having to move the 'scope. The 40 mm Paragon is used as a finder EP and for objects that require a large FOV. The Paragon in my 'scope covers about 1.7° of the visible sky. Therefore, a large portion of objects such as the Veil or M31 can be seen. However the playing field now has been modified, having recently acquired a 20 in. Obsession Dob. Surprisingly it appears that the set of eyepieces have been well chosen, and the only gap now is in the 20–40 mm range. With the increased focal length of the Obsession the effective FOV of the EPs has been reduced along with the increase in magnification. I will likely settle on either a 28 mm UWAN or 26 Nagler.

Microscope Eyepieces on Your Telescope

By Samuel de Roa, Spain

Many observers, particularly new ones, struggle over whether to purchase specialized planetary eyepieces for their telescopes. Over the years, I have found that the best approach has been to recruit microscope eyepieces from the most reputable glass manufacturers, mainly Nikon, Leica/Leitz, Zeiss and Olympus. I thought something like… why wouldn't these eyepieces work in high resolution astronomy work? In doing this, I have come to discover real pearls that would make your dedicated telescope eyepieces pale in terms of performance/price.

As you may know, the eyepiece is not an individual part but makes a full combo with your telescope design. Thus, in order to find my "pearls," I tested their performance in my dedicated telescope using the "slanted-edge" method (for more information you may look at a published article on this method at www.astromart.com or www.cloudynights.com, or just Google "slanted edge Samuel" to find the article). I believe there is a beautiful land of top planetary eyepieces hiding out here in the microscope world. I strongly believe this is a much neglected niche and yet-to-be-discovered place for dedicated planetary glass. It is very exciting, too, to be able to buy top brand glass at a very economical price in the second market world and get that "wow" grin on my face from time to time. Try it and see for yourself! For inexpensive microscope eyepieces and adapters, eBay is an excellent source.

Fig. 5.10 Example of a neglected niche for planetary glass—microscope eyepieces (Image by Samuel de Roa, Spain. Used with permission)

My Thoughts on Eyepieces for General Observing

By Mike Ratcliff, Redlands, CA, USA

You can use simple inexpensive eyepieces at first, when you are thrilled just to see stars at all. I used a 26 mm Plössl, with a relatively narrow 50° apparent field of view, almost exclusively for 2 years. However, adding a wider field, well corrected Tele Vue 19 mm Panoptic (68°) made the sky seem to come alive, and the telescope became a new instrument. I would heartily recommend getting wide fields as your budget allows. With a tight budget, my strategy is to get older, used, and possibly 'cosmetically challenged' eyepieces. My wide-field lineup (now with a 16″ Dob) is a 27 mm Panoptic, and 14 mm, 8.8 mm, and 6.7 mm Meade Series 4000 UWA's.

Orthoscopic (Ortho) eyepieces are very good even though narrower than Plössls (40°). A 12.5 mm Ortho in the 6 in. 'scope was very sharp and showed the shades of gray better on the Moon. Later a 9 mm Ortho was added and delivered quite a surprise. Compared to a 9 mm Nagler Type 1, the Ortho was clearly sharper in the central portion of the field. The same happened later with the 12.5 mm Ortho and a 13 mm Nagler Type 6. Also an 8 mm Tele Vue Plössl seemed amazingly brighter than the 9 mm Nagler and showed a little more in galaxy extents. Since then I have added several Brandons, one Carl Zeiss Jena, and some other Tele Vue Plössls. They can be a bit uncomfortable at small focal lengths.

Another recommendation is a 2× Barlow for increasing magnification with really steady atmospheres. I recently used the 6.7 mm with the 2× Barlow to get 600× to see the central star in M57. Lastly, for public outreaches, I recommend the old standby, a 25 mm or 26 mm Plössl, with a Barlow as needed. Easy to clean, cheap, and comfortable.

Observing Comets and Nebulae: Brilliance and Wide Field Is What We Need First

By Uwe Pilz, Leipzig, Germany

Observing comets and nebulae challenges the observer to perceive very faint and often large filaments extending from these objects. Therefore, the most important task for an eyepiece used for observing comets and nebulae is to conserve the contrast of the image as much as possible so the object appears brilliantly in the field of view. This brilliance is the brightness difference between dark background and the sparsely illuminated areas of the object. Stars in the field and brighter parts of the object tend to scatter light into dark regions. Within the eyepiece, this scattering happens at air-glass surfaces, the edges of the lenses and the internal surface of the barrel. To minimize this scatter would lead us to high quality eyepieces with only one or two elements. But nebulae and comets are often large objects, so an adequate FOV is desirable too.

High quality eyepieces such as Erfles and Königs are the compromise between less optical surfaces and wide field of view, which can extend up to 70°. These types of eyepieces usually contain only three groups of elements; this is only one group more than the simplest suitable eyepiece designs like the Plössl. Erfle and König eyepieces do have two disadvantages: the off-axis sharpness is not as good as other designs, and although their difference is small, having only one extra group, they tend not to be the first choice for planetary observing. Nebulae and comet observers, however, do not need the last bit of sharpness that planetary observers may need. Beside this, the exit pupil distance is quite short for shorter focal length Erfles and Königs; therefore it is not easy to find high quality eyepieces of these designs with a focal length below 15 mm. My collection reaches down to 8 mm, which I use regularly and find the eye position not too uncomfortable. The advantages of the Erfle and König design for comet and nebula observing I find far outweigh any disadvantages.

Fig. 5.11 Telescope Services (TS) 8 mm WA (wide angle low lens count eyepiece) (Image by Uwe Pilz, Leipzig, Germany. Used with permission)

Planetary Observing: Resolution and Contrast Is What We Need First

By Uwe Pilz, Leipzig, Germany

Critically observing planets requires the ability to catch very inconspicuous details on the planet's surface. This is especially true for Jupiter and Saturn, and even harder to achieve for Venus and Mercury. Mars makes the astronomer's life easier in this respect because of its rather high contrast features. The eyepieces used for planetary observing should therefore have the primary task of being able to show all the detail the telescope is able to deliver. Comfort of observation is secondary. Modern eyepieces are usually designed to provide first a comfortable view, with long eye relief, then a huge field of view. But we pay for that comfort and large field of view: much glass, many lenses. Every glass element reflects light and degrades the contrast. Many elements also increase the chance that not every element is centered accurately, thereby reducing sharpness.

I recommend a high quality Plössl eyepiece for planetary observing. They have only two elements and scatter less light than designs with more glass. Centering lenses are also much easier with only two elements to bring in line. The disadvantages are a smaller FOV and a slightly uncomfortable eye relief distance at shorter focal lengths. Quality differences in Plössl eyepieces can be large, however, with cheaper ones usually given with a new instrument as a first eyepiece. What we need are Plössl eyepieces of the highest quality we can achieve: modern anti-reflection coating, blackened lens edges, internal scatter elimination. I found the Meade Series 4000 of very good quality in this respect and use them regularly.

Fig. 5.12 Meade Series 4000 Super Plössl (Image by Uwe Pilz, Leipzig, Germany. Used with permission)

Quick Tip for New Telescope Owners

By Ed Ting, Amherst, NH, USA

One of the most common mistakes I see beginners making is pushing the power too high, too soon. I can't count the number of times someone's come to me for help with their new 'scope, and I find the highest power eyepiece in the focuser. Sometimes all I have to do is swap the low power eyepiece in, and the owner immediately finds that life's gotten easier. So if you're new at this, do yourself a favor and use the lowest power eyepiece (the one with the largest number printed on the side).

Sample Eyepiece Sets for Typical Amateur Telescopes

By Ed Ting, Amherst, NH, USA

I'm often asked to select a set of eyepieces for a particular telescope. There are no hard and fast rules, but you can generally get away with only three or four eyepieces and a Barlow. Your low power eyepiece will be used for framing large objects such as the Double Cluster, or M31 (the Andromeda Galaxy). Medium power is for framing smaller deep space objects such as globular clusters or smaller open clusters. High power will be used sparingly on planets and tight double stars.

A 6″ f/8 Newtonian reflector:
Low power: 32 mm Plössl (38×)
Medium power: 15–20 mm Plössl (61×–81×)
High power: 5–8 mm (152×–243×) wide field design, like Tele Vue's Radian series, or Orthoscopics
2× Barlow lens

An 8″ f/10 Schmidt-Cassegrain:
Low power: 35–40 mm (40×–57×) wide field eyepiece
Medium power: 25 mm Plössl (80×)
High power: 10–12 mm wide field design (166×–200×) or Orthoscopics
2× Barlow lens

A 4″ f/8 refractor:
Low power: 25 mm (32×) Plössl
Medium power: 10 mm (81×) Plössl or wide field design
High power: 5 mm (163×) wide field design, or Orthoscopic
2× Barlow lens

Shallow Sky Sketching Eyepieces

By Thomas McCague, Oak Forest, IL, USA

When I am casually observing and enjoying the view of the nighttime sky with a telescope, I prefer to use long eye relief wide field eyepieces with apparent fields of view greater than 60°. These oculars would include short to long focal lengths with telescopes of differing apertures.

Sketching at the eyepiece presents a different set of choices when selecting oculars. Since I live in a large urban/suburban environment with much light pollution, I must travel substantial distances to reach skies dark enough to glimpse the Milky Way. For this reason I rarely sketch faint deep sky targets from home but instead focus on the shallow sky worlds such as the Moon, Mars, Venus, Jupiter and Saturn. Narrow field eyepieces with these targets are my first choice and preferably ones with short focal lengths to achieve high magnifications. I have three eyepieces I prefer to use under these conditions and all are orthoscopic types. The apparent field of view in all three is at or about 45°. Two are Meade eyepieces at 4 and 6 mm, and the third is a University Optics 9 mm eyepiece. The telescopes I most often use are a 10 in. f/5.7, a 13.1 in. f/6 and a 4.25 in. f/5. All of these 'scopes fit nicely on the same equatorial drive platform, which make sketching at high magnification comfortable.

Should I Buy Complete Sets of Eyepieces or Mix-and-Match?

By Mike Rowles from Crofton, MD, USA

Generally, I don't buy complete sets of any line of eyepiece. We shouldn't assume that all focal lengths in a series will be as good as that series' best examples. I try to determine which are the best of a line—either through reviews or my own experience—and limit myself to those alone. Some focal lengths might not reach the level of performance of other eyepieces in the series. Also, some may work well in your particular 'scope, while others do not. The only exceptions I've made to the "No Sets" rule are the Baader Genuine Orthoscopics and the Pentax XO's. The BGO's are consistently sharp and contrasty, and they all have excellent scatter control and color rendition.

The XO's are excellent—and expensive if you can find them—but there are only two of them: the 5.1 and 2.5 mm. But many lines will have both good and not so good eyepieces. For instance, the Orion Epic ED-II's—the original series—are often maligned for kidney-beaning, or blackouts. However, my own experience with the 22 mm is that it gives a sharp, flat, bright view with good color, and that it binoviews well. VERNONscope Brandons are highly esteemed eyepieces for double stars, lunar and planetary work, but the 4 mm has not been well received. The Pentax XW's are valued for their high light transmission and sharp images. Nevertheless, many observers have complained that the longer focal lengths show too much field curvature. If you've already ordered an entire series of eyepieces, go ahead and try them all out. If some don't work so well in your 'scope, you can always sell those specific eyepieces and keep the ones that do work. Maybe you'll be lucky, and they will all perform well for you.

Some Tips for Viewing the Universe Through an Eyepiece

By Steve Stapf, Crofton, MD, USA

I have many eyepieces, some are my favorites, but they are not necessarily known as the best eyepiece. When looking into the cosmos with a telescope there are those few rare occasions the atmosphere seems to open up with an incredible view for a few seconds/minutes. When this happens the view is stored in your brain for a life time and the eyepiece in the telescope becomes a favorite.

It seems that some eyepieces handle the atmospheric conditions differently due to coatings or type of glass or lens configuration, and those few eyepieces have exceptional clarity at the time of perfect viewing. I have many eyepieces, but if I could start over again I would buy the shorter focal length's Pentax XW's first. With the Pentax you don't have to strain your eye to view through it, so you can view longer. You can also see more detail when the atmosphere becomes still, by being able to view at the eyepiece longer.

Another trick I have learned is that when looking at the night sky through an eyepiece and telescope you need dark-adapted eyes, and while utilizing peripheral vision, the minutest details can be seen. However, regardless of all the tricks one can use the key factors for viewing the night sky are: having the largest telescope aperture possible, having good performing eyepieces, viewing during still atmospheric conditions under a dark sky, and finally having a good astronomy chair at your telescope.

The DIY Eyepiece

By Dr. Tim Wetherell, Canberra, Australia

Naturally enough commercial eyepiece manufacturers create products that will suit the majority of 'scopes out there—they'd be foolish if they didn't. But what if you don't have a "standard" 'scope? I found myself in that position with an 8 in. f9 refractor I recently built. The biggest, widest eyepieces on the market such as the 31 mm Nagler are superb, but they still only give you a quite limited field of view—about 1.4° in this case. If you own a big classic refractor or a very large Newtonian then it's even worse. This is a pity because there's nothing inherent in the optics of a big telescope that prevents it from achieving a much wider view; in fact just the opposite. A 'scope like a big refractor has the capacity to produce stunning coma free wide views. The problem lies with the eyepieces.

Physics dictates that the widest available field of view is limited by the field stop of the eyepiece, which in turn is limited by its physical diameter. The upshot of this is that if you have a 3″ eyepiece then you can achieve a much wider view than you can with a 2 in. eyepiece. The good news is that many modern refractors have really big focusers intended for use with CCD cameras, etc. There are a few commercially available 3″ eyepieces, but not many, and those are quite expensive because they're all small run items. So, in search of wider views I set about making one of my own.

The result—a 65 mm 70° that gives me a field of view of about 2.5°. After a bit of messing about, I managed to achieve a fairly flat field of view, good eye relief and relatively low aberrations, making this one of my favorite eyepieces.

I guess the point of the story in this context is that amateur astronomy is supposed to be fun, and playing around with a bunch of lenses to see what you can achieve is a lot of fun. You probably do need to be Al Nagler to create a truly great eyepiece, but Joe Blogs can with a bit of study and experimentation, create one that's perhaps not perfect but as good as many eyepieces out there. And what's more he or she can learn a lot and have fun doing so.

Fig. 5.13 Homemade large format eyepiece (Image by Dr. Tim Wetherell, Canberra, Australia. Used with permission)

The TV Ethos Series of Eyepieces: Virtues Beyond Just 100° of Panorama

By Chris Mohr, Raleigh, NC, USA

Discussion about the Tele Vue Ethos series of eyepieces naturally tends to focus on their stunningly panoramic 100° AFOV, but the series has another important advantage. Their key performance characteristics, listed below, are remarkably consistent across all eight focal lengths in the series.

1. The size of the eye lens is identical for every focal length of Ethos: a generous 30 mm in diameter (measured with calipers), a fact which you can easily see from a top-perspective group photo of the Ethos family. By comparison, the eye

lens size of the various members of the 82° AFOV Nagler line varies, even across some eyepieces of the same nominal "type." For example, the 31T5 Nagler has the same diameter eye lens as do the Ethos (30 mm), but its sister 26T5 Nagler has a 25 mm diameter eye lens, while all Type 6 Naglers have a 19 mm diameter eye lens (as does the 68° AFOV 24 mm Panoptic).

Obviously, the eye's perception of the apparent field of view of an eyepiece depends on key internal optical design factors and not eyepiece lens size. And in fact, if you experiment with looking through a hand-held 31T5 Nagler and a 13T6 Nagler at a brightly illuminated sheet of white paper filling the entire FOV, your perception of the relative width of the AFOV will be identical.

Nevertheless, under more realistic night sky viewing conditions where you're not looking at a bland, blank canvas, the eye and mind's visual processing finds it more relaxingly easy to perceive the view as "wide" with a generously "wide" eye lens, but especially to perceive that two eyepieces of different focal-lengths share the same AFOV when their respective eye lenses are of identical size. Try looking through a 17T4 Nagler (with the 30 mm diameter eye lens) and a 13T6 Nagler (with the 19 mm eye lens) back to back; at least for me, it takes a few moments of mental adjustment for my perception of the view through the 13T6 to appear the same apparent width as the 17T4. With the Ethos line, the perceptual adjustment in switching one focal length out for another is absolutely seamless. Any differences are merely the ones expected from proportional changes in magnification, exit pupil, and true field of view.

Fig. 5.14 Consistent eye lens size of the Tele Vue Ethos line of eyepieces (Image courtesy of Chris Mohr, Raleigh, NC, USA)

2. The entire Ethos line can be used in stock condition (without modification) as 2″ eyepieces, including the six hybrid 2 in.–1.25 in. format focal lengths (all those 13 mm and below). This would conveniently eliminate the need to use (and swap in/out) the 2 in.–1.25 in. adapter except for one problem—the inability to use 2 in. filters on the 13 mm, 10 mm, 8 mm, and 6 mm Ethos, because of the downward protrusion of the 1.25 in. format portions. Fortunately, there's a relatively inexpensive (@$30 apiece) solution to this. Tele Vue makes a 2 in. eyepiece barrel extender that readily screws onto the 2 in. portions of the 13 mm, 10 mm,

8 mm, or 6 mm Ethos. These extenders also contain threads for 2 in. filters in the bottom portions thereof and have the additional benefit of making the 13 mm through 6 mm Ethos eyepieces much more aesthetically attractive. These extenders are extremely lightweight, adding only a single ounce to the overall weight of the eyepiece when attached, and so do not introduce any balance problems. They used to come in polished aluminum matching that of the native eyepiece barrel but now come only in anodized black, but on the other hand the black extenders give the Ethos eyepieces an attractive "art deco" look, at least to my eye. The hybrid-format 4.7 and 3.7 mm Ethos come stock with their own unique removable 2 in. polished aluminum adapters, so there's no need to separately purchase a barrel extender for these two.

Fig. 5.15 Chrome versus black 2 in. extenders (Image courtesy of Chris Mohr, Raleigh, NC, USA)

3. The weight across the majority of the Ethos lineup is within a similar-enough range that balance issues in changing from one focal length to another are minimized. Only the 21 mm Ethos is likely to cause significant balance issues for most telescope mount setups (its weight is almost identical to the 31T5 Nagler at 2.25 versus. 2.20 lbs). All six focal lengths from 13 mm and below fall within the range of 0.95 lbs–1.30 lbs (about 37 % or 0.35 lb), a range most mounts can handle with no more than a minor tweak to the balance or tension.

The second-heaviest member of the Ethos family, the 17 mm Ethos, weighs 1.55 lbs, which may require a bit more tweaking of balance when swapping out with one of the relatively lighter focal-length Ethos (e.g., the 8 mm at 0.95 lbs), but both the relative and absolute change in eyepiece weight and potential balance issues are much less than when, for example, swapping out a 17T4 Nagler (1.6 lbs) for a 9 mmT6 Nagler (0.42 lbs), a 3.8× relative change and 1.18 lb absolute change in weight.

4. The Ethos line gives seamlessly uniform optical performance and eye relief (15 mm) across all eight focal lengths, differing only in the expected characteristics of magnification, exit pupil, and true field of view. Despite not being a scaled optical design across different focal lengths, none displays any differences in field curvature (all display relatively "flat" views) from the others, nor are there any differences in coloration (which is relatively "neutral" compared to the "warmer" tones many perceive across the Nagler line). Not only are they seamlessly uniform relative to one another as stand-alone focal-length eyepieces, but when coupled with a Tele Vue Powermate, e.g., a 13 mm Ethos with a 2× Powermate, they seamlessly create a perfect virtual Ethos of half the focal length of the original, e.g., 6.5 mm. The one tactical downside for using the Powermate method for obtaining a shorter focal length is that because of the relatively tall size of Ethos eyepieces, the resulting optical "stack" extending from the focuser can be astoundingly long, especially in a reflector where a Paracorr is also being used. For example, a 13 mm Ethos (with 2 in. barrel extender) in a Paracorr in a 2 in. 2× Powermate extends 10 in. out from the focuser (and up to 14 in. from the tube or upper truss cage of the telescope, depending on focuser model). The 13 mm Ethos used in 1.25 in. mode with a 1.25 in. 2.5× Powermate in a Paracorr looks somewhat precarious (due to the relatively thin diameter of the 2.5× Powermate), in addition to the length the stack sticks out of the focuser.

Fig. 5.16 13 mm Ethos with Tele Vue Paracorr and 2× Powermate in Dobsonian's focuser (Image courtesy of Chris Mohr, Raleigh, NC, USA)

5. The available focal lengths in the Ethos line come at nearly uniformly spaced 1.25–1.33× increments in focal length and magnification. This, coupled with the relative uniformity in performance and weight, makes it much more straightforward to build a satisfactorily comprehensive partial set for the user's needs, or to incrementally build the entire set, while minimizing awkward gaps. Either the 21E/13E/8E or the 17E/10E/6E make for excellent "minimalist" sets of such expensive eyepieces, and the 21/13/8 combo is especially well suited to be extended into other particularly useful focal lengths by addition of a 2 in. 2× Powermate.

Try Before You Buy

By Patrick O'Neil, Alamogordo, NM, USA

The best advice where eyepiece selection is concerned would be to attend star parties and talk with other amateurs. Most all of your fellow amateur astronomers are very kind and are more than willing to allow you to view through their equipment. Being able to view through their equipment will permit you to get a better "feel" for what you like to observe and with what kinds of equipment. Once you get this experience with different equipment, then make sure to buy eyepieces specifically for the telescope you intend to use. If you have a fast Newtonian telescope, then opt for eyepieces that offer nice outer field of view correction. If you have a slow focal ratio refractor, then these can bear almost any kind of eyepiece optical design well. My personal favorite eyepieces are the Meade 4000 Series 4-element "Plössls." These are an excellent value, and I've used them in 'scopes with focal ratios ranging from f/3.9–f/17 with satisfying results. They offer soft rubber eyeguards, no undercuts in the barrels, blackened housing tops around the lens for less irritating reflections, and well defined field-stops. In my opinion the only negative feature of this line is the somewhat uncomfortable eye relief on the 9.7 and 6.4 mm focal lengths. The Meade 4000 Series 4-element Plössls can be purchased as a kit that includes case, filters, 2× Barlow, and all the different eyepiece focal lengths (minus the 26 and 20 mm), all for around $200. This line offers focal lengths of 40 mm, 32 mm, 26 mm, 20 mm, 15 mm, 12.4 mm, 9.7 mm, and 6.4 mm and these can provide a lifetime of enjoyment.

Using Zoom Eyepieces Successfully

By Andreas Braun, Braunschweig, Germany

Zoom eyepieces (variable focal length oculars) are a valuable addition to the commonly used fixed focal length eyepieces. Zooms are useful for general viewing by replacing several fixed focal length eyepieces, but their real strength will become apparent when used for high power work where frequent adjustments of the magnification in steps as small as possible are required to enable the user to choose just the right magnification to cope with the target and the seeing conditions.

High magnifications are those where the resolving power of the telescope is utilized. This denotes exit pupils from about 1.5 mm down to about 0.6 mm (17×/in. of aperture to 42×/in. of aperture). For low power work the zooming facility can be handy as well, but it is not similarly important. This application area—where magnification steps of 2× are sufficient—can often be better used with fixed focal length wide-field eyepieces. Typical focal ranges of zoom eyepieces are 24–8 mm, 21–7 mm, 18–9 mm, 6–3 mm, and 4–2 mm. With a focal range of 21–7 mm, for example, a zoom eyepiece can only provide sufficiently small exit pupils if the telescope's focal ratio is around f/10 or higher. For the large number of telescopes with f/8 to f/4 this zoom's genuine focal range will not enable real high power work. Now, a Barlow (1.8×–2.5×) can be favorably employed to sufficiently reduce the focal lengths of the zoom and thus enable almost any telescope to be used for high magnification applications where the fine-tuning ability is of importance. A short focal length zoom (6–3 mm, 4–2 mm) can also be a good alternative. All zoom eyepieces have a limited AFOV at the long focal length end (typically 40°, exceptionally also 60°). This is inherent to the design (exception—constant 50° for the 6–3 mm and 4–2 mm zooms).

What Is the Best Eyepiece?

By William Rose, Larkspur, CO, USA

"What is the best eyepiece?" is a frequent question I'm asked. The age-old adage, 'the one you use the most,' is still the best answer but doesn't address the real question. Typically the question being asked is, "What is the best eyepiece for me to buy?" This is a more difficult question to answer, involving a number of factors, and changes with experience. Only you can answer this question for yourself. The best we can do is pass on what we've learned through the years and hope it will help you avoid some of the mistakes and pitfalls we've experienced.

To understand the question and arrive at a useful answer, you need to appreciate the factors involved in optical viewing astronomy. You should understand that no two people "see" exactly the same. Different telescope and eyepiece designs are better suited for different astronomical targets. What are the viewing conditions? What do you enjoy viewing? What is your experience? What is your budget? Everyone has different answers to these questions, but they determine the answer to the primary question, "What eyepiece should I buy?"

Many amateur astronomers don't appreciate how their eyes and experience influence what they "see." Usually there's little we can do to change our vision. Understand that you may not see the same detail someone else does. What you "see" is dependent on the entire path from the target to your brain. Whatever your vision, everyone can improve their viewing ability through experience. The ability to visually assess detail is primarily, but not solely, dependent on the optical path to your eye. No matter what your equipment, the more time you spend at the eyepiece learning to discern details, structure, and information the better you will become at "seeing."

Recognize that viewing conditions often limit what you see. Parts of the optical path can be altered. Changing location to reduce light pollution and the column of air you're looking through or purchasing a better telescope and/or eyepiece may help. Little can be done about the winds aloft. You can increase your understanding of how conditions affect the optical path by walking out every night there's a clear sky and looking at the stars. Naked eye or with binoculars, start watching how the stars twinkle, etc. This will help you decide when the limiting factor of your optical path is the telescope-eyepiece combination or the conditions. Sometimes you'll find excellent viewing when you're looking between clouds. Barring light pollution, you can see a great deal with your naked eye or a pair of binoculars while helping you learn the skies and increase your experience. This knowledge will help you decide what eyepiece to purchase.

Different telescope and eyepiece designs tend to work better for given astronomical targets. This doesn't mean you shouldn't use whatever you have for observing everything. Just understand why a fast reflector with a wide field of view eyepiece will usually not perform as well as a refractor of similar size with an orthoscopic eyepiece when trying to split doubles.

Don't think that you shouldn't purchase a quality eyepiece because you own a less than perfect telescope. If you stay with amateur astronomy you will own several telescopes and eyepieces. Most people experienced in the hobby have learned that telescopes may come and go, but a favorite eyepiece they enjoy is seldom sold. Both can be a lifetime investment, but in my experience people are more likely to swap their telescope while keeping a set of eyepieces they like.

Most experienced amateur astronomers can tell you what they enjoy viewing and their budget. Many newcomers to astronomy want to see everything and don't want to spend anything. You need to decide what you intend to view. If your target is a nebula then a wide angle eyepiece is in order. If you tend toward planetary viewing or splitting doubles then a narrow field of view is acceptable. There are many eyepieces in both categories in all price ranges. Generally, the wider the view and the better the resolution, the more an eyepiece will cost. As with all things there are exceptions and dependencies. Remember the eyepiece is one part of the optical chain so the lowest quality link of that chain determines the overall view.

This brings us to the question of cost and viewing quality. There's an old saying that a $10 telescope works as well as Mt. Palomar during a rainstorm. There are inexpensive gems that work as well as the very best eyepieces a lot of the time. Where a $1,000 eyepiece with a quality telescope will outperform a less expensive setup is during one of those few nights of unbelievably clear and stable viewing. You need to decide what your viewing habits and tolerances are. Do you go out and view several times a week? If you only view once or twice a month the odds of your hitting a night with exceptional viewing conditions are limited. You need to decide where you want to aim on the optics viewing curve of Cost versus Quality.

For the 'average' amateur astronomer a $1,000 telescope and eyepiece will provide 95 %+of the viewing capability of a $5,000 setup on most nights. What's it worth for you to gain that last 5 % of viewing quality a couple of nights a year? To gain small increases above the 95 % capability will cost increasingly more with fewer and fewer nights that you'll see any difference.

Read the forums and articles to prepare for your next eyepiece purchase. Learn to recognize people who have similar viewing preferences and equipment with more experience. Read what they like and dislike; recognize the reasons why. Increase your understanding of the night sky and experience by looking up every night you can. Naked eye, binoculars, 15 min, or whatever will help increase your understanding of how the optical path will affect what you see at the eyepiece. This will all help you choose the best eyepiece for your next purchase.

What Makes a "Favorite" Eyepiece?

By Tamiji Homma, Newbury Park, CA, USA

Another amateur astronomer once said, "for me the balance of quality eyepieces that are easy to view through, while still giving high quality views, is essential." The key word for me is "balance." It is hard to describe/pinpoint what balance really means. I guess it involves so many parameters under particular target, 'scope, sky condition, location (oxygen needed at very high elevation), temperature, humidity, observer's health (eye, mind, physical), reaction from people observing with, expectation, anticipation, etc. I find it is the total experience that brings WOW. Then when this happens you remember the eyepiece because it was a part of that experience and is a reminder of the experience. It becomes a favorite.

What to Do with Your Old, Inexpensive Eyepieces

By Ed Ting, Amherst, NH, USA

People sometimes ask me what to do with their old, inexpensive eyepieces. Use them for solar observing! Looking at the Sun can be a lot of fun, but there is no sense putting your $400 eyepieces in harm's way. I'll issue the usual caution here—don't even try solar observing unless you are sure you have adequate protection for your 'scope; if you don't, a fried eyepiece will be the least of your concerns. Here's a cautionary tale. One day, a friend and I were doing solar observing through a 4.5 in. Celestron reflector. As we were packing up, we accidentally removed the full-aperture solar filter first (this should always be done last), and while carrying the 'scope inside, the opening of the tube passed through the light path of the Sun for an instant. The total exposure was less than one second. When we got inside, we found the elements of the eyepiece had been fused together, and a hole had burned through the eyepiece cap.

Young People Can Own Nice Eyepieces, Too!

By Jeremy K., Baltimore, MD, USA

Young people can own nice eyepieces, too. Many people are often interested in astronomy, but few of them pursue the hobby seriously until significantly later in life. They believe it is an expensive hobby, one that cannot easily be undertaken when they are young because of so many other expenses and responsibilities. Since eyepieces are primarily 1.25 in. or 2 in. barrels, that means they can be used in nearly every telescope you may ever own. The shortest and longest focal length eyepieces may not provide optimal magnifications in every 'scope, but the middle focal lengths will work well in every telescope.

When I first became serious in the hobby while in high school, I purchased a $99 eyepiece kit that contained five Plössls and a 2× Barlow, and I used these eyepieces extensively for several years. When I finally decided to upgrade, I had looked through a number of different eyepieces and noticed a trend. The average quality wide-field eyepieces may be affordable, but all too often I saw them being replaced, over and over, by owners looking to squeeze more out of a moderately priced bracket of eyepieces that just was not there. It was at that point that I decided, the only way to upgrade your eyepiece collection is to slowly buy the best eyepieces you can, because those eyepieces can last decades and never need to be upgraded again. Going with an average eyepiece will only cost you more money each time you upgrade. I purchased a Pentax XW and slowly replaced the focal lengths I used the most with other top of the line eyepieces. It is almost impossible to justify new eyepieces because they would not really be an improvement, and I have saved a considerable amount of money as a result of having a steady collection.

Part II

Desk Reference of Astronomical Eyepieces

Chapter 6

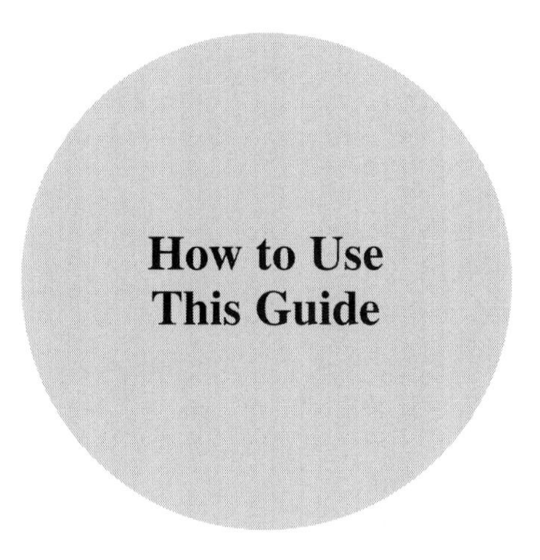

How to Use This Guide

The Desk Reference of Astronomical Eyepieces chapters provide you with alphabetical documentation on over 200 distinct eyepiece lines, both new and used. For each eyepiece line its vital performance characteristics are detailed for every focal length offered. These data include: focal length, apparent field of view, eye relief, field stop diameter, weight, optical lens count and groups, antireflection coating types, and available barrel sizes. The brand maker's Internet URL is also provided (if still available), and as appropriate brief notes indicating any points about the line as well as the other competing eyepiece brands and lines that have similar optical features for the reader to cross reference.

Market Overview

Since the eighteenth century when eyepiece designs began to flourish, the dominant centers of eyepiece manufacture have shifted numerous times. In the eighteenth century the dominant centers for optical manufacturer were of course in Europe, where the telescope was born. Then in the nineteenth century the United States began its venture into optics. An example of this was in 1853 when a German immigrant named John Jacob Bausch set up an optical goods shop in Rochester, New York. That start-up company would then one day grow into what today is known as Bausch & Lomb. Today, the eyepiece manufacturing industry is substantially dominated by Japan, Taiwan, the People's Republic of China (PRC). While there are still many major eyepiece brandings that are not from companies located in Japan, Taiwan, or China, they still depend on these dominant three centers of manufacture to produce the optical glass elements for their designs. As an example, of the many eyepiece brandings in the United States there are only two brands of

W. Paolini, *Choosing and Using Astronomical Eyepieces*, The Patrick Moore Practical Astronomy Series, DOI 10.1007/978-1-4614-7723-5_6,
© Springer Science+Business Media New York 2013

eyepieces that are still marked "Made in the USA" that rely on in-country sources for the manufacture of their optical elements. These two eyepiece lines are the Edmund Scientific RKE and the VERNONscope Brandon.

As for today's major centers of optical manufacture in Japan, Taiwan, and the PRC, the major companies that manufacture either the optical elements, or in some cases the entire eyepiece for the majority of popular brandings, include:

- Canon Global, Japan
- Carton Optical Industries, Japan
- Hioptic Optics, Japan
- Jinghua Optical Electronics Company, PRC
- Kson Optics-Electronics Company, PRC
- Kumming United-Optics, PRC
- Long Perng Company, Taiwan
- Nikon Corporation (Nippon Kogaku), Japan
- Suzhou Synta Optical Technology Company, PRC
- Vixen Optics Japan

The above list of course does not represent all the companies used by the major sellers of eyepiece brands. For brands such as Celestron, Meade, Tele Vue, and others, the exact optical manufacturing sources for their eyepiece optics is undisclosed. Other brands, such as Explore Scientific, readily states that its eyepieces are produced for it by Jinghua Optical in the PRC. And for a number of these brands, although the lens sources may be out-of-country, the optical designs can still be created by opticians in-country with only the manufacture of the glass to their proprietary designs being outsourced to foreign production companies.

Although some observers may prefer eyepieces made entirely in-country, in today's global economy this is less and less feasible, and modern makers of eyepieces instead use the best sources for individual components of an eyepiece sourced from multiple locations worldwide. To track down the original manufacturer of an eyepiece line that has many brandings, a quick visit to the Internet sites for the above companies will readily identify potential original equipment manufacturers for a number of the popular eyepiece brands. Be cautioned, however, that it is sometimes difficult to separate original manufacturers from resale distributors when reviewing these many sources.

If an observer finds what looks to be multiple re-brandings of a single eyepiece type or style, does this then mean these different apparent re-brandings will all perform the same? The answer to this is no. The reason this can be so is that one brand may specify slight changes to the design or coatings or mechanicals being offered by the original manufacturer to slightly distinguish their branding. As a result, there are times when the performance will differ between what looks like the same eyepiece sold under two different brand names. The best way to determine if two very similar-looking eyepiece brands behave exactly the same or not is not to examine the marketing information but instead to review the comments from other observers who have used these eyepieces in the field.

For the over 50 brands and 200 eyepiece lines that are documented, the following reference information is provided:

- Eyepiece brand name and line name (e.g., Tele Vue—Nagler)
- A photo of some or all of the focal lengths (optional)
- A table of the optical statistics detailing:

    1. Focal length
    2. Apparent field of view (AFOV)
    3. Eye relief
    4. Field stop diameter
    5. Weight
    6. Lens count and groups
    7. Coating type
    8. Barrel size

Note: These statistics are based primarily on marketing literature from the manufacturer or the reseller unless otherwise indicated. When no statistics are readily available, then values are footnoted as either measured (^{M}) or estimated (^{E}) based on general optical design criteria or similar brands. In some cases the entry is "—" indicating no reliable estimate can be made.

Notations throughout for the charts are:

E = estimated
M = measured
FC = fully single-coated
MC = some surfaces multi-coated
FMC = fully multi-coated

- Source: Internet URL of the manufacturer/reseller (*optional—supplied if still available*).
- Commonly Advertised Features: marketing features of the eyepiece line from advertised literature (*optional—supplied if still available*).

Note: Supplementing the data tables in Chap. 7, Appendix 2 in this book, Eyepiece Performance Classes, provides additional insight by categorizing the eyepiece brands and lines that perform very similarly to each other. The eyepieces in the Appendix 2 classes often directly compete with each other, and can sometimes even be re-brandings of the exact same original equipment manufacturer's (OEM) eyepiece. These competition classes therefore represent eyepieces that have very similar apparent fields of view, optical correction, and overall performance, and typically differ only in their external appearance/features, build quality, and price. Before choosing any astronomical eyepiece, cross-referencing the data in this chapter with Appendix 2 is highly recommended to ensure a more thorough examination of what is available in the marketplace.

Chapter 7

Agena to Docter

Agena ED (*Discontinued*)

Fig. 7.1 Agena ED eyepieces in focal lengths of 9.5 mm through 25 mm (© 2013 AgenaAstro at AgenaAstro.com)

W. Paolini, *Choosing and Using Astronomical Eyepieces*, The Patrick Moore Practical Astronomy Series, DOI 10.1007/978-1-4614-7723-5_7,
© Springer Science+Business Media New York 2013

Focal length (mm)	AFOV (degrees)	ER (mm)	Field stop (mm)	Weight (oz)	Design (Elem/Grp)	Coatings	Barrel (inches)
2.3	55	20	2.2[E]	7.0	6	FMC	1.25
3.8	55	20	3.6[E]	7.5	6	FMC	1.25
5.2	55	20	5.0[E]	7.2	6	FMC	1.25
7.5	55	20	7.2[E]	6.7	6	FMC	1.25
9.5	55	20	9.1[E]	6.4	6	FMC	1.25
12.5	55	20	12.0[E]	6.0	6	FMC	1.25
14	55	20	13.4[E]	5.8	6	FMC	1.25
18	55	20	17.3[E]	7.0	6	FMC	1.25
21	55	20	20.2[E]	6.9	6	FMC	1.25
25	55	20	24.0[E]	7.0	6	FMC	1.25

Source: www.agenaastro.com

Made with "extra-low dispersion" (ED) glass. This eyepiece line is within the following competition class of eyepieces with similar specifications and performance: Celestron X-Cel, Orion Epic ED II (older version), Vixen Lanthanum (LV), Vixen NLV.

Agena: Enhanced Wide Angle (EWA)

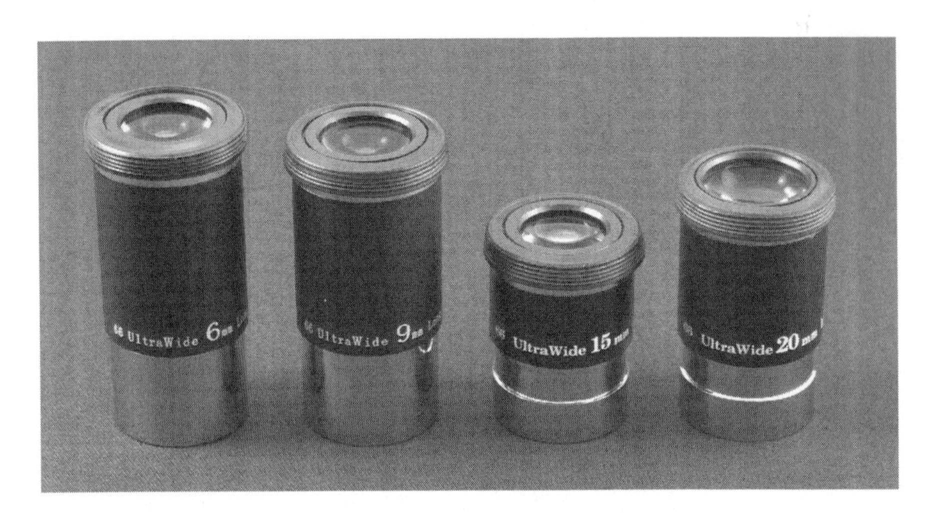

Fig. 7.2 Agena Enhanced Wide Angle 66° eyepieces (© 2013 AgenaAstro at AgenaAstro.com)

Focal length (mm)	AFOV (degrees)	ER (mm)	Field stop (mm)	Weight (oz)	Design (Elem/Grp)	Coatings	Barrel (inches)
6	66	14.8	6.9[E]	4.1	5/3	FMC	1.25
9	66	15	10.4[E]	4.0	6/4	FMC	1.25
15	66	13	17.3[E]	2.8	4/3	FMC	1.25
20	66	18	23.0[E]	3.8	5/3	FMC	1.25

Source: www.agenaastro.com

This eyepiece line is within the following competition class of eyepieces with similar specifications and performance: Orion Expanse, Owl Enhanced Superwide, Sky-Watcher Ultra Wide Angle, Telescope Service SWM Wide Angle, and William Optics WA 66. The 6 and 9 mm focal lengths of the series have a reputation of being excellent performers even in shorter focal ratio telescopes. They are one of the few budget wide fields that perform exceptionally well.

Agena: Mega Wide Angle (MWA)

Fig. 7.3 Agena Mega Wide Angle 100° eyepieces (© 2013 AgenaAstro at AgenaAstro.com)

Focal length (mm)	AFOV (degrees)	ER (mm)	Field stop (mm)	Weight (oz)	Design (Elem/Grp)	Coatings	Barrel (inches)
9	100	16	15.7[E]	14.6	6/4	FMC	1.25/2
16	100	16	27.0[E]	14.5	6/4	FMC	1.25/2

Source: www.agenaastro.com

Has a unique removable 2 in. barrel for use in 1.25 in. focusers. Observer reports the 9 mm focal length as having a fairly good off-axis; however reports for the 16 mm focal length indicate a longer focal ratio telescope is needed for best off-axis performance. This eyepiece line is within the following competition class of eyepieces with similar specifications and performance: Orion GiantView, TMB 100, Zhumell Z100.

Agena: Super Wide Angle

Fig. 7.4 Agena Super Wide Angle 70° eyepieces (© 2013 AgenaAstro at AgenaAstro.com)

Focal length (mm)	AFOV (degrees)	ER (mm)	Field stop (mm)	Weight (oz)	Design (Elem/Grp)	Coatings	Barrel (inches)
10	70	10	12.2[E]	2.9	5/4	FMC	1.25
15	70	13	18.3[E]	3.6	5/4	FMC	1.25
20	70	16	24.4[E]	4.2	5/4	FMC	1.25
26	70	20	31.8[E]	10.2	5/3	FMC	2
32	70	24	39.1[E]	14.4	5/4	FMC	2
38	70	28	46.0[E]	21.3	5/3	FMC	2

Source: www.agenaastro.com

These are parfocal. This eyepiece line is within the following competition class of eyepieces with similar specifications and performance: Astro-Professional SWA, Astro-Tech Wide Field, Garrett SuperWide Angle, GSO Superview, Meade QX, Olivon 70° Wide Angle, Opt Super View, Orion Q70 Super Wide-Field, Sky-Watcher PanaView, Telescope-Service WA, University Optics 70°, William Optics SWAN.

Agena: Ultra Wide Angle (UWA)

Fig. 7.5 Agena Ultra Wide Angle 73°–80° eyepieces (© 2013 AgenaAstro at AgenaAstro.com)

Focal length (mm)	AFOV (degrees)	ER (mm)	Field stop (mm)	Weight (oz)	Design (Elem/Grp)	Coatings	Barrel (inches)
11	74	9	14.2^E	3.4	5/3	FMC	1.25
16	75	12	20.9^E	3.4	6/4	FMC	1.25
15	75	18	19.6^E	9.2	6/4	FMC	2
20	73	20	25.5^E	21.3	7/4	FMC	2
30	80	22	41.9^E	20.1	5/3	FMC	2

Source: www.agenaastro.com

These are parfocal. This eyepiece line is within the following competition class of eyepieces with similar specifications and performance: Apogee Widescan (20 and 30 mm), Astrobuffet 1RPD (30 mm), BW Optik Ultrawide (30 mm), Kokusai Kohki Widescan I/II/III (20 and 30 mm), Moonfish Ultrawide (30 mm), Olivion 80° Ultra Wide Angle, Owl Astronomy Knight Owl Ultrawide Angle, Sky-Watcher Ultra Wide Angle, Surplus Shed Wollensak (30 mm), TAL Ultra Wide Angle, University Optics 80°, University Optics Widescan I/II/III (20 and 30 mm).

Agena: Wide Angle

Fig. 7.6 Agena Wide Angle 60°–65° eyepieces (© 2013 AgenaAstro at AgenaAstro.com)

Focal length (mm)	AFOV (degrees)	ER (mm)	Field stop (mm)	Weight (oz)	Design (Elem/Grp)	Coatings	Barrel (inches)
8	60	9	8.4^E	1.5^E	4/3	FMC	1.25
12	60	10	12.6^E	2^E	4/3	FMC	1.25
17	65	16	19.3^E	3^E	4/3	FMC	1.25
20	65	20	22.7^E	3^E	4/3	FMC	1.25

Source: www.agenaastro.com

This eyepiece line is within the following competition class of eyepieces with similar specifications and performance: Astro-Tech Series 6 Economy Wide Field, Burgess Wide Angle, Owl Advanced Wide Angle.

Antares: Classic Erfle

Fig. 7.7 The Speers Antares 52 mm Erfle (Image courtesy of Tony Miller, Stoney Creek, Ontario, Canada)

Focal length (mm)	AFOV (degrees)	ER (mm)	Field stop (mm)	Weight (oz)	Design (Elem/Grp)	Coatings	Barrel (inches)
30	72	20	37.7[E]	12.25	8/5	FMC	2
32	65	20	36.3[E]	10	6/3	FMC	2
32	70	20	39.1[E]	9.5	6/3	FMC	2
40	65	20	45.4[E]	11.5	6/3	FMC	2
52	50	–	45.4[E]	–	6/3	FMC	2

Source: www.antaresoptical.com

Antares: Elite Plössl

Focal length (mm)	AFOV (degrees)	ER (mm)	Field stop (mm)	Weight (oz)	Design (Elem/Grp)	Coatings	Barrel (inches)
5	52	6^E	4.5^E	4.0	7/4	FMC	1.25
7.5	52	5^E	6.8^E	2.5	5/3	FMC	1.25
10	52	6^E	9.1^E	2.8	5/3	FMC	1.25
15	52	10^E	13.6^E	2.8	5/3	FMC	1.25
20	52	13^E	18.2^E	3.5	5/3	FMC	1.25
25	52	17^E	22.7^E	4.0^E	5/3	FMC	1.25

Source: www.antaresoptical.com

These are "Made in Japan" and are parfocal. This eyepiece line is within the following competition class of eyepieces with similar specifications and performance: Baader Eudiascopic, Bresser 60° Plössl, Celestron Ultima, Kasai Astroplan, Meade Series 4000 Super Plössl (pre-1994, smooth-sided, 5-elements), Meade Series 5000 Super Plössl, Orion Ultrascopic, Parks Gold Series Plössl, Takahashi LE. Note—the Omcon Ultima and Tuthill Super Plössl lines were also competing brands, but these lines are not detailed.

Antares: Ortho (*Discontinued*)

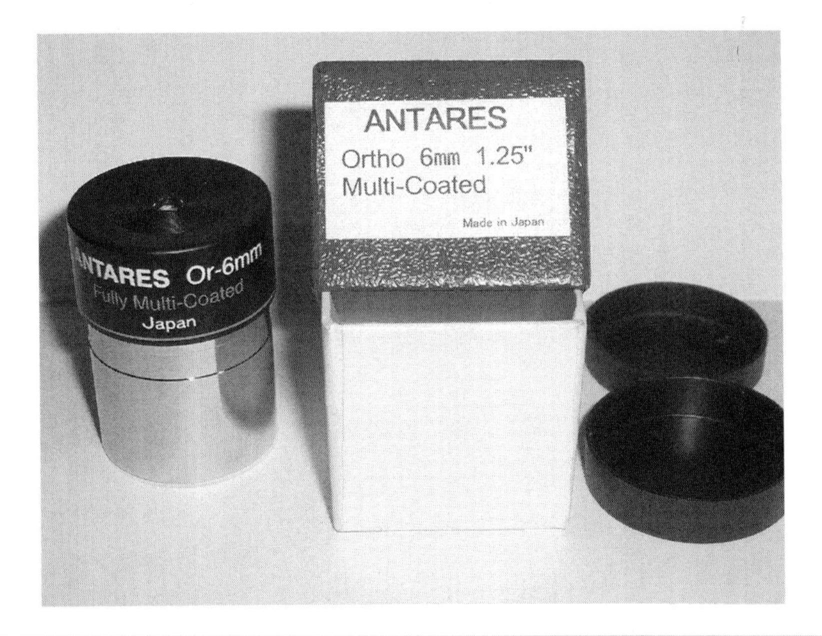

Fig. 7.8 6 mm Antares Ortho (Image courtesy of Andy Howie, Paisley, Scotland)

Focal length (mm)	AFOV (degrees)	ER (mm)	Field stop (mm)	Weight (oz)	Design (Elem/Grp)	Coatings	Barrel (inches)
5	40	4	3.4	<3[E]	4/2	FMC	1.25
6	40	4.9	4.2	<3[E]	4/2	FMC	1.25
7	40	6	4.9	<3[E]	4/2	FMC	1.25
9	40	7.6	6.3	<3[E]	4/2	FMC	1.25
12.5	40	10.4	9.2	<3[E]	4/2	FMC	1.25
18	40	15.2	12.5	<4[E]	4/2	FMC	1.25

Source: www.antaresoptical.com

Very similar in performance and form factor to the much heralded Baader Genuine Orthos and the University Optics HD Orthos. An excellent planetary performer. This eyepiece line is within the following competition class of eyepieces with similar specifications and performance: Apogee Super Abbe Orthoscopic, Baader Planetarium Classic Ortho, Baader Planetarium Genuine Abbe Ortho, Cave Orthostar Orthoscopic, Celestron Ortho, Edmund Scientific Ortho, Kokusai Kohki Abbe Ortho, Kson Super Ortho, Masuyama Orthoscopic, Meade Research Grade Ortho, Meade Series II Orthoscopic, Nikon Ortho, Siebert Optics Star Splitter/Super Star Splitter, Takahashi Ortho, Telescope Service Ortho, Unitron Ortho, University Optics Abbe HD Orthoscopic, University Optics Abbe Volcano Orthoscopic, University Optics O.P.S. Orthoscopic Planetary Series, University Optics Super Abbe Orthoscopic, VERNONscope Brandon.

Antares: Plössl

Fig. 7.9 The Antares Plössls with .096 barrels in 6 mm, 10 mm, 17 mm, and 25 mm (Image courtesy of Doug Bailey, Franklinton, LA, USA)

Focal length (mm)	AFOV (degrees)	ER (mm)	Field stop (mm)	Weight (oz)	Design (Elem/Grp)	Coatings	Barrel (inches)
6	52	4^E	5.4^E	1.5	4/2	FMC	1.25
7.5	52	5^E	6.8^E	1.5	4/2	FMC	1.25
10	52	7^E	9.1^E	1.8	4/2	FMC	1.25
12.5	52	8^E	11.3^E	2.0	4/2	FMC	1.25
15	52	10^E	13.6^E	2.8	4/2	FMC	1.25
17	52	11^E	15.4^E	2.8	4/2	FMC	1.25
20	52	13^E	18.2^E	3.3	4/2	FMC	1.25
25	52	17^E	22.7^E	4.5	4/2	FMC	1.25
32	52	20^E	29.0^E	4.5	4/2	FMC	1.25
40	40^E	20^E	27.9^E	5.3	4/2	FMC	1.25

Source: www.antaresoptical.com

This eyepiece line is within the following competition class of eyepieces with similar specifications and performance: Astro-Professional Plössl, Astro-Tech High Grade Plössl, Astro-Tech Value Line Plössl, Bresser 52° Super Plössl, Carton Plössl, Celestron Omni, Celestron Silvertop Plössl, Clavé Plössl, Coronado CeMax, Edmund Scientific Plössl, Garrett Optical Plössl, GSO Plössl, GTO Plössl, Long Perng Plössl, Meade Series 3000 Plössl, Meade Series 4000 Super Plössl, Olivon Plössl, Opt Plössl, Orion HighLight Plössl, Orion Sirius Plössl, Owl Black Night Plössl, Parks Silver Series, Sky-Watcher SP-Series Super Plössl, Smart Astronomy Sterling Plössl, TAL—Symmetrical Super Plössl, Telescope Service Plössl, Telescope Service Super Plössl, Tele Vue Plössl, Vixen NPL.

Antares: Speers-WALER Series (Wide Angle Long Eye Relief)

Fig. 7.10 Speers WALER 7.2 and 17 mm eyepieces (Eyepieces courtesy of www.handsonoptics.com. Image by the author)

Focal length (mm)	AFOV (degrees)	ER (mm)	Field stop (mm)	Weight (oz)	Design (Elem/Grp)	Coatings	Barrel (inches)
SERIES I							
5–8	82–89	–	–	–	9	FMC	1.25
7.5	82	12	10.7^E	10.8	8/5	FMC	1.25
10	82	12	14.3^E	11.0	8/5	FMC	1.25
14	82	12	20.0^E	10.5	8/5	FMC	1.25
18	82	–	25.8^E	–	9	FMC	1.25
SERIES II							
8.5–12	84–89	12	11.5–14.9	22.4	8	FMC	1.25
4.9	82	12	7.0^E	10.8	8	FMC	1.25
7.2	82	12	10.3^E	11.5	8	FMC	1.25
9.4	82	16	13.5^E	11.5	8	FMC	1.25
13.4	80	12	18.7^E	12.5	9	FMC	1.25
17	78	19	23.1^E	12	9	FMC	1.25

Source: www.antaresoptical.com

The Speers WALER eyepieces provide excellently corrected wide fields of view. Very much a “sleeper” in the 82° AFOV class of eyepieces. This eyepiece line is within the following competition class of eyepieces with similar specifications and performance: Astro-Professional UWA, Celestron Axiom LX, Celestron Luminos, Docter UWA, Explore Scientific 82° Nitrogen Purged, Meade Series 4000 Ultra Wide Angle, Meade Series 5000 Ultra Wide Angle, Orion MegaView Ultra-Wide, Sky-Watcher Nirvana UWA, Sky-Watcher Sky Panorama, Tele Vue Nagler, Williams Optics UWAN.

Antares: W70

Fig. 7.11 Antares 25 mm W70 eyepiece with bolt case (Eyepieces courtesy of David Ittner, Newark, CA, USA)

Focal length (mm)	AFOV (degrees)	ER (mm)	Field stop (mm)	Weight (oz)	Design (Elem/Grp)	Coatings	Barrel (inches)
4.3	70	10	5.0[E]	7[E]	6	FMC	1.25
6	70	14	7.0[E]	7[E]	8	FMC	1.25
9.7	70	15	11.3[E]	7.5	8	FMC	1.25
15	70	10	17.5[E]	3.5	4	FMC	1.25
25	70	11	29.5	4.5	6	FMC	1.25
34	70	15	39.8[E]	9.5	6	FMC	2.00

Source: www.antaresoptical.com

APM: UWA Planetary

Fig. 7.12 APM 7 mm UWA Planetary (Image courtesy of Andy Howie, Paisley, Scotland)

Focal length (mm)	AFOV (degrees)	ER (mm)	Field stop (mm)	Weight (oz)	Design (Elem/Grp)	Coatings	Barrel (inches)
2.5	58	14	2.4^E	5	5/3	FMC	1.25
3.2	58	14	3.1^E	5	5/3	FMC	1.25
4	58	14	3.9^E	5	5/3	FMC	1.25
5	58	14	4.9^E	5	5/3	FMC	1.25
6	58	14	5.8^E	5	5/3	FMC	1.25
7.5	58	14	7.3^E	5	5/3	FMC	1.25
9	58	14	8.7^E	5	6/4	FMC	1.25

Source: www.apm-telescopes.de

This eyepiece line is within the following competition class of eyepieces with similar specifications and performance: Burgess/TMB Planetary, Olivon Wide Angled Plössl, Owl Astronomy High Resolution Planetary, Telescope Service Planetary HR, TMB Planetary II.

Apogee: Super Abbe Orthoscopic (*Discontinued*)

Fig. 7.13 The complete line of Apogee Super Abbe Ortho eyepieces (Image courtesy of William Rose, Larkspur, CO, USA)

Focal length (mm)	AFOV (degrees)	ER (mm)	Field stop (mm)	Weight (oz)	Design (Elem/Grp)	Coatings	Barrel (inches)
4.8	46	4^E	3.9^E	$<3^E$	4/2	FMC	1.25
7.7	46	6^E	6.5	$<3^E$	4/2	FMC	1.25
10.5	46	9^E	8.4^E	$<3^E$	4/2	FMC	1.25
16.8	46	14^E	13.5^E	$<4^E$	4/2	FMC	1.25
24	46	20^E	20.5	$<4^E$	4/2	FMC	1.25

Source: www.apogeeinc.com

This eyepiece line is within the following competition class of eyepieces with similar specifications and performance: Antares Ortho, Baader Planetarium Classic Ortho, Baader Planetarium Genuine Abbe Ortho, Cave Orthostar Orthoscopic, Celestron Ortho, Edmund Scientific Ortho, Kokusai Kohki Abbe Ortho, Kson Super Ortho, Masuyama Orthoscopic, Meade Research Grade Ortho, Meade Series II Orthoscopic, Nikon Ortho, Siebert Optics Star Splitter/Super Star Splitter,

Takahashi Ortho, Telescope Service Ortho, Unitron Ortho, University Optics Abbe HD Orthoscopic, University Optics Abbe Volcano Orthoscopic, University Optics O.P.S. Orthoscopic Planetary Series, University Optics Super Abbe Orthoscopic, VERNONscope Brandon.

Apogee: Widescan III (*Discontinued*)

Focal length (mm)	AFOV (degrees)	ER (mm)	Field stop (mm)	Weight (oz)	Design (Elem/Grp)	Coatings	Barrel (inches)
20	84	–	25[E]	21[E]	7/4	FMC	2
30	84	–	41[E]	20[E]	5/3	FMC	2

Source: www.apogeeinc.com

This eyepiece line is within the following competition class of eyepieces with similar specifications and performance: Agena Ultra Wide Angle, Astrobuffet 1RPD (30 mm), BW Optik Ultrawide (30 mm), Kokusai Kohki Widescan I/II/III (20 and 30 mm), Moonfish Ultrawide (30 mm), Olivion 80° Ultra Wide Angle, Owl Astronomy Knight Owl Ultrawide Angle, Sky-Watcher Ultra Wide Angle, Surplus Shed Wollensak (30 mm), TAL Ultra Wide Angle, University Optics 80°, University Optics Widescan I/II/III (20 and 30 mm).

Astrobuffet: 1 RPD

Focal length (mm)	AFOV (degrees)	ER (mm)	Field stop (mm)	Weight (oz)	Design (Elem/Grp)	Coatings	Barrel (inches)
15	80	11[E]	21[E]	–	5/3[E]	FMC	2
30	84	22[E]	41[E]	20[E]	5/3	FMC	2

Source: www.astrobuffet.com

This eyepiece line is within the following competition class of eyepieces with similar specifications and performance: Agena Ultra Wide Angle, Apogee Widescan (20 and 30 mm), BW Optik Ultrawide (30 mm), Kokusai Kohki Widescan I/II/III (20 and 30 mm), Moonfish Ultrawide (30 mm), Olivion 80° Ultra Wide Angle, Owl Astronomy Knight Owl Ultrawide Angle, Sky-Watcher Ultra Wide Angle, Surplus Shed Wollensak (30 mm), TAL Ultra Wide Angle, University Optics 80°, University Optics Widescan I/II/III (20 and 30 mm).

Astro-Physics: Super Planetary AP-SPL (*Discontinued*)

Fig. 7.14 The complete AP-SPL eyepiece Line (Image courtesy of William Rose, Larkspur, CO, USA)

Focal length (mm)	AFOV (degrees)	ER (mm)	Field stop (mm)	Weight (oz)	Design (Elem/Grp)	Coatings	Barrel (inches)
4	42	5.0	2.60	2.5	3/2	FMC	1.25
5	42	6.4	3.57	2.5	3/2	FMC	1.25
6	42	7.9	4.37	2.6	3/2	FMC	1.25
8	42	10.7	6.04	2.8	3/2	FMC	1.25
10	42	13.7	7.65	3.0	3/2	FMC	1.25
12	42	16.8	8.91	3.2	3/2	FMC	1.25

Source: www.astro-physics.com

Designed for telescopes as fast as f/4; housing made of non-metallic black Delrin to resist fogging and prevent freezing to skin; barrels made of stainless steel for longevity. This eyepiece line is within the following competition class of eyepieces with similar specifications and performance: Pentax SMC Ortho, Pentax XO, Pentax XP, TMB Aspheric Ortho, TMB Supermonocentric, Zeiss CZJ Ortho, Zeiss ZAO I/ZAO II.

Astro-Professional: EF Flatfield

Focal length (mm)	AFOV (degrees)	ER (mm)	Field stop (mm)	Weight (oz)	Design (Elem/Grp)	Coatings	Barrel (inches)
16	60	17	16.6	4	6	FMC	1.25
19	65	17	21.3	4	5	FMC	1.25
27	53	21	24.4	6	6	FMC	1.25

Source: www.astro-professional-france.fr

This eyepiece line is within the following competition class of eyepieces with similar specifications and performance: Astro-Tech Flat Field, BST Flat Field, Orion Edge-On Flat-Field, Sky-Watcher Extra Flat, Smart Astronomy Extra Flat Field, Telescope-Service Edge-On Flat Field.

Astro-Professional: Long Eye Relief Planetary

Focal length (mm)	AFOV (degrees)	ER (mm)	Field stop (mm)	Weight (oz)	Design (Elem/Grp)	Coatings	Barrel (inches)
3	55	20	2.9[E]	10	7/4	FMC	1.25
5	55	20	4.8[E]	10	7/4	FMC	1.25
6	55	20	5.8[E]	10	7/4	FMC	1.25
9	55	20	8.6[E]	9	7/4	FMC	1.25
12.5	55	20	12.0[E]	8	7/4	FMC	1.25
14.5	55	20	13.9[E]	7	7/4	FMC	1.25
18	55	20	17.3[E]	6	5/3	FMC	1.25

Source: www.astro-professional-france.fr

This eyepiece line is within the following competition class of eyepieces with similar specifications and performance: Astro-Tech Long Eye Relief, Long Perng Long Eye Relief, Orion Edge-On Planetary, Smart Astronomy SA Solar System Long Eye Relief, Stellarvue Planetary, William Optics SPL, Zhumell Z Series.

Astro-Professional: Plössl

Focal length (mm)	AFOV (degrees)	ER (mm)	Field stop (mm)	Weight (oz)	Design (Elem/Grp)	Coatings	Barrel (inches)
4	50	3^E	3.5^E	3^E	4/2	–	1.25
6.5	50	4^E	5.7^E	3^E	4/2	–	1.25
10	50	7^E	8.7^E	3^E	4/2	–	1.25
15	50	10^E	13.1^E	3^E	4/2	–	1.25
20	50	14^E	17.5^E	3^E	4/2	–	1.25
25	50	17^E	21.8^E	3^E	4/2	–	1.25
30	50	20^E	26.2^E	4^E	4/2	–	1.25
40	40	27^E	27.9^E	6^E	4/2	–	1.25

Source: www.astro-professional-france.fr

This eyepiece line is within the following competition class of eyepieces with similar specifications and performance: Antares Plössl, Astro-Tech High Grade Plössl, Astro-Tech Value Line Plössl, Bresser 52° Super Plössl, Carton Plössl, Celestron Omni, Celestron Silvertop Plössl, Clavé Plössl, Coronado CeMax, Edmund Scientific Plössl, Garrett Optical Plössl, GSO Plössl, GTO Plössl, Long Perng Plössl, Meade Series 3000 Plössl, Meade Series 4000 Super Plössl, Olivon Plössl, Opt Plössl, Orion HighLight Plössl, Orion Sirius Plössl, Owl Black Night Plössl, Parks Silver Series, Sky-Watcher SP-Series Super Plössl, Smart Astronomy Sterling Plössl, TAL—Symmetrical Super Plössl, Telescope Service Plössl, Telescope Service Super Plössl, Tele Vue Plössl, Vixen NPL.

Astro-Professional: SWA

Focal length (mm)	AFOV (degrees)	ER (mm)	Field stop (mm)	Weight (oz)	Design (Elem/Grp)	Coatings	Barrel (inches)
10	70	10	12.2^E	2.9	5/4	FMC	1.25
15	70	13	18.3^E	3.6	5/4	FMC	1.25
20	70	16	24.4^E	4.2	5/4	FMC	1.25
26	70	20	31.8^E	10.2	5/3	FMC	2
32	70	24	39.1^E	14.4	5/4	FMC	2
38	70	28	46.0^E	21.3	5/3	FMC	2

Source: www.astro-professional-france.fr

These are parfocal. This eyepiece line is within the following competition class of eyepieces with similar specifications and performance: Agena Super Wide Angle, Astro-Tech Wide Field, Garrett SuperWide Angle, GSO Superview, Meade QX, Olivon 70° Wide Angle, Opt Super View, Orion Q70 Super Wide-Field, Sky-Watcher PanaView, Telescope-Service WA, University Optics 70°, William Optics SWAN.

Astro-Professional: UWA

Focal length (mm)	AFOV (degrees)	ER (mm)	Field stop (mm)	Weight (oz)	Design (Elem/Grp)	Coatings	Barrel (inches)
4	82	12	6.0	200	7/4	FMC	1.25
7	82	12	12.0	200	7/4	FMC	1.25
16	82	12	28.6	200	7/4	FMC	1.25
28	82	18	43.5	1,000	6/4	FMC	2

Source: www.astro-professional-france.fr

These are parfocal. This eyepiece line is within the following competition class of eyepieces with similar specifications and performance: Antares Speers-WALER Series, Celestron Axiom LX, Celestron Luminos, Docter UWA, Explore Scientific 82° Nitrogen Purged, Meade Series 4000 Ultra Wide Angle, Meade Series 5000 Ultra Wide Angle, Orion MegaView Ultra-Wide, Sky-Watcher Nirvana UWA, Sky-Watcher Sky Panorama, Tele Vue Nagler, Williams Optics UWAN.

Astro-Tech: AF Series 70° Field

Fig. 7.15 The Astro-Tech AF Series 70 eyepieces (Image by the author)

Focal length (mm)	AFOV (degrees)	ER (mm)	Field stop (mm)	Weight (oz)	Design (Elem/Grp)	Coatings	Barrel (inches)
3.5	70	17	4.3[E]	17.6	8	FMC	1.25/2.00
5	70	17	6.1[E]	16.8	8	FMC	1.25/2.00
8	70	17	9.8[E]	16.8	8	FMC	1.25/2.00
13	70	17	15.9[E]	16.8	8	FMC	1.25/2.00
17	70	17	20.8[E]	16	8	FMC	1.25/2.00
22	70	17	26.9[E]	17.6	8	FMC	2.00

Source: www.astronomics.com

This eyepiece line is within the following competition class of eyepieces with similar specifications and performance: Olivon 70° Ultra Wide Angle, Telescope Service Expanse ED.

Astro-Tech: Flat Field

Fig. 7.16 The Astro-Tech Flat Field eyepieces (Image by the author)

Focal length (mm)	AFOV (degrees)	ER (mm)	Field stop (mm)	Weight (oz)	Design (Elem/Grp)	Coatings	Barrel (inches)
12	60	11	12.6[E]	4	5	FBMC	1.25
19	65	12	21.6[E]	4	5	FBMC	1.25
27	53	18	25.0[E]	6.4	5	FBMC	1.25

Source: www.astronomics.com

These have twist-up eyeguards and are parfocal. This eyepiece line is within the following competition class of eyepieces with similar specifications and performance: Astro-Professional EF Flatfield, BST Flat Field, Orion Edge-On Flat-Field; Sky-Watcher Extra Flat, Smart Astronomy Extra Flat Field, Telescope-Service Edge-On Flat Field.

Astro-Tech: High Grade Plössl

Focal length (mm)	AFOV (degrees)	ER (mm)	Field stop (mm)	Weight (oz)	Design (Elem/Grp)	Coatings	Barrel (inches)
4	55	2	3.8[E]	2	4/2	FMC	1.25
6	55	4	5.8[E]	2	4/2	FMC	1.25
12.5	55	8	12.0[E]	3	4/2	FMC	1.25
17	55	11	16.3[E]	3.5	4/2	FMC	1.25
20	55	12	19.2[E]	3.5	4/2	FMC	1.25

Source: www.astronomics.com

These eyepieces have a unique distinction over most other 4-element Plössl lines as having an extended AFOV. Almost all other Plössl lines have an AFOV of 50–52°, whereas the Astro-Tech High Grade Plössl, Smart Astronomy Sterling Plössl, and Long Perng Plössl have extended this to an advertised 55° (closer to 57° when measured). This larger AFOV, although only a few degrees more than a standard Plössl, is visually more impressive, conveying the impression of an eyepiece in the next larger AFOV class. For improved off-axis performance compared to typical Plössls, this line also uses unconventional concave lens surfaces instead of the standard flat surfaces on the outward-facing eye lens and field lens. Other 4-element Plössl lines that use concave lens surfaces include the Meade 4000 (current 4-element version), Smart Astronomy Sterling Plössl, and Tele Vue Plössl.

This eyepiece line is within the following competition class of eyepieces with similar specifications and performance: Antares Plössl, Astro-Professional Plössl, Astro-Tech Value Line Plössl, Bresser 52° Super Plössl, Carton Plössl, Celestron Omni, Celestron Silvertop Plössl, Clavé Plössl, Coronado CeMax, Edmund Scientific Plössl, Garrett Optical Plössl, GSO Plössl, GTO Plössl, Long Perng Plössl, Meade Series 3000 Plössl, Meade Series 4000 Super Plössl, Olivon Plössl, Opt Plössl, Orion HighLight Plössl, Orion Sirius Plössl, Owl Black Night Plössl, Parks Silver Series, Sky-Watcher SP-Series Super Plössl, Smart Astronomy Sterling Plössl, TAL—Symmetrical Super Plössl, Telescope Service Plössl, Telescope Service Super Plössl, Tele Vue Plössl, Vixen NPL.

Astro-Tech: Long Eye Relief

Fig. 7.17 The Astro-Tech Long Eye Relief eyepieces (Image by the author)

Focal length (mm)	AFOV (degrees)	ER (mm)	Field stop (mm)	Weight (oz)	Design (Elem/Grp)	Coatings	Barrel (inches)
3	55	20	2.9^{E}	8.6	7	FBMC	1.25
9	55	20	8.6^{E}	7.0	7	FBMC	1.25
12.5	55	20	12.0^{E}	6.4	7	FBMC	1.25

Source: www.astronomics.com

These are parfocal. This eyepiece line is within the following competition class of eyepieces with similar specifications and performance: Astro-Professional Long Eye Relief Planetary, Long Perng Long Eye Relief, Orion Edge-On Planetary, Smart Astronomy SA Solar System Long Eye Relief, Stellarvue Planetary, William Optics SPL, Zhumell Z Series.

Astro-Tech: Paradigm Dual ED

Fig. 7.18 The Astro-Tech Paradigm Dual ED eyepieces (Image courtesy of Bob Ryan, Cincinnati, OH, USA)

Focal length (mm)	AFOV (degrees)	ER (mm)	Field stop (mm)	Weight (oz)	Design (Elem/Grp)	Coatings	Barrel (inches)
5	60	13	5.2^E	7.2	6	FMC	1.25
8	60	13	8.4^E	6.4	6	FMC	1.25
12	60	13	12.6^E	6.4	6	FMC	1.25
15	60	15	15.7^E	6.4	6	FMC	1.25
18	60	13	18.8^E	5.6	6	FMC	1.25
25	60	15	26.2^E	6.4	6	FMC	1.25

Source: www.astronomics.com

These contain two "extra low-dispersion glass elements"; have twist-up eyeguard; and are parfocal. This eyepiece line is a reasonably priced excellent alternative to the standard Plössl, offering similarly excellent performance, a wider AFOV, more comfortable longer eye relief, and improved ergonomic features such as adjustable eyeguards. Highly recommended. This eyepiece line is within the following competition class of eyepieces with similar specifications and performance: BST Explorer ED, Olivon 60° ED Wide Angle, Orion Epic II ED, Pentax XF, Telescope Service NED "ED" Flat Field.

Astro-Tech: Series 6 Economy Wide Field

Fig. 7.19 The Astro-Tech Series 6 Economy eyepieces (Image courtesy of David Elosser, NC, USA)

Focal length (mm)	AFOV (degrees)	ER (mm)	Field stop (mm)	Weight (oz)	Design (Elem/Grp)	Coatings	Barrel (inches)
8	60	8	8.4^{E}	1.5	4/3	FMC	1.25
12	60	8	12.6^{E}	2	4/3	FMC	1.25
17	65	13	19.3^{E}	3	4/3	FMC	1.25

Source: www.astronomics.com

These are parfocal. This eyepiece line is within the following competition class of eyepieces with similar specifications and performance: Agena Wide Angle, Burgess Wide Angle, Owl Advanced Wide Angle.

Astro-Tech: Titan Type II ED Premium 2″ Wide Field

Fig. 7.20 Astro-Tech 40 mm Titan-II ED (Image by the author)

Focal length (mm)	AFOV (degrees)	ER (mm)	Field stop (mm)	Weight (oz)	Design (Elem/Grp)	Coatings	Barrel (inches)
30	68	14	35.6[E]	11	6	FMC	2
35	68	14	41.5[E]	12	6	FMC	2
40	68	14	46.0[E]	18	6	FMC	2

Source: www.astronomics.com

These have an extra-low dispersion glass element, twist up eyeguard, and are parfocal. The 40 mm is the best performing member of the line, offering performance indicative of much more expensive wide fields. Offers a very good price-performance ratio. This eyepiece line is within the following competition class of eyepieces with similar specifications and performance: Sky-Watcher Aero, TMB Paragon.

Fig. 7.21 38 mm Titan, predecessor to the Titan-II series (Image courtesy of Erika Rix, Liberty Hill, TX, USA)

Astro-Tech: Value Line Plössl

Focal length (mm)	AFOV (degrees)	ER (mm)	Field stop (mm)	Weight (oz)	Design (Elem/Grp)	Coatings	Barrel (inches)
4	52	6	3.6[E]	2.6	4/2	FMC	1.25
6	52	5	5.4[E]	2.6	4/2	FMC	1.25
9	52	6	8.2[E]	2.6	4/2	FMC	1.25
12	52	8	10.9[E]	2.9	4/2	FMC	1.25
15	52	13	13.6[E]	3.1	4/2	FMC	1.25
20	52	20	18.2[E]	3.5	4/2	FMC	1.25
25	52	22	22.7[E]	3.6	4/2	FMC	1.25
32	50	20	27.0[E]	4.2	4/2	FMC	1.25
40	40	20	27.0[E]	5.5	4/2	FMC	1.25

Source: www.astronomics.com

This eyepiece line is within the following competition class of eyepieces with similar specifications and performance: Antares Plössl, Astro-Professional Plössl, Astro-Tech High Grade Plössl, Bresser 52° Super Plössl, Carton Plössl, Celestron Omni, Celestron Silvertop Plössl, Clavé Plössl, Coronado CeMax, Edmund Scientific Plössl, Garrett Optical Plössl, GSO Plössl, GTO Plössl, Long Perng Plössl, Meade Series 3000 Plössl, Meade Series 4000 Super Plössl, Olivon Plössl, Opt Plössl, Orion HighLight Plössl, Orion Sirius Plössl, Owl Black Night Plössl, Parks Silver Series, Sky-Watcher SP-Series Super Plössl, Smart Astronomy Sterling Plössl, TAL—Symmetrical Super Plössl, Telescope Service Plössl, Telescope Service Super Plössl, Tele Vue Plössl, Vixen NPL.

Astro-Tech: Wide Field

Focal length (mm)	AFOV (degrees)	ER (mm)	Field stop (mm)	Weight (oz)	Design (Elem/Grp)	Coatings	Barrel (inches)
10	70	7	12.2[E]	2.4	5	FMC	1.25
32	70	17	39.1[E]	16	5	FMC	2
38	70	20	46.0[E]	25	5	FMC	2

Source: www.astronomics.com

These have a twist-up eyeguard and are parfocal. This eyepiece line is within the following competition class of eyepieces with similar specifications and performance: Agena Super Wide Angle, Astro-Professional SWA, Garrett SuperWide Angle, GSO Superview, Meade QX, Olivon 70° Wide Angle, Opt Super View, Orion Q70 Super Wide-Field, Sky-Watcher PanaView, Telescope-Service WA, University Optics 70°, William Optics SWAN.

Baader Planetarium: Classic Ortho/Plössl

Fig. 7.22A Baader Classic Orthos and Plössl with the Q-Turret accessory (© Baader-Planetarium at Baader-Planetarium.com)

Focal length (mm)	AFOV (degrees)	ER (mm)	Field stop (mm)	Weight (oz)	Design (Elem/Grp)	Coatings	Barrel (inches)
6	50	5^{E}	5.2^{E}	$<3^{E}$	4/2	FMC	1.25
10	50	8^{E}	8.7^{E}	$<3^{E}$	4/2	FMC	1.25
18	50	15^{E}	15.7^{E}	$<3^{E}$	4/2	FMC	1.25
32	50	21^{E}	27.0^{E}	$<3^{E}$	4/2	FMC	1.25

Source: http://www.baader-planetarium.com/

These have a unique inset-volcano design, have a wing-type eyeguard, and are parfocal. The concept behind this newest Abbe Ortho line is to create an affordable all-in-one set of high performing eyepieces. As such, they reach their fullest potential as a complete set, which can be purchased bundled with a Barlow and Turret a sort of "Volkseyepiece" set. An excellent performer in the tradition of the Baader Genuine Abbe Ortho, receiving many excellent reviews from observers. I highly regarded new entry into the Abbe Ortho marketplace.

This eyepiece line is within the following competition class of eyepieces with similar specifications and performance: Antares Ortho, Apogee Super Abbe Orthoscopic, Baader Planetarium Genuine Abbe Ortho, Cave Orthostar Orthoscopic, Celestron Ortho, Edmund Scientific Ortho, Kokusai Kohki Abbe Ortho, Kson Super Ortho, Masuyama Orthoscopic, Meade Research Grade Ortho, Meade Series II Orthoscopic, Nikon Ortho, Siebert Optics Star Splitter/Super Star Splitter, Takahashi Ortho, Telescope Service Ortho, Unitron Ortho, University Optics Abbe HD Orthoscopic, University Optics Abbe Volcano Orthoscopic, University Optics O.P.S. Orthoscopic Planetary Series, University Optics Super Abbe Orthoscopic, VERNONscope Brandon.

Fig. 7.22B Baader Classic Orthos/Plössl and Q-Turret gift set (© Baader-Planetarium at Baader-Planeetarium.com)

Baader Planetarium: Eudiascopic

Focal length (mm)	AFOV (degrees)	ER (mm)	Field stop (mm)	Weight (oz)	Design (Elem/Grp)	Coatings	Barrel (inches)
3.8	50	5^E	3.3^E	4^E	7/4	FMC	1.25
5	50	6^E	4.4^E	4^E	7/4	FMC	1.25
7.5	50	5^E	6.5^E	3^E	5/3	FMC	1.25
10	50	6^E	8.7^E	3^E	5/3	FMC	1.25
15	50	10^E	13.1^E	3^E	5/3	FMC	1.25
20	50	13^E	17.5^E	4^E	5/3	FMC	1.25
25	50	17^E	21.8^E	4^E	5/3	FMC	1.25
30	50	21^E	26.2^E	6^E	5/3	FMC	1.25
35	45	25^E	27.0^E	7^E	5/3	FMC	1.25

Source: www.baader-planetarium.com

These have coatings that meet U. S. military standards and are supplied with a removable "winged" rubber eyeguard. This eyepiece line is within the following competition class of eyepieces with similar specifications and performance: Antares Elite, Bresser 60° Plössl, Celestron Ultima, Kasai Astroplan, Meade Series 4000 Super Plössl (pre-1994, smooth sided, 5-elements), Meade Series 5000 Super Plössl, Orion Ultrascopic, Parks Gold Series Plössl, Takahashi LE. *Note:* The Omcon Ultima and Tuthill Super Plössl lines were also competing brands, but these lines are not detailed.

Baader Planetarium: Genuine Abbe Ortho (*Discontinued*)

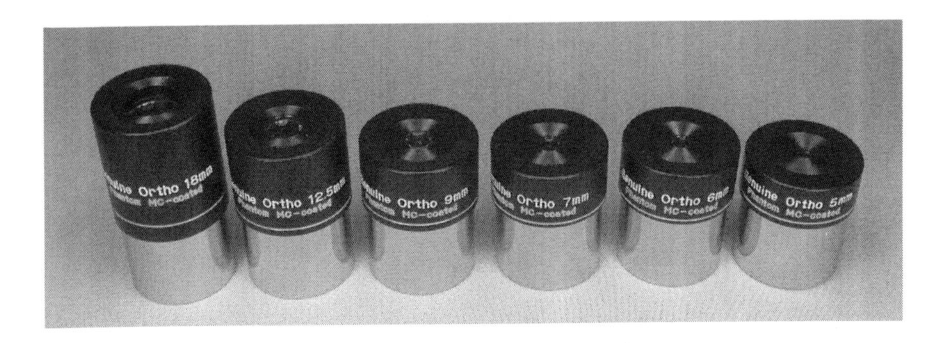

Fig. 7.23 The complete line of Baader Genuine Abbe Ortho eyepieces (Image courtesy of William Rose, Larkspur, CO, USA)

Focal length (mm)	AFOV (degrees)	ER (mm)	Field stop (mm)	Weight (oz)	Design (Elem/Grp)	Coatings	Barrel (inches)
5	40	4	3.4	$<3^{E}$	4/2	FMC	1.25
6	40	4.9	4.2	$<3^{E}$	4/2	FMC	1.25
7	40	6	4.9	$<3^{E}$	4/2	FMC	1.25
9	40	7.6	6.3	$<3^{E}$	4/2	FMC	1.25
12.5	40	10.4	9.2	$<3^{E}$	4/2	FMC	1.25
18	40	15.2	12.5	$<4^{E}$	4/2	FMC	1.25

Source: www.baader-planetarium.com

These have special Phantom Group™ multi-coatings with reflectivity of less than 0.2 % at each air-to-glass surface and are parfocal. This eyepiece line, with the University Optics HD Orthos, have the reputation as being the finest performing Abbe Ortho for their price class. Not quite in the ultimate performance league of the best-of-the-best, like the Zeiss Abbe Ortho (ZAO), but also costing six times less. Always a highly recommended eyepiece line by amateur astronomers, and often called the poor man's ZAO!

This eyepiece line is within the following competition class of eyepieces with similar specifications and performance: Antares Ortho, Apogee Super Abbe Orthoscopic, Baader Planetarium Classic Ortho, Cave Orthostar Orthoscopic, Celestron Ortho, Edmund Scientific Ortho, Kokusai Kohki Abbe Ortho, Kson Super Ortho, Masuyama Orthoscopic, Meade Research Grade Ortho, Meade Series II Orthoscopic, Nikon Ortho, Siebert Optics Star Splitter/Super Star Splitter, Takahashi Ortho, Telescope Service Ortho, Unitron Ortho, University Optics Abbe HD Orthoscopic, University Optics Abbe Volcano Orthoscopic, University Optics O.P.S. Orthoscopic Planetary Series, University Optics Super Abbe Orthoscopic, VERNONscope Brandon.

Baader Planetarium: Hyperion/Hyperion Aspheric

Fig. 7.24 Baader Hyperion eyepieces with fine tuning rings (Image courtesy of William Rose, Larkspur, CO, USA)

Focal length (mm)	AFOV (degrees)	ER (mm)	Field stop (mm)	Weight (oz)	Design (Elem/Grp)	Coatings	Barrel (inches)
3.5	68	20	2.4^{E}	14.6	8/5	FMC	1.25/2.00
5	68	20	3.5^{E}	14.7	8/5	FMC	1.25/2.00
8	68	20	5.6^{E}	13.5	8/5	FMC	1.25/2.00
10	68	20	7.0^{E}	13.1	8/5	FMC	1.25/2.00
13	68	20	9.1^{E}	13.8	8/5	FMC	1.25/2.00
17	68	20	11.9^{E}	13.9	8/5	FMC	1.25/2.00
21	65	20	18.3^{E}	13.7	8/5	FMC	1.25/2.00
24	68	17	29.0	11.6	6	FMC	1.25/2.00
31	72	18	32.0	13.8	6	FMC	1.25/2.00
36	72	20	31.5/44	14.7	6	FMC	1.25/2.00

Source: www.baader-planetarium.com

Standard Hyperions

These have special Phantom Group™ multi-coatings with reflectivity of less than 0.2 % at each air-to-glass surface; use high grade high index water-white glasses; use optional fine-tuning rings to change their focal length (see table below); and have eye lens barrel M43 and M54 threads for direct coupling to cameras. This eyepiece line is within the following competition class of eyepieces with similar specifications and performance: Orion Lanthanum Superwide, Orion Stratus, Vixen Lanthanum Superwide (LVW).

Hyperion	New focal length when fine tuning ring added		
	14 mm	28 mm	14 and 28 mm
Eyepiece	Ring	Ring	Rings
21 mm	17.6	15.5	14.0
17 mm	13.1	10.8	9.2
13 mm	10.8	9.2	8.1
8 mm	6.0	5.0	4.3
5 mm	4.0	3.2	2.6
3.5 mm	2.5	2.1	1.8

Aspheric Hyperions

World's first eyepiece to incorporate aspheric lens surfaces; Phantom Group™ multi-coatings are optimized for each of the glasses used and are threaded to accept the full line of Hyperion Digital T-Rings for direct attachment to almost any camera.

BST: Explorer ED

Focal length (mm)	AFOV (degrees)	ER (mm)	Field stop (mm)	Weight (oz)	Design (Elem/Grp)	Coatings	Barrel (inches)
5	60	13	5.2[E]	7.2	6	FMC	1.25
8	60	13	8.4[E]	6.4	6	FMC	1.25
12	60	13	12.6[E]	6.4	6	FMC	1.25
15	60	15	15.7[E]	6.4	6	FMC	1.25
18	60	13	18.8[E]	5.6	6	FMC	1.25
25	60	15	26.2[E]	6.4	6	FMC	1.25

These use two extra low-dispersion lenses and are parfocal. This eyepiece line is within the following competition class of eyepieces with similar specifications and performance: Astro-Tech Paradigm Dual ED, Olivon 60° ED Wide Angle, Orion Epic II ED, Pentax XF, Telescope Service NED "ED" Flat Field.

BST: Flat Field

Focal length (mm)	AFOV (degrees)	ER (mm)	Field stop (mm)	Weight (oz)	Design (Elem/Grp)	Coatings	Barrel (inches)
8	60	9.5	9.6	2.6	7/4	FMC	1.25
12	60	15	15.2	4.0	7/4	FMC	1.25
16	60	19	16.5	3.5	6/4	FMC	1.25
19	65	18.5	21.2	4.1	5/3	FMC	1.25
27	53	23	24.4	6.5	5/4	FMC	1.25

These have twist-up eyeguards. This eyepiece line is within the following competition class of eyepieces with similar specifications and performance: Astro-Professional EF Flatfield, Astro-Tech Flat Field, Orion Edge-On Flat-Field, Sky-Watcher Extra Flat, Smart Astronomy Extra Flat Field, Telescope-Service Edge-On Flat Field.

Bresser: 52° Super Plössl

Fig. 7.25 Bresser 6.5 mm Super Plössl (Image courtesy of Uwe Pilz, Leipzig, Germany)

Focal length (mm)	AFOV (degrees)	ER (mm)	Field stop (mm)	Weight (oz)	Design (Elem/Grp)	Coatings	Barrel (inches)
6.5	52	6.5	5.4	1.8	4/2	FMC	1.25
9.5	52	7.8	8	2.5	4/2	FMC	1.25
12.5	52	7.8	10.5	2.9	4/2	FMC	1.25
15	52	12	13	2.7	4/2	FMC	1.25
20	52	15.8	18	3.1	4/2	FMC	1.25
26	52	18.3	21.8	3.7	4/2	FMC	1.25
32	52	26.9	26	3.6	4/2	FMC	1.25
40	44	28.4	27.5	5.2	4/2	FMC	1.25
56	52	42.5	46.6	19.2	4/2	FMC	2

This eyepiece line is within the following competition class of eyepieces with similar specifications and performance: Antares Plössl, Astro-Professional Plössl, Astro-Tech High Grade Plössl, Astro-Tech Value Line Plössl, Carton Plössl, Celestron Omni, Celestron Silvertop Plössl, Clavé Plössl, Coronado CeMax, Edmund Scientific Plössl, Garrett Optical Plössl, GSO Plössl, GTO Plössl, Long Perng Plössl, Meade Series 3000 Plössl, Meade Series 4000 Super Plössl, Olivon Plössl, Opt Plössl, Orion HighLight Plössl, Orion Sirius Plössl, Owl Black Night Plössl, Parks Silver Series, Sky-Watcher SP-Series Super Plössl, Smart Astronomy Sterling Plössl, TAL—Symmetrical Super Plössl, Telescope Service Plössl, Telescope Service Super Plössl, Tele Vue Plössl, Vixen NPL.

Bresser: 60° Plössl

Focal length (mm)	AFOV (degrees)	ER (mm)	Field stop (mm)	Weight (oz)	Design (Elem/Grp)	Coatings	Barrel (inches)
5.5	60	7	5.8	2.6	5/3	FMC	1.25
9	60	6.8	9.5	2.5	5/3	FMC	1.25
14	60	10.4	14.7	3.1	5/3	FMC	1.25
20	60	14.8	20.9	3.6	5/3	FMC	1.25
26	60	19.3	27.3	6.3	5/3	FMC	1.25
32	60	22.2	33.6	12.8	5/3	FMC	2
40	60	28.2	42.2	19.1	6/4	FMC	2

These are parfocals. This eyepiece line is within the following competition class of eyepieces with similar specifications and performance: Antares Elite, Baader Eudiascopic, Celestron Ultima, Kasai Astroplan, Meade Series 4000 Super Plössl (pre-1994, smooth-sided, 5-elements), Meade Series 5000 Super Plössl, Orion Ultrascopic, Parks Gold Series Plössl, Takahashi LE. *Note:* The Omcon Ultima and Tuthill Super Plössl lines are also competing brands, but these lines are not detailed.

Burgess: Planetary (*Discontinued*)

Fig. 7.26 Burgess/TMB Planetary eyepieces in 5 mm, 6 mm, 7 mm, and 9 mm (Image courtesy of William Rose, Larkspur, CO, USA)

Focal length (mm)	AFOV (degrees)	ER (mm)	Field stop (mm)	Weight (oz)	Design (Elem/Grp)	Coatings	Barrel (inches)
2.5	60	16	2.6^{E}	5	6	FMC	1.25
3.2	60	16	3.4^{E}	5	6	FMC	1.25
5	60	16	5.2^{E}	5	6	FMC	1.25
6	60	16	6.3^{E}	5	6	FMC	1.25
7	60	16	7.3^{E}	5	6	FMC	1.25
8	60	16	8.4^{E}	5	6	FMC	1.25
9	60	16	9.4^{E}	5	6	FMC	1.25

Source: www.burgessoptical.com

These have a twist-up eyeguard. This eyepiece line is within the following competition class of eyepieces with similar specifications and performance: APM UWA Planetary, Olivon Wide Angled Plössl, Owl Astronomy High Resolution Planetary, Telescope Service Planetary HR, TMB Planetary II.

Burgess: Wide Angle (*Discontinued*)

Focal length (mm)	AFOV (degrees)	ER (mm)	Field stop (mm)	Weight (oz)	Design (Elem/Grp)	Coatings	Barrel (inches)
8	60	10	9	2.4	4/3	FMC	1.25
12	60	12	13	2.9	4/3	FMC	1.25
17	65	17	19	3.6	4/3	FMC	1.25
20	67	20	21	3.6	4/3	FMC	1.25

Source: www.burgessoptical.com

This eyepiece line is within the following competition class of eyepieces with similar specifications and performance: Agena Wide Angle, Astro-Tech Series 6 Economy Wide Field, Owl Advanced Wide Angle.

BW Optic: Ultrawide (*Discontinued*)

Focal length (mm)	AFOV (degrees)	ER (mm)	Field stop (mm)	Weight (oz)	Design (Elem/Grp)	Coatings	Barrel (inches)
30	84	–	41[E]	20[E]	5/3	FMC	2

This eyepiece line is within the following competition class of eyepieces with similar specifications and performance: Agena Ultra Wide Angle, Apogee Widescan (20 and 30 mm), Astrobuffet 1RPD (30 mm), Kokusai Kohki Widescan I/II/III (20 and 30 mm), Moonfish Ultrawide (30 mm), Olivion 80° Ultra Wide Angle, Owl Astronomy Knight Owl Ultrawide Angle, Sky-Watcher Ultra Wide Angle, Surplus Shed Wollensak (30 mm), TAL Ultra Wide Angle, University Optics 80°, University Optics Widescan I/II/III (20 and 30 mm).

Carton: Plössl (*Discontinued*)

Fig. 7.27 Carton 8 mm Plössl (Image courtesy of William Rose, Larkspur, CO, USA)

Focal length (mm)	AFOV (degrees)	ER (mm)	Field stop (mm)	Weight (oz)	Design (Elem/Grp)	Coatings	Barrel (inches)
8	50	5[E]	7.0[E]	<3[E]	4/2	FC[E]	1.25
12	50	8[E]	10.5[E]	<3[E]	4/2	FC[E]	1.25
15	50	10[E]	13.1[E]	<3[E]	4/2	FC[E]	1.25

Source: www.carton-opt.co.jp

This eyepiece line is within the following competition class of eyepieces with similar specifications and performance: Antares Plössl, Astro-Professional Plössl, Astro-Tech High Grade Plössl, Astro-Tech Value Line Plössl, Bresser 52° Super Plössl, Celestron Omni, Celestron Silvertop Plössl, Clavé Plössl, Coronado CeMax, Edmund Scientific Plössl, Garrett Optical Plössl, GSO Plössl, GTO Plössl, Long Perng Plössl, Meade Series 3000 Plössl, Meade Series 4000 Super Plössl, Olivon Plössl, Opt Plössl, Orion HighLight Plössl, Orion Sirius Plössl, Owl Black Night Plössl, Parks Silver Series, Sky-Watcher SP-Series Super Plössl, Smart Astronomy Sterling Plössl, TAL—Symmetrical Super Plössl, Telescope Service Plössl, Telescope Service Super Plössl, Tele Vue Plössl, Vixen NPL.

Cave: Orthostar Orthoscopic (*Discontinued*)

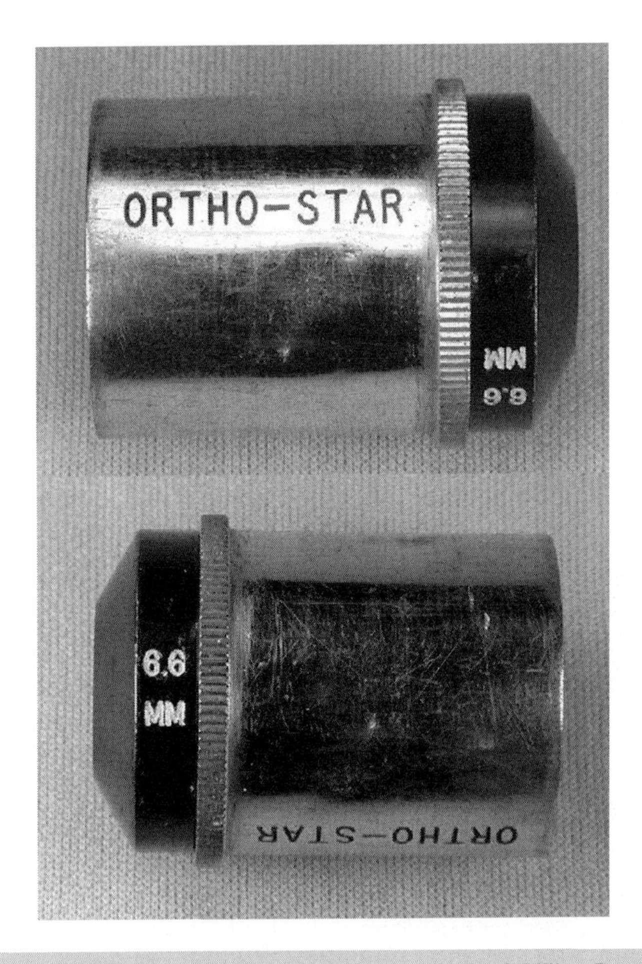

Fig. 7.28 A 6.6 mm Cave Orthostar eyepiece (Image courtesy of William Rose, Larkspur, CO, USA)

Focal length (mm)	AFOV (degrees)	ER (mm)	Field stop (mm)	Weight (oz)	Design (Elem/Grp)	Coatings	Barrel (inches)
6.6	42^E	5^E	4.8^E	3^E	4/2	FC^E	1.25
10	42^E	8^E	7.3^E	3^E	4/2	FC^E	1.25
16	42^E	13^E	11.7^E	3^E	4/2	FC^E	1.25
20	42^E	16^E	14.7^E	3^E	4/2	FC^E	1.25
26.6	42^E	21^E	19.5^E	3^E	4/2	FC^E	1.25

These are parfocal. This eyepiece line is within the following competition class of eyepieces with similar specifications and performance: Antares Ortho, Apogee Super Abbe Orthoscopic, Baader Planetarium Classic Ortho, Baader Planetarium Genuine Abbe Ortho, Celestron Ortho, Edmund Scientific Ortho, Kokusai Kohki Abbe Ortho, Kson Super Ortho, Masuyama Orthoscopic, Meade Research Grade Ortho, Meade Series II Orthoscopic, Nikon Ortho, Siebert Optics Star Splitter/Super Star Splitter, Takahashi Ortho, Telescope Service Ortho, Unitron Ortho, University Optics Abbe HD Orthoscopic, University Optics Abbe Volcano Orthoscopic, University Optics O.P.S. Orthoscopic Planetary Series, University Optics Super Abbe Orthoscopic, VERNONscope Brandon.

Celestron: Axiom (*Discontinued*)

Fig. 7.29 Celestron 34 mm Axiom eyepiece (Image courtesy of Mike Sutherland, Beaverton, OR, USA)

Focal length (mm)	AFOV (degrees)	ER (mm)	Field stop (mm)	Weight (oz)	Design (Elem/Grp)	Coatings	Barrel (inches)
15	70	7	18.3E	–	7	FMC	1.25
19	70	10	23.2E	–	7	FMC	1.25
23	70	10	28.1E	–	7	FMC	1.25
19	70	10	23.2E	–	7	FMC	2
34	70	16	41.5E	–	7	FMC	2
40	70	21	46.0E	–	–	FMC	2
50	52	38	45.4E	–	–	FMC	2

Source: www.celestron.com

These have extra large pop-up eyeguards to block unwanted stray light. The manufacturer's statistics on weight are no longer available.

Celestron: Axiom LX (Discontinued)

Fig. 7.30 Celestron 23 mm Axiom LX (Image courtesy of Hernando Bautista, Manila, Philippines)

Focal length (mm)	AFOV (degrees)	ER (mm)	Field stop (mm)	Weight (oz)	Design (Elem/Grp)	Coatings	Barrel (inches)
7	82	13	10	12	6–8	FMC	1.25
10	82	13	15	12	6–8	FMC	1.25
15	82	13	21	12	6–8	FMC	1.25
19	82	13	27	16.2	6–8	FMC	2
23	82	17	34	16.9	6–8	FMC	2
31	82	21	44	48	6–8	FMC	2

Source: www.celestron.com

These have an adjustable eyeguard. This eyepiece line is within the following competition class of eyepieces with similar specifications and performance: Antares Speers-WALER Series, Astro-Professional UWA, Celestron Luminos, Docter UWA, Explore Scientific 82° Nitrogen Purged, Meade Series 4000 Ultra Wide

Angle, Meade Series 5000 Ultra Wide Angle, Orion MegaView Ultra-Wide, Sky-Watcher Nirvana UWA, Sky-Watcher Sky Panorama, Tele Vue Nagler, Williams Optics UWAN.

Celestron: E-Lux (*Discontinued*)

Fig. 7.31 Celestron 25 mm E-Lux (Image courtesy of Rob Guasto, Sound Beach, NY, USA)

Focal length (mm)	AFOV (degrees)	ER (mm)	Field stop (mm)	Weight (oz)	Design (Elem/Grp)	Coatings	Barrel (inches)
6	50	5	5.2^{E}	$<3^{E}$	4/2	FC	1.25
10	50	7	8.7^{E}	$<4^{E}$	4/2	FC	1.25
25	50	22	21.8^{E}	$<5^{E}$	4/2	FC	1.25
40	43	31	30.0^{E}	$<7^{E}$	4/2	FC	1.25
26	56	20	25.4^{E}	11.8	3/2	FMC	2
32	56	20	31.3^{E}	12.5	3/2	FMC	2
40	50	20	34.9^{E}	16.0	3/2	FMC	2

Source: www.celestron.com

The 1.25″ models of this eyepiece line is within the following competition class of eyepieces with similar specifications and performance: Antares Plössl, Astro-Professional Plössl, Astro-Tech High Grade Plössl, Bresser 52° Super Plössl, Carton Plössl, Celestron Omni, Celestron Silvertop Plössl, Clavé Plössl, Coronado CeMax, Edmund Scientific Plössl, Garrett Optical Plössl, GSO Plössl, GTO Plössl, Long Perng Plössl, Meade Series 3000 Plössl, Meade Series 4000 Super Plössl, Olivon Plössl, Opt Plössl, Orion HighLight Plössl, Orion Sirius Plössl, Owl Black Night Plössl, Parks Silver Series, Sky-Watcher SP-Series Super Plössl, Smart Astronomy Sterling Plössl, TAL—Symmetrical Super Plössl, Telescope Service Plössl, Telescope Service Super Plössl, Tele Vue Plössl, Vixen NPL.

The 2″ models of this eyepiece line are within the following competition class of eyepieces with similar specifications and performance (Kellners and Reverse Kellners, and RKE—Rank-Kaspereit-Erfle): Celestron E-Lux (2 in. models only), Celestron Kellner, Criterion Kellner, Edmund Scientific RKE, GSO Kellner, Kokusai Kohki Kellner, Orion DeepView, Orion E-Series, Russell Optics (2 in. 52 and 60 mm only), Sky-Watcher Kellner, Sky-Watcher Super MA Series, Telescope Service RK, Unitron Kellner.

Celestron: Erfle (*Discontinued*)

Fig. 7.32A Celestron Erfles in 16 mm, 20 mm, 24 mm, 28 mm, and 32 mm (Image courtesy of Blake Andrews, USA)

Focal length (mm)	AFOV (degrees)	ER (mm)	Field stop (mm)	Weight (oz)	Design (Elem/Grp)	Coatings	Barrel (inches)
16	50	9.6	16	6.2	5/3 or 6/3	FC	1.25
20	52	11	20	6.2	5/3 or 6/3	FC	1.25
25	53	15[E]	22	6.2	5/3 or 6/3	FC	1.25
28	53	17[E]	25	6.2[E]	5/3 or 6/3	FC	1.25
32	50	19[E]	28	6.2[E]	5/3 or 6/3	FC	1.25

Source: www.celestron.com

As with an Erfle or similar optical design, these require longer focal length telescopes for a clean performing off-axis. This eyepiece line is within the following competition class of eyepieces with similar specifications and performance: Kokusai Kohki Erfle, University Optics Super Erfle.

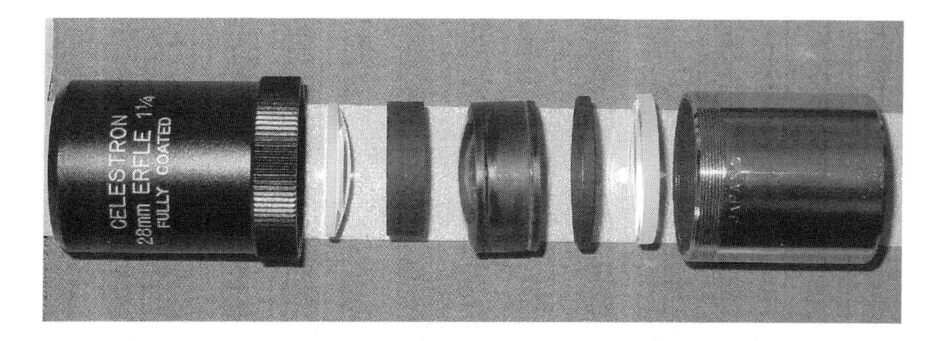

Fig. 7.32B The Celestron 28 mm with lenses exposed. Note that the 1-3-1 design is not an Erfle but may instead be closer to the Göerz design invented by Christian von Hofe; U. S. Patent 1,759,529 (Image by the author)

Celestron: Kellner (*Discontinued*)

Fig. 7.33 Celestron Erfles in 16 mm, 20 mm, 24 mm, 28 mm, and 32 mm (Image courtesy of Blake Andrews, USA)

Focal length (mm)	AFOV (degrees)	ER (mm)	Field stop (mm)	Weight (oz)	Design (Elem/Grp)	Coatings	Barrel (inches)
6	42	4.8	4.4[E]	3.0	3/2	FC	1.25
9	44	7.2	6.9[E]	3.7	3/2	FC	1.25
12	40	9.6	8.4[E]	4.2	3/2	FC	1.25
18	47	9.6	14.8[E]	3.5	3/2	FC	1.25
20	47	15	16.4[E]	3.5	3/2	FC	1.25
25	45	16	19.6[E]	3.7	3/2	FC	1.25

Source: www.celestron.com

Notice that each of the Celestron Kellner's pictured each have a different optical mark on the barrel indicating the likelihood that they were each made by different optical manufacturers. This eyepiece line is within the following competition class of eyepieces with similar specifications and performance (Kellners and Reverse Kellners, and RKE—Rank-Kaspereit-Erfle): Celestron E-Lux (2 in. models only), Edmund Scientific RKE, GSO Kellner, Kokusai Kohki Kellner, Orion DeepView, Orion E-Series, Russell Optics (2 in. 52 and 60 mm only), Sky-Watcher Kellner, Sky-Watcher Super MA Series, Telescope Service RK, Unitron Kellner.

Celestron: Luminos

Fig. 7.34 The Celestron Luminos series (© 2013 Celestron—www.celestron.com)

Focal length (mm)	AFOV (degrees)	ER (mm)	Field stop (mm)	Weight (oz)	Design (Elem/Grp)	Coatings	Barrel (inches)
7	82	12	15	12	7	FMC	1.25
10	82	12	17	12	7	FMC	1.25
15	82	17	26	12	7	FMC	1.25
19	82	20	30	16.2	6	FMC	2
23	82	20	37	32	6	FMC	2
31	82	27	47	40	6	FMC	2

Source: www.celestron.com

These are parfocal. This eyepiece line is within the following competition class of eyepieces with similar specifications and performance: Antares Speers-WALER Series, Astro-Professional UWA, Celestron Axiom LX, Docter UWA, Explore Scientific 82° Nitrogen Purged, Meade Series 4000 Ultra Wide Angle, Meade Series 5000 Ultra Wide Angle, Orion MegaView Ultra-Wide, Sky-Watcher Nirvana UWA, Sky-Watcher Sky Panorama, Tele Vue Nagler, Williams Optics UWAN.

Celestron: Omni

Fig. 7.35 The Celestron Omni series (© 2013 Celestron—www.celestron.com)

Focal length (mm)	AFOV (degrees)	ER (mm)	Field stop (mm)	Weight (oz)	Design (Elem/Grp)	Coatings	Barrel (inches)
4	52	6	3.6[E]	2.5	4/2	FMC	1.25
6	52	5	5.4[E]	2.5	4/2	FMC	1.25
9	52	6	8.2[E]	2.5	4/2	FMC	1.25
12.5	52	8	10.9[E]	2.6	4/2	FMC	1.25
15	52	13	13.6[E]	3.2	4/2	FMC	1.25
20	52	20	18.2[E]	3.5	4/2	FMC	1.25
25	52	22	22.7[E]	3.5	4/2	FMC	1.25
32	52	22	27.0[E]	4.2	4/2	FMC	1.25
40	43	31	27.0[E]	6.5	4/2	FMC	1.25

Source: www.celestron.com

This eyepiece line is within the following competition class of eyepieces with similar specifications and performance: Antares Plössl, Astro-Professional Plössl, Astro-Tech High Grade Plössl, Astro-Tech Value Line Plössl, Bresser 52° Super Plössl, Carton Plössl, Celestron Silvertop Plössl, Clavé Plössl, Coronado CeMax, Edmund Scientific Plössl, Garrett Optical Plössl, GSO Plössl, GTO Plössl, Long Perng Plössl, Meade Series 3000 Plössl, Meade Series 4000 Super Plössl, Olivon Plössl, Opt Plössl, Orion HighLight Plössl, Orion Sirius Plössl, Owl Black Night Plössl, Parks Silver Series, Sky-Watcher SP-Series Super Plössl, Smart Astronomy Sterling Plössl, TAL—Symmetrical Super Plössl, Telescope Service Plössl, Telescope Service Super Plössl, Tele Vue Plössl, Vixen NPL.

Celestron: Ortho (*Discontinued*)

Fig. 7.36 Celestron Orthos in 4–25 mm focal lengths (Image courtesy of Blake Andrews, USA)

Focal length (mm)	AFOV (degrees)	ER (mm)	Field stop (mm)	Weight (oz)	Design (Elem/Grp)	Coatings	Barrel (inches)
4	41	3.5	2.9E	2.8	4/2	FC	1.25
5	43	4	3.8E	3.0	4/2	FC	1.25
6	43	4.8	4.5E	3.2	4/2	FC	1.25
7	42	5.6	5.1E	3.2	4/2	FC	1.25
9	42	7.2	6.6E	3.2	4/2	FC	1.25
12.5	44	10	9.6E	3.4	4/2	FC	1.25
18	46	14.4	14.5E	3.4	4/2	FC	1.25
25	47	20	20.5E	3.5	4/2	FC	1.25

Source: www.celestron.com

This eyepiece line is within the following competition class of eyepieces with similar specifications and performance: Antares Ortho, Apogee Super Abbe Orthoscopic, Baader Planetarium Classic Ortho, Baader Planetarium Genuine Abbe Ortho, Cave Orthostar Orthoscopic, Edmund Scientific Ortho, Kokusai Kohki

Abbe Ortho, Kson Super Ortho, Masuyama Orthoscopic, Meade Research Grade Ortho, Meade Series II Orthoscopic, Nikon Ortho, Siebert Optics Star Splitter/ Super Star Splitter, Takahashi Ortho, Telescope Service Ortho, Unitron Ortho, University Optics Abbe HD Orthoscopic, University Optics Abbe Volcano Orthoscopic, University Optics O.P.S. Orthoscopic Planetary Series, University Optics Super Abbe Orthoscopic, VERNONscope Brandon.

Celestron: Silvertop Plössl (*Discontinued*)

Fig. 7.37 The complete Celestron Silvertop Plössl collection (Image courtesy of Steven Cotton, Florida, USA)

Focal length (mm)	AFOV (degrees)	ER (mm)	Field stop (mm)	Weight (oz)	Design (Elem/Grp)	Coatings	Barrel (inches)
7.5	46	6	6.0[E]	–	4/2	MC	1.25
10	46	7	8.0[E]	–	4/2	MC	1.25
15	46	11	12.0[E]	–	4/2	MC	1.25
17	47	14	13.9[E]	–	4/2	MC	1.25
22	48	17	18.4[E]	–	4/2	MC	1.25
26	49	20	22.2[E]	–	4/2	MC	1.25
30	50	28	26.2[E]	–	4/2	MC	1.25
36	35	22	22.0[E]	–	4/2	FC/MC	1.25
45	33	25	25.9[E]	–	4/2	MC	1.25
50	43	37	37.5[E]	–	4/2	MC	2

Source: www.celestron.com

The Celestron Silvertop Plössls have an excellent reputation and are one of the more sought after lines of discontinued eyepieces offering a distinctive look, feel, and performance. Many observers characterize their performance on par with the much renowned vintage Tele Vue Smoothside Plössls. Note that some versions of these share the same Circle-NJ optical mark that appeared on the early generations of many Tele Vue eyepiece lines. This eyepiece line is within the following competition class of eyepieces with similar specifications and performance: Antares Plössl, Astro-Professional Plössl, Astro-Tech High Grade Plössl, Astro-Tech Value Line Plössl, Bresser 52° Super Plössl, Carton Plössl, Celestron Omni, Clavé Plössl, Coronado CeMax, Edmund Scientific Plössl, Garrett Optical Plössl, GSO Plössl, GTO Plössl, Long Perng Plössl, Meade Series 3000 Plössl, Meade Series 4000 Super Plössl, Olivon Plössl, Opt Plössl, Orion HighLight Plössl, Orion Sirius Plössl, Owl Black Night Plössl, Parks Silver Series, Sky-Watcher SP-Series Super Plössl, Smart Astronomy Sterling Plössl, TAL—Symmetrical Super Plössl, Telescope Service Plössl, Telescope Service Super Plössl, Tele Vue Plössl, Vixen NPL.

Fig. 7.38 View of the Silvertop eye lenses (Image courtesy of Steven Cotton, FL, USA)

Fig. 7.39 Promotional case with orange cap containers for the Silvertops that was bundled with some telescopes (Image courtesy of Steven Cotton, FL, USA)

Fig. 7.40 Rare Silvertop with the Circle-NJ optical mark (see Tele Vue Plössls for more details) (Image courtesy of Steven Cotton, FL, USA)

Celestron: Ultima (*Discontinued*)

Fig. 7.41 Rare 45 mm Celestron Ultima (Image by the author)

Focal length (mm)	AFOV (degrees)	ER (mm)	Field stop (mm)	Weight (oz)	Design (Elem/Grp)	Coatings	Barrel (inches)
5	50	4	4.4[E]	4	5/3	FMC	1.25
7.5	51	5	6.7[E]	4	5/3	FMC	1.25
10	51	9	8.9[E]	5	5/3	FMC	1.25
12.5	51	13	11.1[E]	6	5/3	FMC	1.25
18	51	18	16.0[E]	7	5/3	FMC	1.25
30	50	21	26.2[E]	8	5/3	FMC	1.25
35	49	25	27.0[E]	8	5/3	FMC	1.25
42	36	32	25.7[E]	10	5/3	FMC	1.25
45	51[E]	29[E]	40[E]	14[E]	5/3	FMC	2
60	43[E]	32[E]	45[E]	20[E]	5/3	FMC	2
80	32[E]	42[E]	45[E]	27[E]	5/3	FMC	2

Source: www.celestron.com

The 35 mm focal length of this line offers a distinctive view as the image can sometimes seems to float somewhat in front of your eye (similar, but not as pronounced as the Edmund 28 mm RKE). Makes a great addition to any eyepiece collection. This eyepiece line is within the following competition class of eyepieces with similar specifications and performance: Antares Elite, Baader Eudiascopic,

Bresser 60° Plössl, Kasai Astroplan, Meade Series 4000 Super Plössl (pre-1994, smooth-sided, 5-elements), Meade Series 5000 Super Plössl, Orion Ultrascopic, Parks Gold Series Plössl, Takahashi LE. *Note:* The Omcon Ultima and Tuthill Super Plössl lines are also competing brands, but these lines are not detailed.

Celestron: Ultima LX

Fig. 7.42 The Celestron Ultima LX series (© 2013 Celestron—www.celestron.com)

Focal length (mm)	AFOV (degrees)	ER (mm)	Field stop (mm)	Weight (oz)	Design (Elem/Grp)	Coatings	Barrel (inches)
5	70	16	6.1^E	22	8	FMC	1.25/2
8	70	16	9.8^E	21	8	FMC	1.25/2
13	70	16	15.9^E	21	8	FMC	1.25/2
17	70	16	20.8^E	20	8	FMC	1.25/2
22	70	16	26.9^E	19	6	FMC	2
32	70	16	39.1^E	17	6	FMC	2

Source: www.celestron.com

These have an extendable twist-up guard and a shock-resistant rubber covering.

Celestron: X-Cel (*Discontinued*)

Focal length (mm)	AFOV (degrees)	ER (mm)	Field stop (mm)	Weight (oz)	Design (Elem/Grp)	Coatings	Barrel (inches)
2.3	55	20	2.2[E]	<6[E]	6	FMC	1.25
5	55	20	4.8[E]	<6[E]	6	FMC	1.25
8	55	20	7.7[E]	<6[E]	6	FMC	1.25
10	55	20	9.6[E]	<6[E]	6	FMC	1.25
12.5	55	20	12.0[E]	<6[E]	6	FMC	1.25
18	55	20	17.3[E]	<6[E]	6	FMC	1.25
21	55	20	20.2[E]	<6[E]	6	FMC	1.25
25	55	20	24.0[E]	<6[E]	6	FMC	1.25

Source: www.celestron.com

These use extra low-dispersion glass and are parfocal. This eyepiece line is within the following competition class of eyepieces with similar specifications and performance: Agena ED, Orion Epic ED II (older version), Vixen Lanthanum (LV), Vixen NLV.

Celestron: X-Cel LX

Fig. 7.43 The Celestron X-Cel LX series (© 2013 Celestron—www.celestron.com)

Focal length (mm)	AFOV (degrees)	ER (mm)	Field stop (mm)	Weight (oz)	Design (Elem/Grp)	Coatings	Barrel (inches)
2.3	60	16	2.4[E]	8	6	FMC	1.25
5	60	16	5.2[E]	7	6	FMC	1.25
7	60	16	7.3[E]	7	6	FMC	1.25
9	60	16	9.4[E]	7	6	FMC	1.25
12	60	16	12.6[E]	7	6	FMC	1.25
18	60	16	18.8[E]	7	6	FMC	1.25
25	60	16	26.2[E]	7	6	FMC	1.25

Source: www.celestron.com

The Celestron X-Cel LX eyepiece line often receives very good observer reports, better than many of the other lines available in this AFOV class. It has all the hallmarks of a "sleeper" in that it is not often discussed on the online astronomy forums, but it fairly consistently receives very high marks from observers. These are parfocal.

Clavé: Plössl (*Discontinued*)

Fig. 7.44 The Clavé Plössls (Image courtesy of William Rose, Larkspur, CO, USA)

Focal length (mm)	AFOV (degrees)	ER (mm)	Field stop (mm)	Weight (oz)	Design (Elem/Grp)	Coatings	Barrel (inches)
3	48	2[E]	2.7	–	4/2	MC	1.25
4	48	3[E]	3.6	–	4/2	MC	1.25
5	48	3[E]	4.4	–	4/2	MC	1.25
6	48	4[E]	5.3	–	4/2	MC	1.25
8	48	5[E]	7.1	–	4/2	MC	1.25
10	48	7[E]	8.9	–	4/2	MC	1.25
12	48	8[E]	10.7	–	4/2	MC	1.25
16	48	10[E]	14.2	–	4/2	MC	1.25
20	48	13[E]	17.8	–	4/2	MC	1.25
25	48	16[E]	22.3	–	4/2	MC	1.25
30	48	20[E]	26.8	–	4/2	MC	1.25
35	42	23[E]	26.8	–	4/2	MC	1.25

Source: www.kinoptik.com

Clavé eyepieces were manufactured by Est. S. R. Clavé between 1955 and 1985. In 1985, KINOPTIK purchased Clavé and continued to manufacture the Clavé Plössls from the same facility in Paris until the late 1990s. During their production, they developed a reputation as a top line eyepiece, including for planetary observing. Tele Vue used the Clavé Plössl as the standard to beat when they developed their patented Tele Vue Plössl design. The manufacturer's information on the weight of each eyepiece is no longer readily available.

This eyepiece line is within the following competition class of eyepieces with similar specifications and performance: Antares Plössl, Astro-Professional Plössl, Astro-Tech High Grade Plössl, Astro-Tech Value Line Plössl, Bresser 52° Super Plössl, Carton Plössl, Celestron Omni, Celestron Silvertop Plössl, Coronado CeMax, Edmund Scientific Plössl, Garrett Optical Plössl, GSO Plössl, GTO Plössl, Long Perng Plössl, Meade Series 3000 Plössl, Meade Series 4000 Super Plössl, Olivon Plössl, Opt Plössl, Orion HighLight Plössl, Orion Sirius Plössl, Owl Black Night Plössl, Parks Silver Series, Sky-Watcher SP-Series Super Plössl, Smart Astronomy Sterling Plössl, TAL—Symmetrical Super Plössl, Telescope Service Plössl, Telescope Service Super Plössl, Tele Vue Plössl, Vixen NPL.

Fig. 7.45 The Clavé color filters (Image courtesy of William Rose, Larkspur, CO, USA)

Coronado: CeMax

Focal length (mm)	AFOV (degrees)	ER (mm)	Field stop (mm)	Weight (oz)	Design (Elem/Grp)	Coatings	Barrel (inches)
12	52	9	10.5	–	≈4	Special	1.25
18	52	12	18	–	≈4	Special	1.25
25	47	13	19.5	–	≈4	Special	1.25

Source: www.meade.com

These are marketed specifically for the Coronado Personal Solar Telescope (PST) as enhanced contrast eyepieces. Observer reports that these perform on par with a quality Plössl for nighttime astronomical observing. The manufacturer does not list the weights of these eyepieces.

This eyepiece line is within the following competition class of eyepieces with similar specifications and performance: Antares Plössl, Astro-Professional Plössl, Astro-Tech High Grade Plössl, Astro-Tech Value Line Plössl, Bresser 52° Super Plössl, Carton Plössl, Celestron Omni, Celestron Silvertop Plössl, Clavé Plössl (discontinued), Edmund Scientific Plössl, Garrett Optical Plössl, GSO Plössl, GTO Plössl, Long Perng Plössl, Meade Series 3000 Plössl, Meade Series 4000 Super

Plössl, Olivon Plössl, Opt Plössl, Orion HighLight Plössl, Orion Sirius Plössl, Owl Black Night Plössl, Parks Silver Series, Sky-Watcher SP-Series Super Plössl, Smart Astronomy Sterling Plössl, TAL—Symmetrical Super Plössl, Telescope Service Plössl, Telescope Service Super Plössl, Tele Vue Plössl, Vixen NPL.

Couture: Ball Singlet

Fig. 7.46 The Couture Ball eyepieces (Image courtesy of Steve Couture, Belle Mead, NJ, USA)

Focal length (mm)	AFOV (degrees)	ER (mm)	Field stop (mm)	Weight (oz)	Design (Elem/Grp)	Coatings	Barrel (inches)
2.9	15	0.9	None	<2	1/1	None	1.25
4.4	15	1.4	None	<2	1/1	None	1.25
5.9	15	1.9	None	<2	1/1	None	1.25
7.3	15	2.4	None	<2	1/1	None	1.25

This is not a production eyepiece, but an amateur astronomer-made eyepiece whose on-axis can compete with some of the best high contrast planetary eyepieces. See Chap. 5 for the essay by Stephen Couture detailing how this eyepiece was constructed. Best when used for critical on-axis planetary observing due to its very small AFOV. Although sphere singlet eyepieces may show a larger AFOV, the portion of the AFOV that remains aberration-free for planetary observing is generally between 10° and 20°, depending on the focal ratio of the telescope used. While the eye relief for these eyepieces is short, due to their narrow AFOV and the typical practice of mounting the ball lens so that a portion of it protrudes above the housing, their eye relief can still feel no tighter than a typical Abbe Ortho of similar focal length.

This eyepiece line is within the following competition class of eyepieces with similar specifications and performance: Siebert Planesphere.

Criterion: Ortho/Kellner/A.R. (Achromatic Ramsden) (*Discontinued*)

Fig. 7.47 Criterion eyepieces (Image courtesy of William Rose, Larkspur, CO, USA)

Focal length (mm)	AFOV (degrees)	ER (mm)	Field stop (mm)	Weight (oz)	Design (Elem/Grp)	Coatings	Barrel (inches)
4	42[E]	3[E]	2.9[E]	<3[E]	4/2-Ortho	Enhanced	1.25
6	42[E]	5[E]	4.4[E]	<3[E]	4/2-Ortho	Enhanced	1.25
9	45[E]	4[E]	7.1[E]	<3[E]	3/2-Kellner	Enhanced	1.25
12.7	45[E]	6[E]	10.0[E]	<3[E]	3/2-Kellner	Enhanced	1.25
16.3	65[E]	5[E]	18.5[E]	<3[E]	5/3-Erfle	Enhanced	1.25
18	45[E]	8[E]	14.1[E]	<3[E]	3/2-Kellner	Enhanced	1.25
18	40[E]	2[E]	12.6[E]	<3[E]	2/2-Huygen	Enhanced	1.25
30	45[E]	14[E]	23.6[E]	<3[E]	3/2-Kellner	Enhanced	1.25
50	30[E]	5[E]	26.2[E]	<4[E]	3/1-Hastings	Enhanced	1.25
50	40[E]	5[E]	34.9[E]	<4[E]	2/2-Ramsden	Enhanced	1.25

This eyepiece line is within the following competition class of eyepieces with similar specifications and performance (Kellners and Reverse Kellners, and RKE—Rank-Kaspereit-Erfle): Celestron E-Lux (2 in. models only), Celestron Kellner, Edmund Scientific RKE, GSO Kellner, Kokusai Kohki Kellner, Orion DeepView, Orion E-Series, Russell Optics (2 in. 52 and 60 mm only), Sky-Watcher Kellner, Sky-Watcher Super MA Series, Telescope Service RK, Unitron Kellner.

Denkmeier: D21/D14

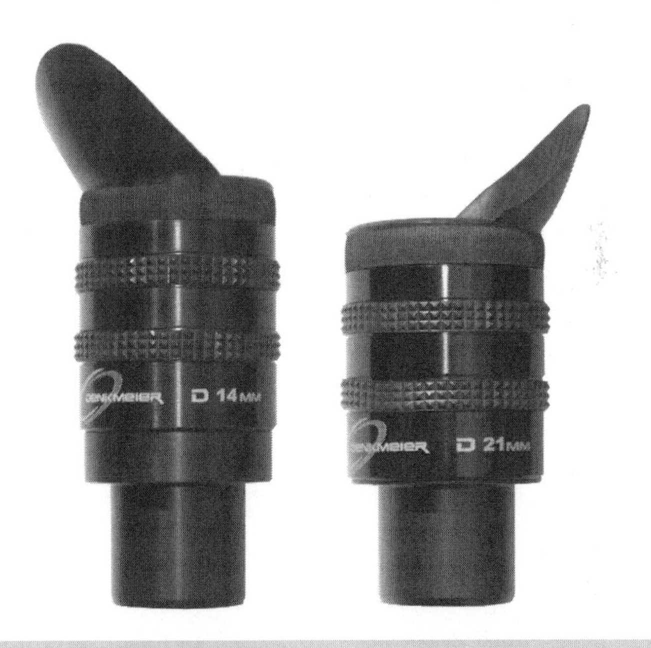

Fig. 7.48 The 14 and 21 mm Denkmeier eyepieces (© 2013 Denkmeier Optical, Inc.—www.denkmeier.com)

Focal length (mm)	AFOV (degrees)	ER (mm)	Field stop (mm)	Weight (oz)	Design (Elem/Grp)	Coatings	Barrel (inches)
14	65	20	15.9[E]	8	6/5	FMC	1.25
21	65	20	23.8[E]	8	6/4	FMC	1.25

Source: www.denkmeier.com

The barrels and housings of this line are made of 100% anodized aluminum to reduce weight. The D14 has six elements in five groups (e.g., a 1-2-1-1-1 configuration). The last two singlets are contained in the barrel and use high index glass. The D21 uses a doublet group in the barrel instead of two singlets (e.g., a 1-2-1-2 configuration). The elements in the housing of the D21, although grouped the same as the D14, are a different optical prescription.

The D14 and D21 optical design was thoroughly field tested prior to production under "real" observing conditions by the creator of the line, Russ Lederman, using his own large Dobsonian. These tests examined edge sharpness, ghosting, reflections, on-axis sharpness, and color correction both on- and off-axis. Targets used for the testing were Jupiter, Mars, Saturn, and the Moon. Stellar diffraction patterns were also examined for astigmatism. Star clusters were used to verify no pincushion distortion was in the design and that and edge correction was visually very good for their designed 65° AFOV.

Docter: UWA

Fig. 7.49 A 12.5 mm Docter 84° eyepiece (Image courtesy of Tamiji Homma, Newbury Park, CA, USA)

Focal length (mm)	AFOV (degrees)	ER (mm)	Field stop (mm)	Weight (oz)	Design (Elem/Grp)	Coatings	Barrel (inches)
12.5	84	18	19.2	525g	8/5	FMC	1.25

Source: www.docter-germany.de

This eyepiece maintains an outstanding reputation. Considered by some experienced observers as the best 12.5 mm eyepiece made and without peer. Waterproof.

This eyepiece line is within the following competition class of eyepieces with similar specifications and performance: Antares Speers-WALER Series, Astro-Professional UWA, Celestron Axiom LX, Celestron Luminos, Explore Scientific 82° Nitrogen Purged, Meade Series 4000 Ultra Wide Angle, Meade Series 5000 Ultra Wide Angle, Orion MegaView Ultra-Wide, Sky-Watcher Nirvana UWA, Sky-Watcher Sky Panorama, Tele Vue Nagler, Williams Optics UWAN.

Chapter 8

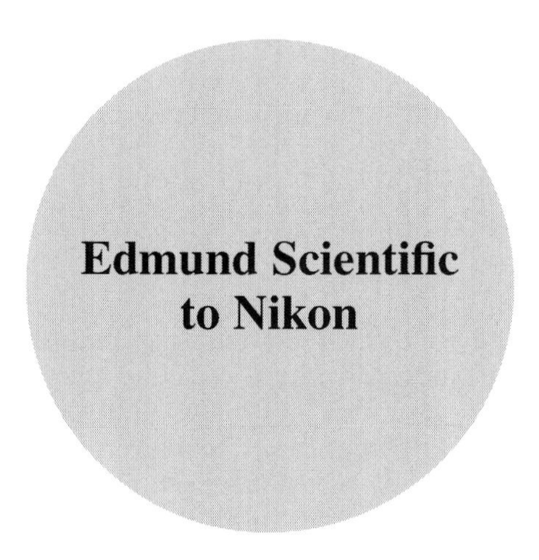

Edmund Scientific to Nikon

Edmund Scientific: Ortho

Fig. 8.1 Edmund 12.5 mm Ortho (Eyepiece courtesy of www.handsonoptics.com. Image by the author)

W. Paolini, *Choosing and Using Astronomical Eyepieces*, The Patrick Moore Practical Astronomy Series, DOI 10.1007/978-1-4614-7723-5_8,
© Springer Science+Business Media New York 2013

Focal length (mm)	AFOV (degrees)	ER (mm)	Field stop (mm)	Weight (oz)	Design (Elem/Grp)	Coatings	Barrel (inches)
4	41	3.2	2.8	2.8	4/2	FC	1.25
6	43	4.9	4.2	3.2	4/2	FC	1.25
12.5	44	10.4	8.5	3.4	4/2	FC	1.25
18	46	15.2	13.5	3.4	4/2	FC	1.25
25	47	22.2	17.5	3.5	4/2	FC	1.25

Source: www.edmundoptics.com

This eyepiece line is within the following competition class of eyepieces with similar specifications and performance: Antares Ortho, Apogee Super Abbe Orthoscopic, Baader Planetarium Classic Ortho, Baader Planetarium Genuine Abbe Ortho, Cave Orthostar Orthoscopic, Celestron Ortho, Kokusai Kohki Abbe Ortho, Masuyama Orthoscopic, Meade Research Grade Ortho, Meade Series II Orthoscopic, Nikon Ortho, Siebert Optics Star Splitter/Super Star Splitter, Takahashi Ortho, Telescope Service Ortho, Unitron Ortho, University Optics Abbe HD Orthoscopic, University Optics Abbe Volcano Orthoscopic, University Optics O.P.S. Orthoscopic Planetary Series, University Optics Super Abbe Orthoscopic, VERNONscope Brandon.

Edmund Scientific: Plössl

Fig. 8.2 Edmund 28 mm Plössl (*right*) compared to Edmund 28 RKE (*left*) (Image by the author)

Focal length (mm)	AFOV (degrees)	ER (mm)	Field stop (mm)	Weight (oz)	Design (Elem/Grp)	Coatings	Barrel (inches)
8	60	9.5[M]	8.4[E]	3[E]	7/4	–	1.25
12	50	8[E]	10.5[E]	3[E]	4/2	–	1.25
15	50	10[E]	13.1[E]	3[E]	4/2	–	1.25
21	50	14[E]	18.3[E]	3[E]	4/2	–	1.25
28	50	19[E]	24.4[E]	3[E]	4/2	–	1.25

Source: www.scientificsonline.com

An excellent performing Plössl eyepiece. Brightness less than expected as reported by observers, however they also report them as excellent for planetary observing as well as for binoviewers. The 28 mm shows a "floating image" effect that the Edmund 28 mm RKE is famous for, only milder.

This eyepiece line is within the following competition class of eyepieces with similar specifications and performance: Antares Plössl, Astro-Professional Plössl, Astro-Tech High Grade Plössl, Astro-Tech Value Line Plössl, Bresser 52° Super Plössl, Carton Plössl, Celestron Omni, Celestron Silvertop Plössl, Clavé Plössl, Coronado CeMax, Garrett Optical Plössl, GSO Plössl, GTO Plössl, Long Perng Plössl, Meade Series 3000 Plössl, Meade Series 4000 Super Plössl, Olivon Plössl, Opt Plössl, Orion HighLight Plössl, Orion Sirius Plössl, Owl Black Night Plössl, Parks Silver Series, Sky-Watcher SP-Series Super Plössl, Smart Astronomy Sterling Plössl, TAL—Symmetrical Super Plössl, Telescope Service Plössl, Telescope Service Super Plössl, Tele Vue Plössl, Vixen NPL.

Edmund Scientific: RKE

Fig. 8.3 Edmund RKE (Rank-Kaspereit-Erfle) eyepieces (Image courtesy of John W., MA, USA)

Focal length (mm)	AFOV (degrees)	ER (mm)	Field stop (mm)	Weight (oz)	Design (Elem/Grp)	Coatings	Barrel (inches)
8	45	8.2	6.6	<3[E]	3/2	FC	1.25
12	45	10.7	9.7	<3[E]	3/2	FC	1.25
15	45	13.4	11.9	<3[E]	3/2	FC	1.25
21	45	18.8	17.3	<3[E]	3/2	FC	1.25
28	45	24.5	23.3	<3[E]	3/2	FC	1.25

Source: www.edmundoptics.com

Edmund Scientific has a long history in the astronomy community, with products dating back to the 1950s (including the 28 mm pre-RKE pictured below). The 28 mm RKE is the most talked about eyepiece of the series, as it provides a view where the image seems to float in space above the eyepiece. For some observers the long eye relief of the 28 mm RKE can make holding the correct eye position difficult; however many feel the unique view it provides more than overcomes this eye positioning difficulty. Best off-axis performance in telescopes with medium length focal ratios.

The "RKE" trademark was first used in commerce on October 14, 1977, by Edmund Scientific, and what these letters stand for has never been fully revealed in any marketing documentation made readily available by Edmund Scientific. As a result, amateur astronomers over the decades have conjectured it as meaning Reverse Kellner eyepiece or possibly Rank Kellner eyepiece, since Dr. David Rank of Edmund Scientific was the inventor. The latter term is also what the marketing and engineering departments of Edmund Scientific commonly say is the meaning to the "best of their knowledge." However, in an amendment to their trademark application on January 16, 1979, Edmund Scientific did reveal that their RKE trademark stands for Rank, Kaspereit, Erfle, the three designs from which Dr. David Rank derived the new design. The most notable member of this family is the 28 mm RKE. This eyepiece has gained a very large and loyal following over the decades. The most unusual aspect of this eyepiece is how the eyepiece seems to vanish from view when observing and the image can literally appear as if it is floating in space above the focuser. If generally needs a longer focal length for the off-axis to be well behaved, but given the uniqueness of its view this is hardly a detriment. For many observers a very valued member of their astronomical gear. The Edmund RKE is one of the very few eyepieces remaining that are offered as "Made in the USA."

Fig. 8.4 The most talked about focal length of the Edmund RKE Series—28 mm RKE (Image by the author)

Fig. 8.5 Edmund 1-1/8 in. eyepiece that predates the current 28 mm RKE with elements exposed. Note the design of this earlier version appears to be a true non-symmetrical Plössl (Image by the author)

This eyepiece line is within the following competition class of eyepieces with similar specifications and performance (Kellners and Reverse Kellners, and RKE—Rank-Kaspereit-Erfle): Celestron E-Lux (2 in. models only), Celestron Kellner, Criterion Kellner, GSO Kellner, Kokusai Kohki Kellner, Orion DeepView, Orion E-Series, Russell Optics (2 in. 52 and 60 mm only), Sky-Watcher Kellner, Sky-Watcher Super MA Series, Telescope Service RK, Unitron Kellner.

Explore Scientific: 68° Nitrogen Purged (ES68 N2)

Fig. 8.6 Complete set of Explore Scientific 68° eyepieces (Image courtesy of John W., MA, USA)

Focal length (mm)	AFOV (degrees)	ER (mm)	Field stop (mm)	Weight (oz)	Design (Elem/Grp)	Coatings	Barrel (inches)
16	68	11.9	19.0^{E}	5.5	6/4	FMC	1.25
20	68	15.3	23.7^{E}	8.7	6/4	FMC	1.25
24	68	18.4	27.0^{E}	11.6	6/4	FMC	1.25
28	68	21.6	33.2^{E}	16	6/4	FMC	2
34	68	26.4	40.4^{E}	24	6/4	FMC	2
40	68	31	46.0^{E}	35	6/4	FMC	2

Source: www.explorescientific.com

These have a 15-layer enhanced multi-layer deposition coating; are waterproofed with argon gas; are individually serial numbered; and have a lifetime warranty. Considered by many as an excellent performing ultra wide-field that keeps pace well with even the premium brands. This eyepiece line is within the following competition class of eyepieces with similar specifications and performance: Meade 4000 SWA, Meade 5000 SWA, Nikon NAV-SW, Pentax XL, Pentax XW, Tele Vue Delos, and Tele Vue Panoptic.

Explore Scientific: 82° Nitrogen Purged (ES82 N2)

Fig. 8.7 Complete set of Explore Scientific 82° eyepieces (Image courtesy of John W., MA, USA)

Focal length (mm)	AFOV (degrees)	ER (mm)	Field stop (mm)	Weight (oz)	Design (Elem/Grp)	Coatings	Barrel (inches)
4.7	82	14	6.7[E]	7.5	7/4	FMC	1.25
6.7	82	14	9.6[E]	8	7/4	FMC	1.25
8.8	82	14	12.6[E]	8	7/4	FMC	1.25
11	82	15	15.7[E]	10	7/4	FMC	1.25
14	82	15	20.0[E]	9	7/4	FMC	1.25
18	82	13	25.8[E]	14	6/4	FMC	2
24	82	17	34.3[E]	25.6	6/4	FMC	2
30	82	21	42.9[E]	35.2	6/4	FMC	2

Source: www.explorescientific.com

These are waterproofed with nitrogen gas; are individually serial numbered; and have a lifetime warranty. Considered by many as an excellent performing ultra wide-field that keeps pace with even the premium brands.

This eyepiece line is within the following competition class of eyepieces with similar specifications and performance: Antares Speers-WALER Series, Astro-Professional UWA, Celestron Axiom LX, Celestron Luminos, Docter UWA, Meade Series 4000 Ultra Wide Angle, Meade Series 5000 Ultra Wide Angle, Orion MegaView Ultra-Wide, Sky-Watcher Nirvana UWA, Sky-Watcher Sky Panorama, Tele Vue Nagler, Williams Optics UWAN.

Fig. 8.8 Example of an earlier style of the 82° series without nitrogen purging (Image by the author)

Explore Scientific: 100° Nitrogen Purged (ES100 N2)

Fig. 8.9 The 9 mm, 14 mm, and 20 mm Explore Scientific 100° eyepieces (Image courtesy of John W., MA, USA)

Focal length (mm)	AFOV (degrees)	ER (mm)	Field stop (mm)	Weight (oz)	Design (Elem/Grp)	Coatings	Barrel (inches)
9	100	12.5	15.7[E]	21	9	FMC	2
14	100	14.5	24.4[E]	30	9	FMC	2
20	100	14.5	34.9[E]	34	9	FMC	2
25	100	–	43.6[E]	40	8	FMC	2
30	100	–	52.4[E]	128*	–	FMC	3

Source: www.explorescientific.com

These use low dispersion and high refractive index optical glasses; are waterproofed with nitrogen gas; are individually serial numbered; and have a lifetime warranty. This eyepiece line is within the following competition class of eyepieces with similar specifications and performance: Nikon NAV-HW, Tele Vue Ethos. The 30 mm eyepiece is the soon to be released with a 3 in. barrel eyepiece. *Weight is estimated at 128 ounces including diagonal.

Explore Scientific: 120° Nitrogen Purged (ES100 N2)

Fig. 8.10 The 9 mm Explore Scientific 120° zeries. TMB supermonocentric shown for size comparison (Image courtesy of Tamiji Homma, Newbury Park, CA, USA)

Focal length (mm)	AFOV (degrees)	ER (mm)	Field stop (mm)	Weight (oz)	Design (Elem/Grp)	Coatings	Barrel (inches)
9	120	13	18.8[E]	48	12/8	FMC	2

Source: www.explorescientific.com

The first production 120° AFOV eyepiece in the consumer marketplace. Initial reaction from observers has been very positive. These are o-ring sealed and argon-purged to be 100 % waterproof; use dense crown, light crown, dense flint, and lanthanum rare Earth optical glasses; have15-layer multi-coatings on all air/glass surfaces; designed for use with telescopes having focal ratios down to f/4 and below; are individually serial numbered; and have a lifetime warranty. This eyepiece is the only 120° AFOV eyepiece on the market; as such it has no direct competition as of late 2012.

Galland/Gailand/Galoc: Ortho/Erfle/König (*Discontinued*)

Fig. 8.11 Galoc eyepieces and Barlow (Image courtesy of Jim Rosenstock, Fort Washington, MD, USA)

Focal length (mm)	AFOV (degrees)	ER (mm)	Field stop (mm)	Weight (oz)	Design (Elem/Grp)	Coatings	Barrel (inches)
4	42[E]	3[E]	2.9[E]	<3[E]	4/2-Ortho	FC	1.25
7	42[E]	6[E]	5.1[E]	<3[E]	4/2-Ortho	FC	1.25
10	42[E]	8[E]	7.3[E]	<3[E]	4/2-Ortho	FC	1.25
16.3	65[E]	5[E]	18.5[E]	<3[E]	5/3-Erfle	FC	1.25
28	42[E]	22[E]	20.5[E]	<3[E]	4/2-Ortho	FC	1.25

Garrett Optical: Orthoscopic (*Discontinued*)

Focal length (mm)	AFOV (degrees)	ER (mm)	Field stop (mm)	Weight (oz)	Design (Elem/Grp)	Coatings	Barrel (inches)
4	41	3.5	2.9[E]	2.8	4/2	FC	1.25
5	43	4	3.8[E]	3.0	4/2	FC	1.25
6	43	4.8	4.5[E]	3.2	4/2	FC	1.25
7	42	5.6	5.1[E]	3.2	4/2	FC	1.25
9	42	7.2	6.6[E]	3.2	4/2	FC	1.25
12.5	44	10	9.6[E]	3.4	4/2	FC	1.25
18	46	14.4	14.5[E]	3.4	4/2	FC	1.25
25	47	20	20.5[E]	3.5	4/2	FC	1.25

Source: www.garrettoptical.com

Garrett Optical: Plössl

Focal length (mm)	AFOV (degrees)	ER (mm)	Field stop (mm)	Weight (oz)	Design (Elem/Grp)	Coatings	Barrel (inches)
10	52	7	9.1[E]	<3[E]	4/2	FMC	1.25
12.5	52	8	11.3[E]	<3[E]	4/2	FMC	1.25
15	52	13	13.6[E]	<3[E]	4/2	FMC	1.25
20	52	20	18.2[E]	<3[E]	4/2	FMC	1.25
25	52	17	22.7[E]	<3[E]	4/2	FMC	1.25
30	52	22	27.0[E]	<4[E]	4/2	FMC	1.25

Source: www.garrettoptical.com

This eyepiece line is within the following competition class of eyepieces with similar specifications and performance: Antares Plössl, Astro-Professional Plössl, Astro-Tech High Grade Plössl, Astro-Tech Value Line Plössl, Bresser 52° Super Plössl, Carton Plössl, Celestron Omni, Celestron Silvertop Plössl, Clavé Plössl, Coronado CeMax, Edmund Scientific Plössl, GSO Plössl, GTO Plössl, Long Perng

Plössl, Meade Series 3000 Plössl, Meade Series 4000 Super Plössl, Olivon Plössl, Opt Plössl, Orion HighLight Plössl, Orion Sirius Plössl, Owl Black Night Plössl, Parks Silver Series, Sky-Watcher SP-Series Super Plössl, Smart Astronomy Sterling Plössl, TAL—Symmetrical Super Plössl, Telescope Service Plössl, Telescope Service Super Plössl, Tele Vue Plössl, Vixen NPL.

Garrett Optical: SuperWide Angle

Focal length (mm)	AFOV (degrees)	ER (mm)	Field stop (mm)	Weight (oz)	Design (Elem/Grp)	Coatings	Barrel (inches)
10	72	10	12.6[E]	2.9	5/4	FMC	1.25
15	72	13	18.8[E]	3.6	5/4	FMC	1.25
20	72	16	25.1[E]	4.2	5/4	FMC	1.25
26	71	20	32.2[E]	12	5/3	FMC	2
32	72	24	40.2[E]	14	5/4	FMC	2
38	69	28	45.8[E]	21	5/3	FMC	2

Source: www.garrettoptical.com

These are parfocal. This eyepiece line is within the following competition class of eyepieces with similar specifications and performance: Agena Super Wide Angle, Astro-Professional SWA, Astro-Tech Wide Field, GSO Superview, Meade QX, Olivon 70° Wide Angle, Opt Super View, Orion Q70 Super Wide-Field, Sky-Watcher PanaView, Telescope-Service WA, University Optics 70°, William Optics SWAN.

GSO (Guan Sheng Optical): Kellner

Focal length (mm)	AFOV (degrees)	ER (mm)	Field stop (mm)	Weight (oz)	Design (Elem/Grp)	Coatings	Barrel (inches)
26	65	20	29.5[E]	9.3	3/2	FMC	2
32	65	20	36.3[E]	10.5	3/2	FMC	2
40	56	20	39.1[E]	10.5	3/2	FMC	2

Source: www.gs-telescope.com

This eyepiece line is within the following competition class of eyepieces with similar specifications and performance (Kellners and Reverse Kellners, and RKE—Rank-Kaspereit-Erfle): Celestron E-Lux (2 in. models only), Celestron Kellner, Criterion Kellner, Edmund Scientific RKE, Kokusai Kohki Kellner, Orion DeepView, Orion E-Series, Russell Optics (2 in. 52 and 60 mm only), Sky-Watcher Kellner, Sky-Watcher Super MA Series, Telescope Service RK, Unitron Kellner.

GSO (Guan Sheng Optical): Super Plössl

Fig. 8.12 The 40 mm GSO Super Plössl (Image courtesy of William Rose, Larkspur, CO, USA)

Focal length (mm)	AFOV (degrees)	ER (mm)	Field stop (mm)	Weight (oz)	Design (Elem/Grp)	Coatings	Barrel (inches)
4	52	6	3.6^E	2.6	4/2	FMC	1.25
6	52	5	5.4^E	1.7	4/2	FMC	1.25
9	52	6	8.2^E	1.7	4/2	FMC	1.25
12	52	8	10.9^E	2.9	4/2	FMC	1.25
15	52	13	13.6^E	3.1	4/2	FMC	1.25
20	52	20	18.2^E	3.5	4/2	FMC	1.25
25	52	22	22.7^E	3.6	4/2	FMC	1.25
32	52	22	27.0^E	4.2	4/2	FMC	1.25

Source: www.gs-telescope.com

This eyepiece line is within the following competition class of eyepieces with similar specifications and performance: Antares Plössl, Astro-Professional Plössl, Astro-Tech High Grade Plössl, Astro-Tech Value Line Plössl, Bresser 52° Super Plössl, Carton Plössl, Celestron Omni, Celestron Silvertop Plössl, Clavé Plössl, Coronado CeMax, Edmund Scientific Plössl, Garrett Optical Plössl, GTO Plössl,

Long Perng Plössl, Meade Series 3000 Plössl, Meade Series 4000 Super Plössl, Olivon Plössl, Opt Plössl, Orion HighLight Plössl, Orion Sirius Plössl, Owl Black Night Plössl, Parks Silver Series, Sky-Watcher SP-Series Super Plössl, Smart Astronomy Sterling Plössl, TAL—Symmetrical Super Plössl, Telescope Service Plössl, Telescope Service Super Plössl, Tele Vue Plössl, Vixen NPL.

GSO (Guan Sheng Optical): Superview

Fig. 8.13 Complete set of GSO Superview eyepieces (Image courtesy of John W., MA, USA)

Focal length (mm)	AFOV (degrees)	ER (mm)	Field stop (mm)	Weight (oz)	Design (Elem/Grp)	Coatings	Barrel (inches)
15	68	13	17.8[E]	3.8	4/3	FMC	1.25
20	68	18	23.7[E]	5.7	5/3	FMC	1.25
30	68	22	35.6[E]	12.9	5/3	FMC	2
42	65	30	46.0[E]	13	5/3	FMC	2
50	60	35	52.4[E]	13.6	5/3	FMC	2

Source: www.gs-telescope.com

This eyepiece line is within the following competition class of eyepieces with similar specifications and performance: Agena Super Wide Angle, Astro-Professional SWA, Astro-Tech Wide Field, Garrett SuperWide Angle, Meade QX, Olivon 70° Wide Angle, Opt Super View, Orion Q70 Super Wide-Field, Sky-Watcher PanaView, Telescope-Service WA, University Optics 70°, William Optics SWAN.

GTO: Plössl

Focal length (mm)	AFOV (degrees)	ER (mm)	Field stop (mm)	Weight (oz)	Design (Elem/Grp)	Coatings	Barrel (inches)
4	52	4	3.6[E]	<3[E]	4/2	FMC	1.25
6	52	5	5.4[E]	<3[E]	4/2	FMC	1.25
9	52	7	8.2[E]	<3[E]	4/2	FMC	1.25
12	52	12	10.9[E]	<3[E]	4/2	FMC	1.25
15	52	12	13.6[E]	<3[E]	4/2	FMC	1.25
25	52	20	22.7[E]	<3[E]	4/2	FMC	1.25
32	52	24	29.0[E]	<4[E]	4/2	FMC	1.25

This eyepiece line is within the following competition class of eyepieces with similar specifications and performance: Antares Plössl, Astro-Professional Plössl, Astro-Tech High Grade Plössl, Astro-Tech Value Line Plössl, Bresser 52° Super Plössl, Carton Plössl, Celestron Omni, Celestron Silvertop Plössl, Clavé Plössl, Coronado CeMax, Edmund Scientific Plössl, Garrett Optical Plössl, GSO Plössl, Long Perng Plössl, Meade Series 3000 Plössl, Meade Series 4000 Super Plössl, Olivon Plössl, Opt Plössl, Orion HighLight Plössl, Orion Sirius Plössl, Owl Black Night Plössl, Parks Silver Series, Sky-Watcher SP-Series Super Plössl, Smart Astronomy Sterling Plössl, TAL—Symmetrical Super Plössl, Telescope Service Plössl, Telescope Service Super Plössl, Tele Vue Plössl, Vixen NPL.

GTO: Proxima

Fig. 8.14 A 31 mm Proxima eyepiece (Eyepieces courtesy of www.handsonoptics.com. Image by the author)

Focal length (mm)	AFOV (degrees)	ER (mm)	Field stop (mm)	Weight (oz)	Design (Elem/Grp)	Coatings	Barrel (inches)
31	71	20	38.4^{E}		5	MC	2

Not recommended for faster than f/5 optical systems.

GTO: Wide Field

Focal length (mm)	AFOV (degrees)	ER (mm)	Field stop (mm)	Weight (oz)	Design (Elem/Grp)	Coatings	Barrel (inches)
32	65	22	36.3^{E}	12^{E}	–	–	2
40	55	22	38.4^{E}	13^{E}	–	–	2
50	50	22	43.6^{E}	14^{E}	–	–	2

The manufacturer's statistics on optical design and coatings are no longer available.

I.R. Poyser: Plössl and Adapted Military

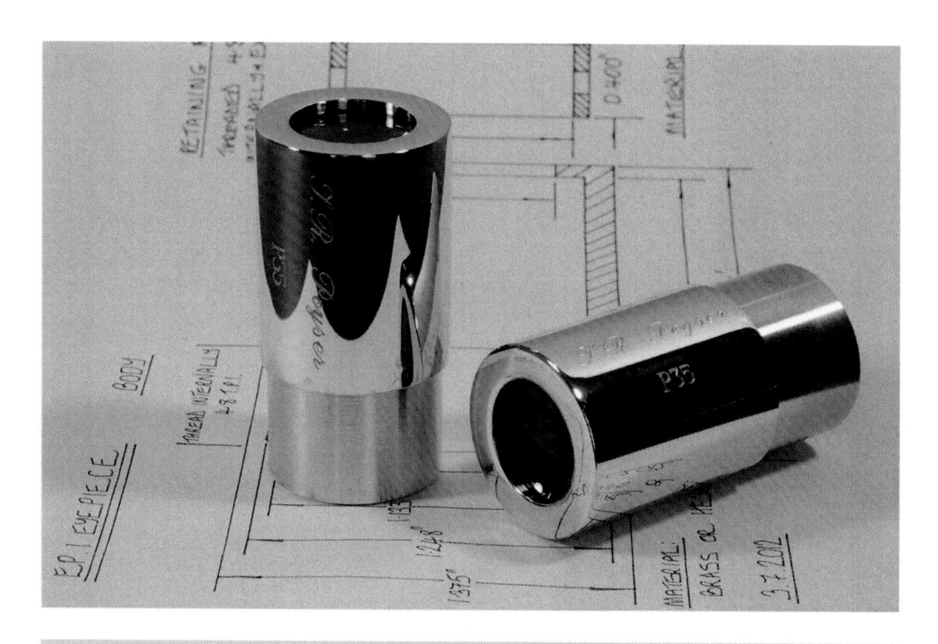

Fig. 8.15 I. R. Poyser 35 mm brass Plössl (© I.R.Poyser Telescope Makers. Image courtesy of Ian Poyser, Ystrad Meurig, Wales, U. K.—www.irpoyser.co.uk)

Fig. 8.16 I. R. Poyser 37 mm wide field (© I.R.Poyser Telescope Makers. Image courtesy of Ian Poyser, Ystrad Meurig, Wales, U. K.—www.irpoyser.co.uk)

Focal length (mm)	AFOV (degrees)	ER (mm)	Field stop (mm)	Weight (oz)	Design (Elem/Grp)	Coatings	Barrel (inches)
35	55	24[E]	21.2	7.2	4/2	None	1.25
37	55–60[E]	–	–	11	–	FMC	1.25 or 2

Source: www.irpoyser.co.uk

These eyepieces are made in the U.K. The 35 mm eyepiece is made in brass and can be optionally ordered with aluminum housings. The optics are available unmounted. The body is made of polished brass machined from a single piece of metal. The optics are not anti-reflection coated to provide the purest color image possible. Optics are British made and cemented elements use Canadian balsam. The optical components are easily removable without the need for special tools to aid cleaning. The 37 mm eyepiece is made from military optics and performs well in both long and short focal ratio telescopes.

Kasai: Astroplan (AP)

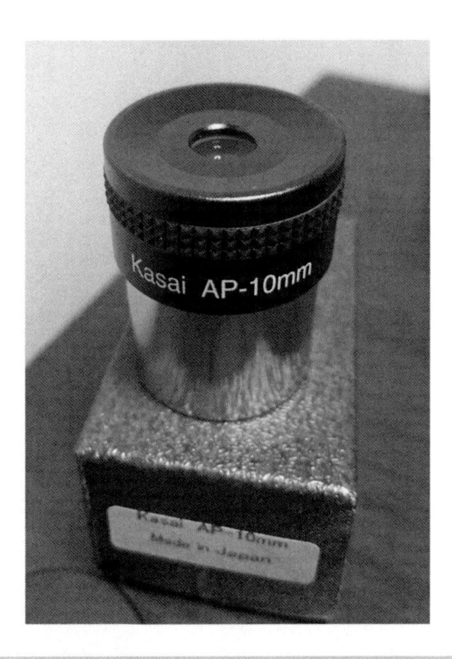

Fig. 8.17 The Kasai 10 mm Astroplan (AP) eyepiece (Image courtesy of Christoph Bosshard, Zürich, Switzerland)

Focal length (mm)	AFOV (degrees)	ER (mm)	Field stop (mm)	Weight (oz)	Design (Elem/Grp)	Coatings	Barrel (inches)
5	50	4^{E}	4.4^{E}	4^{E}	5/3	FMC	1.25
7.5	50	6^{E}	6.5^{E}	4^{E}	5/3	FMC	1.25
10	50	8^{E}	8.7^{E}	5^{E}	5/3	FMC	1.25
12.5	50	7.2^{E}	10.9^{E}	6^{E}	5/3	FMC	1.25
15	50	11^{E}	13.1^{E}	7^{E}	5/3	FMC	1.25
20	50	17^{E}	17.5^{E}	7^{E}	5/3	FMC	1.25

Source: www.kasai-trading.jp; www.hutech.com

The optics of this line are based on the Carl Zeiss 2-1-2 Super Plössl design and made by the company that produced the famous Masuyama eyepieces. This eyepiece line is within the following competition class of eyepieces with similar specifications and performance: Antares Elite, Baader Eudiascopic, Bresser 60° Plössl, Celestron Ultima, Meade Series 4000 Super Plössl (pre-1994, smooth sided, five-elements), Meade Series 5000 Super Plössl, Orion Ultrascopic, Parks Gold Series Plössl, Takahashi LE. Note—the Omcon Ultima and Tuthill Super Plössl lines are also competing brands, but these lines are not detailed.

Kokusai Kohki: Abbe Orthos (*Discontinued*)

Fig. 8.18 The entire line of Kokusai Kohki Abbe Ortho eyepieces (Image courtesy of Glen Moulton, Lyra Optic, Lancashire, U. K.—www.LyraOptic.co.uk)

Focal length (mm)	AFOV (degrees)	ER (mm)	Field stop (mm)	Weight (oz)	Design (Elem/Grp)	Coatings	Barrel (inches)
4	41	3.5	2.9^E	2.8	4/2	FC	1.25
5	43	4	3.8^E	3.0	4/2	FC	1.25
6	43	4.8	4.5^E	3.2	4/2	FC	1.25
7	42	5.6	5.1^E	3.2	4/2	FC	1.25
9	42	7.2	6.6^E	3.2	4/2	FC	1.25
12.5	44	10	9.6^E	3.4	4/2	FC	1.25
18	46	14.4	14.5^E	3.4	4/2	FC	1.25
25	47	20	20.5^E	3.5	4/2	FC	1.25

Source: www.kkohki.com

Along with the University Optics Volcano Orthos, these eyepieces have been considered over the years the standard for planetary viewing, offering performance well above their price point and having a volcano housing design making the short eye relief of the shorter focal lengths much easier to use. This eyepiece line is within the following competition class of eyepieces with similar specifications and performance: Antares Ortho, Apogee Super Abbe Orthoscopic, Baader Planetarium Classic Ortho, Baader Planetarium Genuine Abbe Ortho, Cave Orthostar Orthoscopic, Celestron Ortho, Edmund Scientific Ortho, Masuyama Orthoscopic, Meade Research Grade Ortho, Meade Series II Orthoscopic, Nikon Ortho, Siebert Optics Star Splitter/Super Star Splitter, Takahashi Ortho, Telescope Service Ortho, Unitron Ortho, University Optics Abbe HD Orthoscopic, University Optics Abbe Volcano Orthoscopic, University Optics O.P.S. Orthoscopic Planetary Series, University Optics Super Abbe Orthoscopic, VERNONscope Brandon.

Kokusai Kohki: Erfle (*Discontinued*)

Fig. 8.19 Kokusai Kohki 16 and 20 mm Erfles (Image courtesy of Malcolm Neo, Singapore)

Focal length (mm)	AFOV (degrees)	ER (mm)	Field stop (mm)	Weight (oz)	Design (Elem/Grp)	Coatings	Barrel (inches)
16	60	9.6	16.8[E]	6.2	5/3	MC	1.25
20	62	11	21.6[E]	6.2	5/3	MC	1.25
25	62	21	27.0[E]	6.2	5/3	MC	1.25

Source: www.kkohki.com

This eyepiece line is within the following competition class of eyepieces with similar specifications and performance: Celestron Erfle, University Optics Super Erfle.

Kokusai Kohki: Kellner (*Discontinued*)

Fig. 8.20 Kokusai Kohki Kellners in 6 mm, 9 mm, and 12 mm focal lengths (Image courtesy of Andy Howie, Paisley, Scotland)

Focal length (mm)	AFOV (degrees)	ER (mm)	Field stop (mm)	Weight (oz)	Design (Elem/Grp)	Coatings	Barrel (inches)
6	41	4.8	4.3^E	3.0	3/2	MC	1.25
9	44	7.2	6.9^E	3.7	3/2	MC	1.25
12	40	9.6	8.4^E	4.3	3/2	MC	1.25
18	47	9.6	14.8^E	3.5	3/2	MC	1.25
20	47	15	16.4^E	3.5	3/2	MC	1.25
25	45	16	19.6^E	3.7	3/2	MC	1.25

Source: www.kkohki.com

This eyepiece line is within the following competition class of eyepieces with similar specifications and performance (Kellners and Reverse Kellners, and RKE—Rank-Kaspereit-Erfle): Celestron E-Lux (2 in. models only), Celestron Kellner, Criterion Kellner, Edmund Scientific RKE, GSO Kellner, Orion DeepView, Orion E-Series, Russell Optics (2 in. 52 and 60 mm only), Sky-Watcher Kellner, Sky-Watcher Super MA Series, Telescope Service RK, Unitron Kellner.

Kokusai Kohki: Widescan III (*Discontinued*)

Focal length (mm)	AFOV (degrees)	ER (mm)	Field stop (mm)	Weight (oz)	Design (Elem/Grp)	Coatings	Barrel (inches)
20	84	–	25[E]	21[E]	7/4	FMC	2
30	84	–	41[E]	20[E]	5/3	FMC	2

Source: www.kkohki.com

This eyepiece line is within the following competition class of eyepieces with similar specifications and performance: Agena Ultra Wide Angle, Apogee Widescan (20 and 30 mm), Astrobuffet 1RPD (30 mm), BW Optik Ultrawide (30 mm), Moonfish Ultrawide (30 mm), Olivion 80° Ultra Wide Angle, Owl Astronomy Knight Owl Ultrawide Angle, Sky-Watcher Ultra Wide Angle, Surplus Shed Wollensak (30 mm), TAL—Ultra Wide Angle, University Optics 80°, University Optics Widescan I/II/III (20 and 30 mm).

Kson: Super Abbe Orthoscopic

Fig. 8.21 The complete line of KSON Super Abbe eyepieces (© KSON Optics-Electronics Company—www.ksonoptics.com)

Focal length (mm)	AFOV (degrees)	ER (mm)	Field stop (mm)	Weight (oz)	Design (Elem/ Grp)	Coatings	Barrel (inches)
4.8	46	4[E]	3.9[E]	<3[E]	4/2	FMC	1.25
7.7	46	6[E]	6.5	<3[E]	4/2	FMC	1.25
10.5	46	9[E]	8.4[E]	<3[E]	4/2	FMC	1.25
16.8	46	14[E]	13.5[E]	<4[E]	4/2	FMC	1.25
24	46	20[E]	20.5	<4[E]	4/2	FMC	1.25

Source: www.ksonoptics.com

The Kson Super Abbe Orthos are identical in form and available focal lengths as the University Optics Super Abbe Orthoscopic. This eyepiece line is within the following competition class of eyepieces with similar specifications and performance: Antares Ortho, Apogee Super Abbe Orthoscopic, Baader Planetarium Classic Ortho, Baader Planetarium Genuine Abbe Ortho, Cave Orthostar Orthoscopic, Celestron Ortho, Edmund Scientific Ortho, Kokusai Kohki Abbe Ortho, Masuyama Orthoscopic, Meade Research Grade Ortho, Meade Series II Orthoscopic, Nikon Ortho, Siebert Optics Star Splitter/Super Star Splitter, Takahashi Ortho, Telescope Service Ortho, Unitron Ortho, University Optics Abbe HD Orthoscopic, University Optics Abbe Volcano Orthoscopic, University Optics O.P.S. Orthoscopic Planetary Series, University Optics Super Abbe Orthoscopic, VERNONscope Brandon.

Leitz: Ultra Wide (30 mm 88°)

Fig. 8.22 The Leitz 30 mm 88° eyepiece (Image courtesy of William Rose, Larkspur, CO, USA)

Focal length (mm)	AFOV (degrees)	ER (mm)	Field stop (mm)	Weight (oz)	Design (Elem/Grp)	Coatings	Barrel (inches)
30	88	15	46^{E}	20^{E}	6	–	2

This eyepiece line is within the following competition class of eyepieces with similar specifications and performance: Astro-Professional 28 mm UWA, Leitz 30 mm Ultra Wide, Orion 28 mm MegaView Ultra-Wide, Sky-Watcher 28 mm Nirvana UWA, Tele Vue 31 mm Nagler, and Williams Optics 28 mm UWAN.

Long Perng: 68° Wide Angle

Focal length (mm)	AFOV (degrees)	ER (mm)	Field stop (mm)	Weight (oz)	Design (Elem/Grp)	Coatings	Barrel (inches)
3.5	68	20	4.2^{E}	13^{E}	8/6	FMC	1.25
5	68	20	5.9^{E}	12^{E}	8/6	FMC	1.25
7	68	20	8.3^{E}	11^{E}	8/6	FMC	1.25
9	68	20	10.7^{E}	10^{E}	8/6	FMC	1.25
12	68	20	14.2^{E}	10^{E}	8/6	FMC	1.25
14.5	68	20	17.2^{E}	10^{E}	8/6	FMC	1.25
18	68	20	21.4^{E}	10^{E}	8/6	FMC	1.25

Source: www.longperng.com.tw

This eyepiece line is within the following competition class of eyepieces with similar specifications and performance: Long Perng 68° Wide Angle and Orion Premium 68° Long Eye Relief.

Long Perng: Long Eye Relief

Fig. 8.23 Long Perng Long Eye Relief eyepieces in 3 mm, 9 mm, 5 mm, and 14.5 mm focal lengths (Photograph reprinted with authorization of Long Perng, Taiwan)

Focal length (mm)	AFOV (degrees)	ER (mm)	Field stop (mm)	Weight (oz)	Design (Elem/Grp)	Coatings	Barrel (inches)
3	55	20	2.9[E]	10	7/4	FMC	1.25
5	55	20	4.8[E]	10	7/4	FMC	1.25
6	55	20	5.8[E]	10	7/4	FMC	1.25
9	55	20	8.6[E]	9	7/4	FMC	1.25
12.5	55	20	12.0[E]	8	7/4	FMC	1.25
14.5	55	20	13.4[E]	7	7/4	FMC	1.25
18	55	20	17.3[E]	6	5/3	FMC	1.25

Source: www.longperng.com.tw

This eyepiece line is within the following competition class of eyepieces with similar specifications and performance: Astro-Professional Long Eye Relief Planetary, Astro-Tech Long Eye Relief, Long Perng Long Eye Relief, Orion Edge-On Planetary, Smart Astronomy SA Solar System Long Eye Relief, Stellarvue Planetary, William Optics SPL, Zhumell Z Series.

Long Perng: Plössl

Fig. 8.24 The complete line of Long Perng Plössls (Photograph reprinted with authorization of Long Perng, Taiwan)

Focal length (mm)	AFOV (degrees)	ER (mm)	Field stop (mm)	Weight (oz)	Design (Elem/Grp)	Coatings	Barrel (inches)
4	55	2.4	3.8^{E}	2.1	4/2	FMC	1.25
6	55	3.6	8.6^{E}	2.2	4/2	FMC	1.25
12.5	55	7.5	12.0^{E}	3.0	4/2	FMC	1.25
17	55	10.2	16.3^{E}	3.4	4/2	FMC	1.25
20	55	12	19.2^{E}	3.4	4/2	FMC	1.25
25	55	15	24.0^{E}	4.7	4/2	FMC	1.25
30	55	18	28.8^{E}	13.8	4/2	FMC	2
40	55	24	38.4^{E}	13.3	4/2	FMC	2

Source: www.longperng.com.tw

These are fully multicoated with 40-layer high transmission/anti-reflection optical coatings. They have a unique distinction over most other four-element Plössl lines as having an extended AFOV. Almost all other Plössl lines have an AFOV of 50–52°, whereas the Astro-Tech High Grade Plössl, Smart Astronomy Sterling

Plössl, and Long Perng Plössl have extended this to an advertised 55° (closer to 57° when measured). This larger AFOV, although only a few degrees more than a standard Plössl, is visually more impressive, conveying the impression of an eyepiece in the next larger AFOV class.

This eyepiece line is within the following competition class of eyepieces with similar specifications and performance: Antares Plössl, Astro-Professional Plössl, Astro-Tech High Grade Plössl, Astro-Tech Value Line Plössl, Bresser 52° Super Plössl, Carton Plössl, Celestron Omni, Celestron Silvertop Plössl, Clavé Plössl, Coronado CeMax, Edmund Scientific Plössl, Garrett Optical Plössl, GSO Plössl, GTO Plössl, Meade Series 3000 Plössl, Meade Series 4000 Super Plössl, Olivon Plössl, Opt Plössl, Orion HighLight Plössl, Orion Sirius Plössl, Owl Black Night Plössl, Parks Silver Series, Sky-Watcher SP-Series Super Plössl, Smart Astronomy Sterling Plössl, TAL—Symmetrical Super Plössl, Telescope Service Plössl, Telescope Service Super Plössl, Tele Vue Plössl, Vixen NPL.

Masuyama: Masuyama (*Discontinued*)

Fig. 8.25 Masuyama 7.5 mm, 15 mm, 25 mm, and 35 mm eyepieces (Image courtesy of Carol Anderson, Helena, MT, USA)

Focal length (mm)	AFOV (degrees)	ER (mm)	Field stop (mm)	Weight (oz)	Design (Elem/Grp)	Coatings	Barrel (inches)
5	52	3.9	4.5[E]	2.3	5/3	FMC	1.25
7.5	52	5	6.8[E]	2.3	5/3	FMC	1.25
10	52	5.9	9.1[E]	2.8	5/3	FMC	1.25
15	52	9.1	13.6[E]	2.8	5/3	FMC	1.25
20	52	12.7	18.2[E]	4.1	5/3	FMC	1.25
25	52	16.7	22.7[E]	4.6	5/3	FMC	1.25
25W	65	14.6	27.0[E]	9.7	6/4	FMC	1.25/1.43
30	52	19.2	27.0[E]	9.3	5/3	FMC	1.25/1.43
35	52	23.5	27.0[E]	10.1	5/3	FMC	1.25/1.43
45	50	32.5	39.3[E]	20.6	5/3	FMC	2/1.43
60	46	46.7	47.0[E]	29.8	5/3	FMC	2
80	34	50.3	47.0[E]	38.8	4/2	FMC	2
100	46	34	80.3[E]	60.0	3/2	FMC	4

Source: Ohi Optical Manufacturing Co., Ltd, Japan

The Masuyamas are a legendary eyepiece line with very distinctive styling and an outstanding reputation. The longer focal lengths in particular are often reported to have phenomenal contrast with velvet black background fields of view.

This eyepiece line is within the following competition class of eyepieces with similar specifications and performance: Antares Elite, Baader Eudiascopic, Bresser 60° Plössl, Celestron Ultima, Meade Series 4000 Super Plössl (pre-1994, smooth sided, five-elements), Meade Series 5000 Super Plössl, Orion Ultrascopic, Parks Gold Series Plössl, Takahashi LE. Note—the Omcon Ultima and Tuthill Super Plössl lines are also competing brands, but these lines are not detailed.

Masuyama: Orthoscopic

Focal length (mm)	AFOV (degrees)	ER (mm)	Field stop (mm)	Weight (oz)	Design (Elem/Grp)	Coatings	Barrel (inches)
4	42	3.35	2.9[E]	<3[E]	4/2	FMC	1.25
5	42	4.05	3.7[E]	<3[E]	4/2	FMC	1.25
6	42	4.91	4.4[E]	<3[E]	4/2	FMC	1.25
7	42	6.07	5.1[E]	<3[E]	4/2	FMC	1.25
9	42	7.61	6.6[E]	<3[E]	4/2	FMC	1.25
12.5	42	10.41	9.2[E]	<3[E]	4/2	FMC	1.25
18	42	15.21	13.2[E]	<3[E]	4/2	FMC	1.25
25	42	22.22	18.3[E]	<4[E]	4/2	FMC	1.25
40	39[E]	34[E]	27.0[E]	<4[E]	4/2	FMC	1.25

Source: Ohi Optical Manufacturing Co., Ltd, Japan

The Masuyama Orthos have been announced on the manufacturer's website but have not been released as of the publication of this book.

This eyepiece line is expected to compete with the following other eyepiece lines: Antares Ortho, Apogee Super Abbe Orthoscopic, Baader Planetarium Classic Ortho, Baader Planetarium Genuine Abbe Ortho, Cave Orthostar Orthoscopic, Celestron Ortho, Edmund Scientific Ortho, Kokusai Kohki Abbe Ortho, Kson Ortho, Meade Research Grade Ortho, Meade Series II Orthoscopic, Nikon Ortho, Siebert Optics Star Splitter/Super Star Splitter, Takahashi Ortho, Telescope Service Ortho, Unitron Ortho, University Optics Abbe HD Orthoscopic, University Optics Abbe Volcano Orthoscopic, University Optics O.P.S. Orthoscopic Planetary Series, University Optics Super Abbe Orthoscopic, VERNONscope Brandon.

Meade: Research Grade Ortho and Wide Field (*Discontinued*)

Fig. 8.26 Full collection of Meade Research Grade Orthos and Wide Fields (Image courtesy of Jim Barnett, Petaluma, CA, USA)

Focal length (mm)	AFOV (degrees)	ER (mm)	Field stop (mm)	Weight (oz)	Design (Elem/Grp)	Coatings	Barrel (inches)
4	42	3^{E}	2.9^{E}	$<3^{E}$	4/2	MC	1.25
7	42	6^{E}	5.1^{E}	$<3^{E}$	4/2	MC	1.25
10.5	42	8^{E}	7.7^{E}	$<3^{E}$	4/2	MC	1.25
16.8	42	13^{E}	12.3^{E}	$<3^{E}$	4/2	MC	1.25
28	42	22^{E}	20.5^{E}	$<3^{E}$	4/2	MC	1.25
7	60	4^{E}	7.3^{E}	$<3^{E}$	5/3	MC	1.25
12.4	65	8^{E}	14.1^{E}	$<3^{E}$	5/3	MC	1.25
15.5	65	10^{E}	17.6^{E}	$<3^{E}$	5/3	MC	1.25
20	65	12.7^{M}	22.7^{E}	$<3^{E}$	5/3	MC	1.25
32	65	20^{E}	36.3^{E}	$<3^{E}$	5/3	MC	2

Source: www.meade.com

These use a seven-layer multi-coating and are parfocal. Usually a much sought after vintage eyepiece line with an excellent reputation for planetary observing. This eyepiece line is within the following competition class of eyepieces with similar specifications and performance: Antares Ortho, Apogee Super Abbe Orthoscopic, Baader Planetarium Classic Ortho, Baader Planetarium Genuine Abbe Ortho, Cave Orthostar Orthoscopic, Celestron Ortho, Kokusai Kohki Abbe Ortho, Kson Super Ortho, Masuyama Orthoscopic, Meade Series II Orthoscopic, Nikon Ortho, Siebert Optics Star Splitter/Super Star Splitter, Takahashi Ortho, Telescope Service Ortho, Unitron Ortho, University Optics Abbe HD Orthoscopic, University Optics Abbe Volcano Orthoscopic, University Optics O.P.S. Orthoscopic Planetary Series, University Optics Super Abbe Orthoscopic, VERNONscope Brandon.

Meade: Series II Modified Achromatic MA (*Discontinued*)

Fig. 8.27 The Meade Modified Achromat eyepieces (Image courtesy of Paul D. Webb, Columbia, SC, USA)

Focal length (mm)	AFOV (degrees)	ER (mm)	Field stop (mm)	Weight (oz)	Design (Elem/Grp)	Coatings	Barrel (inches)
9	40	5	6.3[E]	≈3[E]	3	MC	1.25
12	40	8	8.4[E]	≈3[E]	3	MC	1.25
25	40	16	17.5[E]	≈4[E]	3	MC	1.25
40	36	18	25.1[E]	≈6[E]	3	MC	1.25
40WF	40	–	27.0[E]	≈6[E]	–	MC	1.25

Source: www.meade.com

Meade: Series II Orthoscopic (*Discontinued*)

Fig. 8.28 The complete Meade Series II Ortho collection (Image courtesy of Jamie Crona, Fanwood, NJ, USA)

Focal length (mm)	AFOV (degrees)	ER (mm)	Field stop (mm)	Weight (oz)	Design (Elem/Grp)	Coatings	Barrel (inches)
4	45	3^{E}	3.1^{E}	$\approx 3^{E}$	4/2	MC	1.25
6	45	5^{E}	4.7^{E}	$\approx 3^{E}$	4/2	MC	1.25
9	45	7^{E}	7.1^{E}	$\approx 3^{E}$	4/2	MC	1.25
12.5	45	10^{E}	9.8^{E}	$\approx 3^{E}$	4/2	MC	1.25
18	45	14^{E}	14.1^{E}	$\approx 4^{E}$	4/2	MC	1.25
25	45	16	19.6^{E}	$\approx 4^{E}$	4/2	MC	1.25

Source: www.meade.com

Aside from comments on the age of this vintage eyepiece line they often receive very excellent reports from observers and are considered excellent performers for planetary observing. These are also parfocal.

This eyepiece line is within the following competition class of eyepieces with similar specifications and performance: Antares Ortho, Apogee Super Abbe Orthoscopic, Baader Planetarium Classic Ortho, Baader Planetarium Genuine Abbe Ortho, Cave Orthostar Orthoscopic, Celestron Ortho, Edmund Scientific Ortho, Kokusai Kohki Abbe Ortho, Kson Super Ortho, Masuyama Orthoscopic, Meade Research Grade Ortho, Nikon Ortho, Siebert Optics Star Splitter/Super Star Splitter, Takahashi Ortho, Telescope Service Ortho, Unitron Ortho, University Optics Abbe HD Orthoscopic, University Optics Abbe Volcano Orthoscopic, University Optics O.P.S. Orthoscopic Planetary Series, University Optics Super Abbe Orthoscopic, VERNONscope Brandon.

Meade: Series 3000 Plössl (*Discontinued*)

Fig. 8.29 The complete Meade Series 3000 Plössl collection. Older smooth-sided version (front). Newer version with rubber eyeguards and grip panels (back) (Image courtesy of Jamie Crona, Fanwood, NJ, USA)

Focal length (mm)	AFOV (degrees)	ER (mm)	Field stop (mm)	Weight (oz)	Design (Elem/Grp)	Coatings	Barrel (inches)
5	50	4.4^E	4.4^E	≈2^E	4/2	MC	1.25
6.7	50	5.8^E	5.8^E	≈3^E	4/2	MC	1.25
9.5	50	8.3^E	8.3^E	≈3^E	4/2	MC	1.25
16	50	14.0^E	14.0^E	≈3^E	4/2	MC	1.25
25	50	21.8^E	21.8^E	≈4^E	4/2	MC	1.25
40	44	27.0^E	27.0^E	≈4^E	4/2	MC	1.25

Source: www.meade.com

The Meade 3000 Plössls are considered by many observers as one of the consistently better performing vintage Plössls. The 40 mm of the line is considered by some to be one of the better examples of this focal length compared to any Plössl. These are also parfocal.

This eyepiece line is within the following competition class of eyepieces with similar specifications and performance: Antares Plössl, Astro-Professional Plössl, Astro-Tech High Grade Plössl, Astro-Tech Value Line Plössl, Bresser 52° Super Plössl, Carton Plössl, Celestron Omni, Celestron Silvertop Plössl, Clavé Plössl, Coronado CeMax, Edmund Scientific Plössl, Garrett Optical Plössl, GSO Plössl, GTO Plössl, Long Perng Plössl, Meade Series 4000 Super Plössl, Olivon Plössl, Opt Plössl, Orion HighLight Plössl, Orion Sirius Plössl, Owl Black Night Plössl, Parks Silver Series, Sky-Watcher SP-Series Super Plössl, Smart Astronomy Sterling Plössl, TAL—Symmetrical Super Plössl, Telescope Service Plössl, Telescope Service Super Plössl, Tele Vue Plössl, Vixen NPL.

Meade: Series 4000 QX Wide Angle (*Discontinued*)

Fig. 8.30 The complete line of Orion Stratus eyepieces (Image courtesy of Paul Surowiec, LaPorte, IN, USA)

Focal length (mm)	AFOV (degrees)	ER (mm)	Field stop (mm)	Weight (oz)	Design (Elem/Grp)	Coatings	Barrel (inches)
15	70	10	18.3[E]	3.8	5	FMC	1.25
20	70	14	24.4[E]	4.6	5	FMC	1.25
26	70	18	31.8[E]	13.0	5	FMC	2
30	70	21[E]	36.7[E]	13.6	5	FMC	2
36	70	25	44.0[E]	21.2	5	FMC	2

Source: www.meade.com

This eyepiece line is within the following competition class of eyepieces with similar specifications and performance: Agena Super Wide Angle, Astro-Professional SWA, Astro-Tech Wide Field, Garrett SuperWide Angle, GSO Superview, Olivon 70° Wide Angle, Opt Super View, Orion Q70 Super Wide-Field, Sky-Watcher PanaView, Telescope-Service WA, University Optics 70°, William Optics SWAN.

Meade: Series 4000 Super Plössl (Four-Element and Five-Element)

Fig. 8.31 The Meade 4000 Super Plössl (Image courtesy of Meade Instruments Corp.)

Focal length (mm)	AFOV (degrees)	ER (mm)	Field stop (mm)	Weight (oz)	Design (Elem/Grp)	Coatings	Barrel (inches)
6.4	52	2	5.8[E]	3	4/2	FMC	1.25
9.7	52	4	8.8[E]	3	4/2	FMC	1.25
12.4	52	7	11.3[E]	4	4/2	FMC	1.25
15	52	9	13.6[E]	4	4/2	FMC	1.25
20	52	13	18.2[E]	5	4/2	FMC	1.25
26	52	10	23.6[E]	5	4/2	FMC	1.25
32	52	13	27.0[E]	6	4/2	FMC	1.25
40	44	17	27.0[E]	≈6	4/2	FMC	1.25
56	52	21	46.0[E]	22	4/2	FMC	2

Source: www.meade.com

These use a seven-layer multi-coating; are parfocal; and are designed with the latest in optical glass types. For improved off-axis performance as compared to typical Plössls, current Meade 4000 4-element line also uses unconventional concave lens surfaces instead of the standard flat surfaces on the outward-facing eye lens and field lens. Other four-element Plössl lines that use concave lens surfaces include the Astro-Tech High Grade Plössl, Smart Astronomy Sterling Plössl, and Tele Vue Plössl.

This more current four-element version of this eyepiece line is within the following competition class of eyepieces with similar specifications and performance: Antares Plössl, Astro-Professional Plössl, Astro-Tech High Grade Plössl, Astro-Tech Value Line Plössl, Bresser 52° Super Plössl, Carton Plössl, Celestron Omni, Celestron Silvertop Plössl, Clavé Plössl, Coronado CeMax, Edmund Scientific Plössl, Garrett Optical Plössl, GSO Plössl, GTO Plössl, Long Perng Plössl, Meade Series 3000 Plössl, Olivon Plössl, Opt Plössl, Orion HighLight Plössl, Orion Sirius Plössl, Owl Black Night Plössl, Parks Silver Series, Sky-Watcher SP-Series Super Plössl, Smart Astronomy Sterling Plössl, TAL—Symmetrical Super Plössl, Telescope Service Plössl, Telescope Service Super Plössl, Tele Vue Plössl, Vixen NPL.

The older five-element (pseudo-Masuyama) version of this eyepiece line is generally reported as an excellent performing vintage line of eyepieces and is within the following competition class of eyepieces with similar specifications and performance: Antares Elite Plössl, Baader Eudiascopic, Bresser 60° Plössl, Celestron Ultima, Kasai Astroplan, Meade Series 5000 Super Plössl, Orion Ultrascopic, Parks Gold Series Plössl, Takahashi LE. *Note:* The Omcon Ultima and Tuthill Super Plössl lines are also competing brands, but these lines are not detailed.

Fig. 8.32 The complete Meade Series 4000 Plössl collection. Older 5-element smooth-sided version (Image courtesy of Jamie Crona, Fanwood, NJ, USA)

Meade: Series 4000 Super Wide Angle (SWA) (*Discontinued*)

Fig. 8.33 The 13.8 mm, 18 mm, 24.5 mm, 32 mm, and 40 mm Meade 4000 SWAs (Image courtesy of Michel Guévin, Longueuil, Québec, Canada)

Focal length (mm)	AFOV (degrees)	ER (mm)	Field stop (mm)	Weight (oz)	Design (Elem/Grp)	Coatings	Barrel (inches)
13.8	67	10	16.1[E]	6	6/4	FMC	1.25
18	67	14	21.0[E]	5	6/4	FMC	1.25
24.8	67	19	27.0[E]	10	6/4	FMC	1.25
32	67	25[E]	37.4[E]	18	6/4	FMC	2
40	67	31[E]	46.0[E]	25	6/4	FMC	2

Source: www.meade.com

These use a seven-layer multi-coating; are designed for low to medium power use with any telescope having a focal ratio down to f/4; and are parfocal. This eyepiece line is within the following competition class of eyepieces with similar specifications and performance: Explore Scientific 68 Series, Meade 5000 SWA, Nikon NAV-SW, Pentax XL, Pentax XW, Tele Vue Delos, and Tele Vue Panoptic.

Meade: Series 4000 Ultra Wide Angle (UWA) (*Discontinued*)

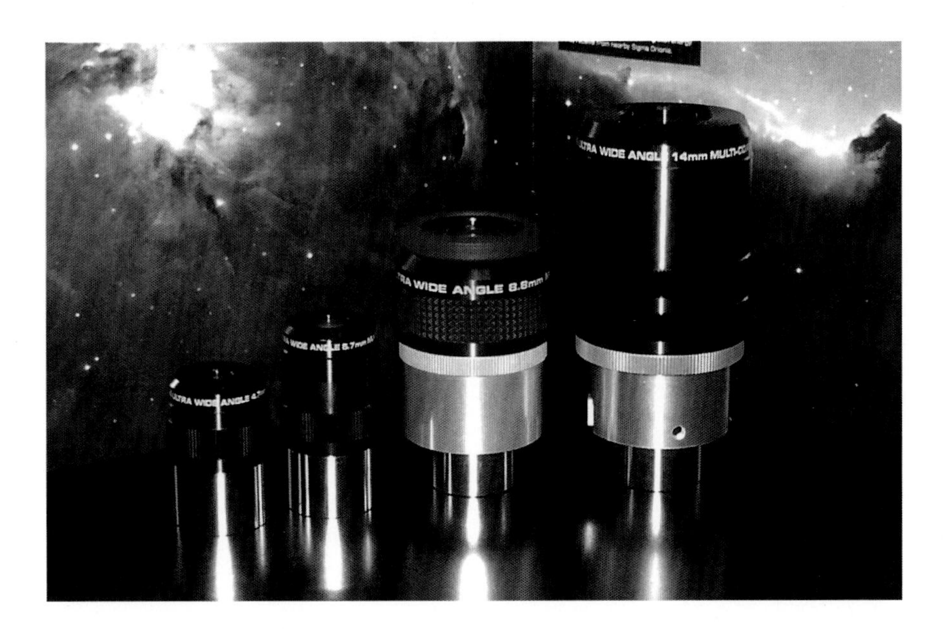

Fig. 8.34 The 4.7 mm, 6.7 mm, 8.8 mm, and 14 mm Meade 4000 UWAs (Image courtesy of Tony Miller, Stoney Creek, Ontario, Canada)

Focal length (mm)	AFOV (degrees)	ER (mm)	Field stop (mm)	Weight (oz)	Design (Elem/Grp)	Coatings	Barrel (inches)
4.7	84	8^E	6.9^E	5	8/5	FMC	1.25
6.7	84	9^E	9.8^E	6	8/5	FMC	1.25
8.8	84	10^E	12.9^E	18	8/5	FMC	1.25/2
14	84	16^E	20.5^E	26	8/5	FMC	1.25/2

Source: www.meade.com

These use a seven-layer multi-coating; the 4.7 and 8.8 mm are parfocal; and all are designed with the latest in optical glass types. The 14 mm of this line, although large, has a reputation of being a highly engaging eyepiece that challenges even the most modern 82° eyepieces. For many, the Meade 4000 UWAs are considered a "classic." As in a 2011 observation report by the author:

> *Taking it out into the field, I found using the 14 mm Meade 4000 UWA a very pleasurable experience. I'm usually not a fan of large eyepieces, like this is, nor of dual skirt eyepieces, but I was enjoying this one. I had a Nagler 13 mm T6 at the ready to do some compares in my Orion XT10 (250 mm) f/4.7 Dobsonian, so was ready for some fun. Overall the field of view in the Meade UWA was very good. Off-axis was not as tight as the Nagler 13 mm T6 but still very good. Lateral color, which is becoming a nuisance of late for me, was surprisingly excellent, being about twice as good as what the Nagler T6 was showing off-axis. Tonal qualities were also much more neutral in the Meade UWA. Once in a while, I did come across some eye positioning sensitivity with the Meade UWA, whereas I was getting none of this from the Nagler T6, but it was minimal. Star fields had a lot more "pop" to them from the Meade UWA than in the Nagler T6, so the visual experience was more interesting because of that between the two. In addition, the Meade UWA's field of view gave a much more engaging feel than the Nagler T6, which was feeling flat by comparison. If it were a "spacewalk" impression contest between them, the Meade UWA would have been the clear winner for me.*

This eyepiece line is within the following competition class of eyepieces with similar specifications and performance: Antares Speers-WALER Series, Astro-Professional UWA, Celestron Axiom LX, Celestron Luminos, Docter UWA, Explore Scientific 82° Nitrogen Purged, Meade Series 5000 Ultra Wide Angle, Orion MegaView Ultra-Wide, Sky-Watcher Nirvana UWA, Sky-Watcher Sky Panorama, Tele Vue Nagler, Williams Optics UWAN.

Meade: Series 5000 HD-60

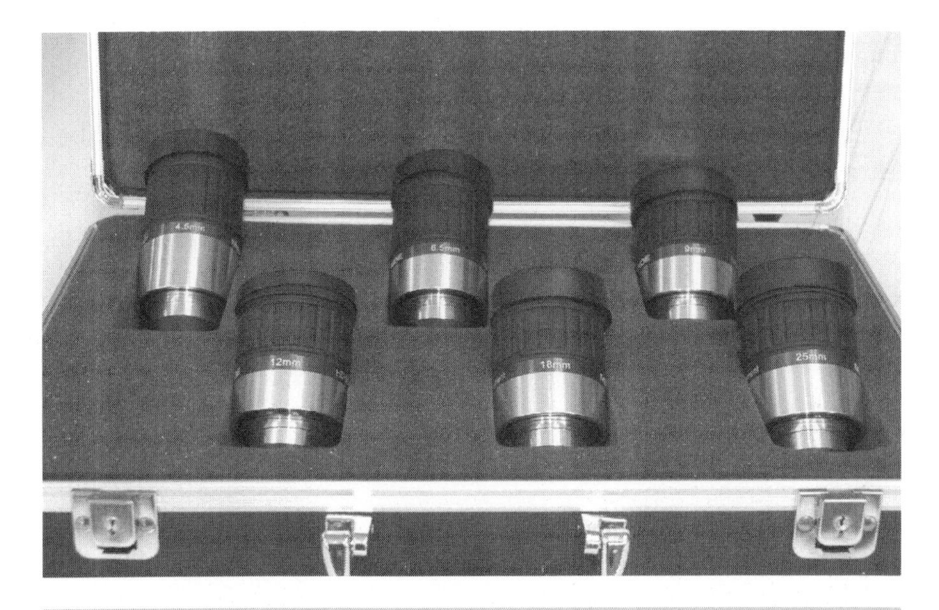

Fig. 8.35 Meade HD-60 eyepieces in presentation case (Eyepieces courtesy of www.handsonoptics.com. Image by the author)

Focal length (mm)	AFOV (degrees)	ER (mm)	Field stop (mm)	Weight (oz)	Design (Elem/Grp)	Coatings	Barrel (inches)
4.5	60	17	4.7[E]	7.2	6	FMC	1.25
6.5	60	17	6.8[E]	6.7	6	FMC	1.25
9	60	17	9.4[E]	6.3	6	FMC	1.25
12	60	17	12.6[E]	6.6	6	FMC	1.25
18	60	17	18.8[E]	7.4	6	FMC	1.25
25	60	17	26.2[E]	7.2	6	FMC	1.25

Source: www.meade.com

Meade: Series 5000 Plössl

Fig. 8.36 The complete Meade Series 5000 Plössls (Image courtesy of Stephen Chen, Dartmouth, Nova Scotia, Canada)

Focal length (mm)	AFOV (degrees)	ER (mm)	Field stop (mm)	Weight (oz)	Design (Elem/Grp)	Coatings	Barrel (inches)
5.5	60	7	5.8^E	2.6	6	FMC	1.25
9	60	6.8	9.4^E	2.5	5	FMC	1.25
14	60	10.4	14.7^E	3.2	5	FMC	1.25
20	60	14.8	20.9^E	4.9	5	FMC	1.25
26	60	19.3	27.0^E	8.2	5	FMC	1.25
32	60	22.4	33.5^E	15	5	FMC	2
40	60	28.2	41.9^E	25.4	5	FMC	2

Source: www.meade.com

These have individually tuned multi-layered coatings; have an adjustable twist-up rubber eyeguard; and are parfocal. This eyepiece line is within the following competition class of eyepieces with similar specifications and performance: Antares Elite, Baader Eudiascopic, Bresser 60° Plössl, Celestron Ultima, Kasai Astroplan, Meade Series 4000 Super Plössl (pre-1994, smooth sided, five-elements), Orion Ultrascopic, Parks Gold Series Plössl, Takahashi LE. *Note:* The Omcon Ultima and Tuthill Super Plössl lines are also competing brands, but these lines are not detailed here.

Meade: Series 5000 Super Wide Angle (SWA)

Fig. 8.37 Meade 5000 SWA eyepieces in 28 mm, 24 mm, 20 mm, and 16 mm (Image courtesy of David Elosser, NC, USA)

Focal length (mm)	AFOV (degrees)	ER (mm)	Field stop (mm)	Weight (oz)	Design (Elem/Grp)	Coatings	Barrel (inches)
16	68	11.9	19.0^E	5.6	6	FMC	1.25
20	68	15.3	23.7^E	9.5	6	FMC	1.25
24	68	18.4	29.7^E	13.1	6	FMC	2
28	68	21.6	33.2^E	18.0	6	FMC	2
34	68	26.4	40.4^E	27.5	6	FMC	2
40	68	31.1	46.0^E	43.7	6	FMC	2

Source: www.meade.com

This eyepiece line is within the following competition class of eyepieces with similar specifications and performance: Explore Scientific 68 Series, Meade 4000 SWA, Nikon NAV-SW, Pentax XL, Pentax XW, Tele Vue Delos, and Tele Vue Panoptic.

Meade: Series 5000 Ultra Wide Angle

Fig. 8.38 The Meade Series 5000 UWA eyepieces (Image courtesy of Meade Instruments Corp.)

Focal length (mm)	AFOV (degrees)	ER (mm)	Field stop (mm)	Weight (oz)	Design (Elem/Grp)	Coatings	Barrel (inches)
4.7	82	13.6	6.7[E]	7.8	7/4	FMC	1.25
5.5	82	13.2	7.9[E]	8.9	6/4	FMC	1.25
6.7	82	15	9.6[E]	7.6	7/4	FMC	1.25
8.8	82	15	12.6[E]	8.5	7/4	FMC	1.25
14	82	15	20.0[E]	8.2	7/4	FMC	1.25
18	82	13	25.8[E]	14.2	6/4	FMC	1.25
20	82	17.6	28.6[E]	26.1	6/4	FMC	2
24	82	17.6	34.3[E]	29.6	6/4	FMC	2
30	82	22	42.9[E]	47.2	6/4	FMC	2

Source: www.meade.com

Another really excellent performing ultra wide-angle eyepiece line, competing well even against more expensive brands. This eyepiece line is within the following competition class of eyepieces with similar specifications and performance: Antares Speers-WALER Series, Astro-Professional UWA, Celestron Axiom LX, Celestron Luminos, Docter UWA, Explore Scientific 82° Nitrogen Purged, Meade Series 4000 Ultra Wide Angle, Orion MegaView Ultra-Wide, Sky-Watcher Nirvana UWA, Sky-Watcher Sky Panorama, Tele Vue Nagler, Williams Optics UWAN.

Meade: XWA

Fig. 8.39 The Meade XWA eyepieces (Image courtesy of Meade Instruments Corp.)

Focal length (mm)	AFOV (degrees)	ER (mm)	Field stop (mm)	Weight (oz)	Design (Elem/Grp)	Coatings	Barrel (inches)
9	100	12.5	15.7^E	24	9/6	FMC	2
14	100	14.5	24.4^E	29	9/6	FMC	2
20	100	14.5	34.9^E	31	9/6	FMC	2

Source: www.meade.com

Designed to work with telescopes having focal ratios as fast as f/4; parfocal.

Moonfish: Ultrawide

Focal length (mm)	AFOV (degrees)	ER (mm)	Field stop (mm)	Weight (oz)	Design (Elem/Grp)	Coatings	Barrel (inches)
30	84	22^E	41^E	20^E	5/3	FMC	2

Source: www.moonfishgroup.com

This eyepiece line is within the following competition class of eyepieces with similar specifications and performance: Agena Ultra Wide Angle, Apogee Widescan (20 and 30 mm), Astrobuffet 1RPD (30 mm), BW Optik Ultrawide (30 mm),

Kokusai Kohki Widescan I/II/III (20 and 30 mm), Olivion 80° Ultra Wide Angle, Owl Astronomy Knight Owl Ultrawide Angle, Sky-Watcher Ultra Wide Angle, Surplus Shed Wollensak (30 mm), TAL Ultra Wide Angle, University Optics 80°, University Optics Widescan I/II/III (20 and 30 mm).

Nikon: NAV-HW (Hyper Wide)

Fig. 8.40 The 12.5 and 17 mm nikon NAV-HW eyepieces. The supplied EIC accessory converts them to 10 and 14 mm (Image courtesy of Tamiji Homma, Newbury Park, CA, USA)

Focal length (mm)	AFOV (degrees)	ER (mm)	Field stop (mm)	Weight (oz)	Design (Elem/Grp)	Coatings	Barrel (inches)
12.5	102	16	22.3^{E}	23.6	10/7	FMC	1.25/2
10	102	16	17.8^{E}	24.3	12/8	FMC	1.25/2
17	102	16	30.3^{E}	33.5	10/7	FMC	2
14	102	16	24.9^{E}	34.9	12/8	FMC	2

Source: www.nikonsportoptics.com

Each eyepiece includes the unique benefit of an "EiC" tele-extender lens, specifically designed for each individual eyepiece, that shortens the focal length and is like getting two eyepieces for the price of one. This eyepiece line is within the following competition class of eyepieces with similar specifications and performance: Explore Scientific 100 Series, Tele Vue Ethos.

Nikon: NAV-SW (Super Wide)

Fig. 8.41 The complete line of Nikon NAV-SW eyepieces (Image courtesy of Tamiji Homma, Newbury Park, CA, USA)

Focal length (mm)	AFOV (degrees)	ER (mm)	Field stop (mm)	Weight (oz)	Design (Elem/Grp)	Coatings	Barrel (inches)
5	72	18	6.3^{E}	10.6	8/6	FMC	1.25
7	72	17	8.8^{E}	10.2	8/6	FMC	1.25
10	72	19	12.6^{E}	10.6	8/6	FMC	1.25
14	72	18	17.6^{E}	9.7	7/5	FMC	1.25
17.5	72	26	22.0^{E}	12.9	7/5	FMC	1.25

Source: www.nikonsportoptics.com

This eyepiece design emphasizes suppression of field curvature and all astigmatism to bring about the maximum improvement off-axis without sacrificing the quality of the image at the center of the field. The rubber eyeguard is painted with a phosphorescent logo that glows slightly to help locate the eyepieces in the dark. Excellent off-axis performer. This eyepiece line is within the following competition class of eyepieces with similar specifications and performance: Explore Scientific 68 Series, Meade 4000 SWA, Meade Series 5000 SWA, Pentax XL, Pentax XW, Tele Vue Delos.

Nikon: Ortho (*Discontinued*)

Focal length (mm)	AFOV (degrees)	ER (mm)	Field stop (mm)	Weight (oz)	Design (Elem/Grp)	Coatings	Barrel (inches)
5	45	4^{E}	3.9^{E}	$<3^{E}$	4/2	FMC	.965
7	45	6^{E}	5.5^{E}	$<3^{E}$	4/2	FMC	.965
9	45	7^{E}	7.1^{E}	$<3^{E}$	4/2	FMC	.965
12.5	45	10^{E}	9.8^{E}	$<3^{E}$	4/2	FMC	.965
18	45	14^{E}	14.1^{E}	$<3^{E}$	4/2	FMC	.965
25	45	16^{E}	19.6^{E}	$<3^{E}$	3/2	FMC	.965
40	45	25^{E}	31.4^{E}	$<3^{E}$	3/2	FMC	.965

Source: www.nikonsportoptics.com

This eyepiece line is within the following competition class of eyepieces with similar specifications and performance: Antares Ortho, Apogee Super Abbe Orthoscopic, Baader Planetarium Classic Ortho, Baader Planetarium Genuine Abbe Ortho, Cave Orthostar Orthoscopic, Celestron Ortho, Edmund Scientific Ortho, Kokusai Kohki Abbe Ortho, Kson Super Ortho, Masuyama Orthoscopic, Meade Research Grade Ortho, Meade Series II Orthoscopic, Siebert Optics Star Splitter/Super Star Splitter, Takahashi Ortho, Telescope Service Ortho, Unitron Ortho, University Optics Abbe HD Orthoscopic, University Optics Abbe Volcano Orthoscopic, University Optics O.P.S. Orthoscopic Planetary Series, University Optics Super Abbe Orthoscopic, VERNONscope Brandon.

Chapter 9

Olivon: 60° ED Wide Angle

Focal length (mm)	AFOV (degrees)	ER (mm)	Field stop (mm)	Weight (oz)	Design (Elem/Grp)	Coatings	Barrel (inches)
5	60	13^{E}	5.2^{E}	7^{E}	6	FMC	1.25
8	60	13^{E}	8.4^{E}	6^{E}	6	FMC	1.25
13	60	13^{E}	12.6^{E}	6^{E}	6	FMC	1.25
15	60	15^{E}	15.7^{E}	6^{E}	6	FMC	1.25
18	60	13^{E}	18.8^{E}	6^{E}	6	FMC	1.25
25	60	15^{E}	26.2^{E}	6^{E}	6	FMC	1.25

Source: www.olivonmanufacturing.com

These use two extra low-dispersion lenses and are parfocal. This eyepiece line is within the following competition class of eyepieces with similar specifications and performance: Astro-Tech Paradigm Dual ED, BST Explorer ED, Orion Epic II ED, Pentax XF, Telescope Service NED "ED" Flat Field.

W. Paolini, *Choosing and Using Astronomical Eyepieces*, The Patrick Moore Practical Astronomy Series, DOI 10.1007/978-1-4614-7723-5_9,
© Springer Science+Business Media New York 2013

Olivon: 70° Wide Angle

Focal length (mm)	AFOV (degrees)	ER (mm)	Field stop (mm)	Weight (oz)	Design (Elem/Grp)	Coatings	Barrel (inches)
26	70	20^{E}	31.8^{E}	10^{E}	5	FMC	2
32	70	24^{E}	39.1^{E}	14^{E}	5	FMC	2
38	70	28^{E}	46.0^{E}	21^{E}	5	FMC	2

Source: www.olivonmanufacturing.com

This eyepiece line is within the following competition class of eyepieces with similar specifications and performance: Agena Super Wide Angle, Astro-Professional SWA, Astro-Tech Wide Field, Garrett SuperWide Angle, GSO Superview, Meade QX, Opt Super View, Orion Q70 Super Wide-Field, Sky-Watcher PanaView, Telescope-Service WA, University Optics 70°, William Optics SWAN.

Olivon: 70° Ultra Wide Angle

Focal length (mm)	AFOV (degrees)	ER (mm)	Field stop (mm)	Weight (oz)	Design (Elem/Grp)	Coatings	Barrel (inches)
3.5	70	17^{E}	4.3^{E}	18^{E}	8^{E}	FMC	1.25/2
5	70	17^{E}	6.1^{E}	17^{E}	8^{E}	FMC	1.25/2
8	70	17^{E}	9.8^{E}	17^{E}	8^{E}	FMC	1.25/2
13	70	17^{E}	15.9^{E}	17^{E}	8^{E}	FMC	1.25/2
17	70	17^{E}	20.8^{E}	16^{E}	8^{E}	FMC	1.25/2

Source: www.olivonmanufacturing.com

This eyepiece line is within the following competition class of eyepieces with similar specifications and performance: Astro-Tech AF Series 70, Telescope Service Expanse ED.

Olivon: 80° Ultra Wide Angle

Focal length (mm)	AFOV (degrees)	ER (mm)	Field stop (mm)	Weight (oz)	Design (Elem/Grp)	Coatings	Barrel (inches)
11	80	9[E]	15.4[E]	3[E]	5/3[E]	FMC	1.25
15	80	12[E]	20.9[E]	3[E]	6/4[E]	FMC	1.25
16	80	18[E]	22.3[E]	9[E]	6/4[E]	FMC	2
20	80	20[E]	27.9[E]	21[E]	7/4[E]	FMC	2
30	80	22[E]	41.9[E]	20[E]	5/3[E]	FMC	2

Source: www.olivonmanufacturing.com

These are parfocal. This eyepiece line is within the following competition class of eyepieces with similar specifications and performance: Agena Ultra Wide Angle, Apogee Widescan (20 and 30 mm), Astrobuffet 1RPD (30 mm), BW Optik Ultrawide (30 mm), Kokusai Kohki Widescan I/II/III (20 and 30 mm), Moonfish Ultrawide (30 mm), Owl Astronomy Knight Owl Ultrawide Angle, Sky-Watcher Ultra Wide Angle, Surplus Shed Wollensak (30 mm), TAL Ultra Wide Angle, University Optics 80°, University Optics Widescan I/II/III (20 and 30 mm).

Olivon: Plössl

Focal length (mm)	AFOV (degrees)	ER (mm)	Field stop (mm)	Weight (oz)	Design (Elem/Grp)	Coatings	Barrel (inches)
4	50	3[E]	3.5[E]	2[E]	4/2	—	1.25
6.3	50	4[E]	5.5[E]	2[E]	4/2	—	1.25
10	50	7[E]	8.7[E]	2[E]	4/2	—	1.25
12.5	50	9[E]	10.9[E]	3[E]	4/2	—	1.25
17	50	12[E]	14.8[E]	3[E]	4/2	—	1.25
20	50	14[E]	17.5[E]	3[E]	4/2	—	1.25
25	50	17[E]	21.8[E]	4[E]	4/2	—	1.25
32	50	22[E]	27.9[E]	5[E]	4/2	—	1.25
40	40	27[E]	27.9[E]	6[E]	4/2	—	1.25

Source: www.olivonmanufacturing.com

The manufacturer's detail on coatings used for these is not specified. This eyepiece line is within the following competition class of eyepieces with similar specifications and performance: Antares Plössl, Astro-Professional Plössl, Astro-Tech High Grade Plössl, Astro-Tech Value Line Plössl, Bresser 52° Super Plössl,

Carton Plössl, Celestron Omni, Celestron Silvertop Plössl, Clavé Plössl, Coronado CeMax, Edmund Scientific Plössl, Garrett Optical Plössl, GSO Plössl, GTO Plössl, Long Perng Plössl, Meade Series 3000 Plössl, Meade Series 4000 Super Plössl, Opt Plössl, Orion HighLight Plössl, Orion Sirius Plössl, Owl Black Night Plössl, Parks Silver Series, Sky-Watcher SP-Series Super Plössl, Smart Astronomy Sterling Plössl, TAL—Symmetrical Super Plössl, Telescope Service Plössl, Telescope Service Super Plössl, Tele Vue Plössl, Vixen NPL.

Olivon: Wide Angled Plössl

Focal length (mm)	AFOV (degrees)	ER (mm)	Field stop (mm)	Weight (oz)	Design (Elem/Grp)	Coatings	Barrel (inches)
2.5	58	16	2.5[E]	5	6	FMC	1.25
3.2	58	16	3.2[E]	5	6	FMC	1.25
4	58	16	4.0[E]	5	6	FMC	1.25
5	58	16	5.1[E]	5	6	FMC	1.25
6	58	16	6.1[E]	5	6	FMC	1.25
7	58	16	7.1[E]	5	6	FMC	1.25
8	58	16	8.1[E]	5	6	FMC	1.25
9	58	16	9.1[E]	5	6	FMC	1.25
15	58	16	15.2[E]	5[E]	6	FMC	1.25
20	58	16	20.2[E]	5[E]	6	FMC	1.25

Source: www.olivonmanufacturing.com

Although the name Plössl is used for this eyepiece line, their design is much more complex than a simple Plössl. Each focal length has an integrated twist-up eyeguard. This eyepiece line is within the following competition class of eyepieces with similar specifications and performance: APM UWA Planetary, Burgess/TMB Planetary, Owl Astronomy High Resolution Planetary, Telescope Service Planetary HR, TMB Planetary II.

Opt: Plössl

Focal length (mm)	AFOV (degrees)	ER (mm)	Field stop (mm)	Weight (oz)	Design (Elem/Grp)	Coatings	Barrel (inches)
4	45	6	3.1[E]	2[E]	4/2	FMC	1.25
6	52	5	5.4[E]	2[E]	4/2	FMC	1.25
9	52	6	8.2[E]	3[E]	4/2	FMC	1.25
15	52	13	13.6[E]	3[E]	4/2	FMC	1.25
20	52	20	18.2[E]	3[E]	4/2	FMC	1.25
25	52	20	22.7[E]	4[E]	4/2	FMC	1.25
32	52	20	27.0[E]	5[E]	4/2	FMC	1.25
40	45	31	27.0[E]	6[E]	4/2	FMC	1.25

Source: www.optcorp.com

This eyepiece line is within the following competition class of eyepieces with similar specifications and performance: Antares Plössl, Astro-Professional Plössl, Astro-Tech High Grade Plössl, Astro-Tech Value Line Plössl, Bresser 52° Super Plössl, Carton Plössl, Celestron Omni, Celestron Silvertop Plössl, Clavé Plössl, Coronado CeMax, Edmund Scientific Plössl, Garrett Optical Plössl, GSO Plössl, GTO Plössl, Long Perng Plössl, Meade Series 3000 Plössl, Meade Series 4000 Super Plössl, Olivon Plössl, Orion HighLight Plössl, Orion Sirius Plössl, Owl Black Night Plössl, Parks Silver Series, Sky-Watcher SP-Series Super Plössl, Smart Astronomy Sterling Plössl, TAL—Symmetrical Super Plössl, Telescope Service Plössl, Telescope Service Super Plössl, Tele Vue Plössl, Vixen NPL.

Opt: Super View

Focal length (mm)	AFOV (degrees)	ER (mm)	Field stop (mm)	Weight (oz)	Design (Elem/Grp)	Coatings	Barrel (inches)
15	68	13	17.8[E]	3.8	4/3	FMC	1.25
20	68	20	23.7[E]	5.7	5/3	FMC	1.25
30	68	20	35.6[E]	12.9	5/3	FMC	2
42	60	20	44.0[E]	13	5/3	FMC	2
50	60	20	46.0[E]	13.6	5/3	FMC	2

Source: www.optcorp.com

This eyepiece line is within the following competition class of eyepieces with similar specifications and performance: Agena Super Wide Angle, Astro-Professional SWA, Astro-Tech Wide Field, Garrett SuperWide Angle, GSO Superview, Meade QX, Olivon 70° Wide Angle, Orion Q70 Super Wide-Field, Sky-Watcher PanaView, Telescope-Service WA, University Optics 70°, William Optics SWAN.

Orion: Deep View

Fig. 9.1 The Orion Deep View eyepieces (© 2013 Orion Telescopes. Reprinted with permission from Orion Telescopes & Binoculars, www.telescope.com)

Focal length (mm)	AFOV (degrees)	ER (mm)	Field stop (mm)	Weight (oz)	Design (Elem/Grp)	Coatings	Barrel (inches)
28	56	20	27	10	3	MC	2
35	56	20	37	12	3	MC	2
42	52	20	38	13	3	MC	2

Source: www.telescope.com

These are parfocal. This eyepiece line is within the following competition class of eyepieces with similar specifications and performance (Kellners and Reverse Kellners, and RKE—Rank-Kaspereit-Erfle): Celestron E-Lux (2″ models only), Celestron Kellner, Criterion Kellner, Edmund Scientific RKE, GSO Kellner, Kokusai Kohki Kellner, Orion E-Series, Russell Optics (2 in. 52 and 60 mm only), Sky-Watcher Kellner, Sky-Watcher Super MA Series, Telescope Service RK, Unitron Kellner.

Orion: E-Series

Fig. 9.2 The Orion E-Series eyepieces (© 2013 Orion Telescopes. Reprinted with permission from Orion Telescopes & Binoculars, www.telescope.com)

Focal length (mm)	AFOV (degrees)	ER (mm)	Field stop (mm)	Weight (oz)	Design (Elem/Grp)	Coatings	Barrel (inches)
3.6	40	5	2.5^E	1	4/2	FC	1.25
6.3	46	8	5.1^E	2	3/2	FC	1.25
10	52	10	9.1^E	2	3/2	FC	1.25
20	52	12	18.2^E	3	3/2	FC	1.25

Source: www.telescope.com

Designed as a beginner's telescope eyepiece line to provide quality views. These are parfocal. This eyepiece line is within the following competition class of eyepieces with similar specifications and performance: Kellners and Reverse Kellners, and RKE—Rank-Kaspereit-Erfle, Celestron E-Lux (2 in. models only), Celestron Kellner, Criterion Kellner, Edmund Scientific RKE, GSO Kellner, Kokusai Kohki Kellner, Orion DeepView, Russell Optics (2″ 52 and 60 mm only), Sky-Watcher Kellner, Sky-Watcher Super MA Series, Telescope Service RK, Unitron Kellner.

Orion: Edge-On Flat-Field

Fig. 9.3 The Orion Edge-On Flat Field eyepieces (© 2013 Orion Telescopes. Reprinted with permission from Orion Telescopes & Binoculars, www.telescope.com)

Focal length (mm)	AFOV (degrees)	ER (mm)	Field stop (mm)	Weight (oz)	Design (Elem/Grp)	Coatings	Barrel (inches)
16	60	17	16.6	4	6	FMC	1.25
19	65	17	21.3	4	5	FMC	1.25
27	53	21	24.4	6	6	FMC	1.25

Source: www.telescope.com

These have a twist-up eyeguard and are parfocal. This eyepiece line is within the following competition class of eyepieces with similar specifications and performance: Astro-Professional EF Flatfield, Astro-Tech Flat Field, BST Flat Field, Sky-Watcher Extra Flat, Smart Astronomy Extra Flat Field, Telescope-Service Edge-On Flat Field.

Orion: Edge-On Planetary

Fig. 9.4 The Orion Edge-On Flat Field Planetary eyepieces (© 2013 Orion Telescopes. Reprinted with permission from Orion Telescopes & Binoculars, www.telescope.com)

Focal length (mm)	AFOV (degrees)	ER (mm)	Field stop (mm)	Weight (oz)	Design (Elem/Grp)	Coatings	Barrel (inches)
3	55	20	2.9[E]	10	7/4	FMC	1.25
5	55	20	4.8[E]	10	7/4	FMC	1.25
6	55	20	5.8[E]	10	7/4	FMC	1.25
9	55	20	8.6[E]	9	7/4	FMC	1.25
12.5	55	20	12.0[E]	8	7/4	FMC	1.25
14.5	55	20	13.9[E]	7	7/4	FMC	1.25

Source: www.telescope.com

These are parfocal. This eyepiece line is within the following competition class of eyepieces with similar specifications and performance: Astro-Professional Long Eye Relief Planetary, Astro-Tech Long Eye Relief, Long Perng Long Eye Relief, Smart Astronomy SA Solar System Long Eye Relief, Stellarvue Planetary, William Optics SPL, Zhumell Z Series.

Orion: Epic ED II (*Discontinued*)

Focal length (mm)	AFOV (degrees)	ER (mm)	Field stop (mm)	Weight (oz)	Design (Elem/Grp)	Coatings	Barrel (inches)
5	58	13	8.1	7	6	FMC	1.25
8	58	13	11.1	6	6	FMC	1.25
12	60	13	15.3	6	6	FMC	1.25
15	60	15	18.0	6	6	FMC	1.25
18	60	13	20.2	6	6	FMC	1.25
25	60	15	25.9	6	6	FMC	1.25

Source: www.telescope.com

These have a twist-up eyeguard and are parfocal. This eyepiece line is within the following competition class of eyepieces with similar specifications and performance: Astro-Tech Paradigm Dual ED, BST Explorer ED, Olivon 60° ED Wide Angle, Pentax XF, Telescope Service NED "ED" Flat Field.

Orion previously had a different style of eyepiece available under this same name. These eyepieces were available in 3.7 mm, 5.1 mm, 9.5 mm, 12.3 mm, 14 mm, 18 mm, 22 mm, and 25 mm, all having a fixed 20 mm of eye relief and a 55° AFOV. This line was within the following competition class of eyepieces with similar specifications and performance: Agena ED, Celestron X-Cel, Vixen Lanthanum (LV), Vixen NLV.

Orion: Expanse Wide-Field

Fig. 9.5 The Orion Expanse Wide-Field eyepieces (© 2013 Orion Telescopes. Reprinted with permission from Orion Telescopes & Binoculars, www.telescope.com)

Focal length (mm)	AFOV (degrees)	ER (mm)	Field stop (mm)	Weight (oz)	Design (Elem/Grp)	Coatings	Barrel (inches)
6	66	14.8	8.0	4	5	MC	1.25
9	66	15.0	15.0	4	6	MC	1.25
15	66	13.0	17.0	3	4	MC	1.25
20	66	18.0	23.5	4	4	MC	1.25

Source: www.telescope.com

The 6 and 9 mm focal lengths of the series have a reputation of being excellent performers even in faster focal ratio telescopes. This eyepiece line is within the following competition class of eyepieces with similar specifications and performance: Agena Enhanced Wide Angle, Owl Enhanced Superwide, Sky-Watcher Ultra Wide Angle, Telescope Service SWM Wide Angle, William Optics WA 66.

Orion: GiantView 100° UltraWide

Focal length (mm)	AFOV (degrees)	ER (mm)	Field stop (mm)	Weight (oz)	Design (Elem/Grp)	Coatings	Barrel (inches)
9	100	13	15.7[E]	15	6/4	FMC	2
16	100	15	27.9[E]	15	6/4	FMC	2

Source: www.telescope.com

Observers report the 9 mm focal length as having a fairly good off-axis; however, reports for the 16 mm focal length indicate a longer focal ratio telescope is needed for best off-axis performance. This eyepiece line is within the following competition class of eyepieces with similar specifications and performance: Agena Mega Wide Angle, TMB 100, Zhumell Z100.

Orion: HighLight Plössl

Fig. 9.6 The Orion HighLight Plössl eyepieces (© 2013 Orion Telescopes. Reprinted with permission from Orion Telescopes & Binoculars, www.telescope.com)

Focal length (mm)	AFOV (degrees)	ER (mm)	Field stop (mm)	Weight oz)	Design (Elem/Grp)	Coatings	Barrel (inches)
6.3	52	4.1	5.0	2.0	4/2	FMC	1.25
7.5	52	4.9	5.8	2.1	4/2	FMC	1.25
10	52	6.5	8.0	2.3	4/2	FMC	1.25
12.5	52	8.1	10.2	2.6	4/2	FMC	1.25
17	52	11.0	14.5[E]	3.2	4/2	FMC	1.25
20	52	13.0	17.1	3.4	4/2	FMC	1.25
25	52	16.9	24.1	3.9	4/2	FMC	1.25
32	52	20.0	28.6	5.2	4/2	FMC	1.25
40	43	20.0	28.5	6.5	4/2	FMC	1.25

Source: www.telescope.com

These have barrels that are safety undercut and precision-machined to a tolerance of +/−0.05 mm from the nominal specification to ensure a good fit into your telescope diagonal or focuser. This eyepiece line is within the following competition class of eyepieces with similar specifications and performance: Antares Plössl, Astro-Professional Plössl, Astro-Tech High Grade Plössl, Astro-Tech Value Line Plössl, Bresser 52° Super Plössl, Carton Plössl, Celestron Omni, Celestron Silvertop Plössl, Clavé Plössl, Coronado CeMax, Edmund Scientific Plössl, Garrett Optical Plössl, GSO Plössl, GTO Plössl, Long Perng Plössl, Meade Series 3000 Plössl, Meade Series 4000 Super Plössl, Olivon Plössl, Opt Plössl, Orion Sirius Plössl, Owl Black

Night Plössl, Parks Silver Series, Sky-Watcher SP-Series Super Plössl, Smart Astronomy Sterling Plössl, TAL—Symmetrical Super Plössl, Telescope Service Plössl, Telescope Service Super Plössl, Tele Vue Plössl, Vixen NPL.

Orion: Lanthanum Superwide (*Discontinued*)

Focal length (mm)	AFOV (degrees)	ER (mm)	Field stop (mm)	Weight (oz)	Design (Elem/Grp)	Coatings	Barrel (inches)
8	65	20	9.1^{E}	17.4	8/5	FMC	1.25/2
13	65	20	14.7^{E}	16.2	8/5	FMC	1.25/2
17	65	20	19.3^{E}	15.5	8/5	FMC	1.25/2
22	65	20	25.0^{E}	15.2	8/5	FMC	1.25/2

Source: www.telescope.com

This eyepiece line is within the following competition class of eyepieces with similar specifications and performance: Baader Hyperion, Orion Stratus, Vixen Lanthanum Superwide (LVW).

Orion: Long Eye Relief

Fig. 9.7 The Orion Long Eye Relief eyepieces (© 2013 Orion Telescopes. Reprinted with permission from Orion Telescopes & Binoculars, www.telescope.com)

Focal length (mm)	AFOV (degrees)	ER (mm)	Field stop (mm)	Weight (oz)	Design (Elem/Grp)	Coatings	Barrel (inches)
3	55	20	2.9^E	10	7	FMC	1.25
5	55	20	4.8^E	10	7	FMC	1.25
6	55	20	5.8^E	10	7	FMC	1.25
9	55	20	8.6^E	9	7	FMC	1.25
12.5	55	20	12.0^E	8	7	FMC	1.25
14.5	55	20	13.4^E	7	7/4	FMC	1.25
18	55	20	17.3^E	6	5/3	FMC	1.25

Source: www.telescope.com

This eyepiece line is within the following competition class of eyepieces with similar specifications and performance: Astro-Professional Long Eye Relief Planetary, Astro-Tech Long Eye Relief, Long Perng Long Eye Relief, Smart Astronomy SA Solar System Long Eye Relief, Stellarvue Planetary, William Optics SPL, Zhumell Z Series.

Orion: MegaView Ultra-Wide

Fig. 9.8 The Orion MegaView eyepieces (© 2013 Orion Telescopes. Reprinted with permission from Orion Telescopes & Binoculars, www.telescope.com)

Focal length (mm)	AFOV (degrees)	ER (mm)	Field stop (mm)	Weight (oz)	Design (Elem/Grp)	Coatings	Barrel (inches)
4	82	12	7.4	8	7	FMC	1.25
7	82	12	14.0	8	7	FMC	1.25
16	82	12	24.0	9	7	FMC	1.25
28	82	18	43.1	35	7	FMC	2

Source: www.telescope.com

These have a twist-up eyeguard and are parfocal. This eyepiece line is within the following competition class of eyepieces with similar specifications and performance: Antares Speers-WALER Series, Astro-Professional UWA, Celestron Axiom LX, Celestron Luminos, Docter UWA, Explore Scientific 82° Nitrogen Purged, Meade Series 4000 Ultra Wide Angle, Meade Series 5000 Ultra Wide Angle, Sky-Watcher Nirvana UWA, Sky-Watcher Sky Panorama, Tele Vue Nagler, Williams Optics UWAN.

Orion: Optiluxe (*Discontinued*)

Fig. 9.9 Orion 40 mm Optiluxe (Image courtesy of Wade Wheeler, Adrian, MI, USA)

Focal length (mm)	AFOV (degrees)	ER (mm)	Field stop (mm)	Weight (oz)	Design (Elem/Grp)	Coatings	Barrel (inches)
32	58	20	32.4[E]	—	4	FMC	2
40	62	20	43.3[E]	—	4	FMC	2
50	45	37	39.3[E]	—	5	FMC	2

Source: www.telescope.com

These are manufactured in Japan. The manufacturer's statistics on weight are no longer available.

Orion: Premium 68° Long Eye Relief

Fig. 9.10 The Orion 7 mm Premium 68° Long Eye Relief eyepiece (© 2013 Orion Telescopes. Reprinted with permission from Orion Telescopes & Binoculars, www.telescope.com)

Focal length (mm)	AFOV (degrees)	ER (mm)	Field stop (mm)	Weight (oz)	Design (Elem/Grp)	Coatings	Barrel (inches)
7	68	20	26.0	11	8/6	FMC	1.25

Source: www.telescope.com

This eyepiece line is within the following competition class of eyepieces with similar specifications and performance: Long Perng 68° Wide Angle.

Orion: Q70 Super Wide-Field

Fig. 9.11 The Orion Q70 eyepieces (© 2013 Orion Telescopes. Reprinted with permission from Orion Telescopes & Binoculars, www.telescope.com)

Focal length (mm)	AFOV (degrees)	ER (mm)	Field stop (mm)	Weight (oz)	Design (Elem/Grp)	Coatings	Barrel (inches)
26	70	20	32.2	12	5	FMC	2
32	70	24	40.0	15	5	FMC	2
38	70	28	45.7	21	5	FMC	2

Source: www.telescope.com

These eyepieces are parfocal. This eyepiece line is within the following competition class of eyepieces with similar specifications and performance: Agena Super Wide Angle, Astro-Professional SWA, Astro-Tech Wide Field, Garrett SuperWide Angle, GSO Superview, Meade QX, Olivon 70° Wide Angle, Opt Super View, Sky-Watcher PanaView, Telescope-Service WA, University Optics 70°, William Optics SWAN.

Orion: Sirius Plössl

Fig. 9.12 The Orion Sirius Plössl eyepieces (© 2013 Orion Telescopes. Reprinted with permission from Orion Telescopes & Binoculars, www.telescope.com)

Focal length (mm)	AFOV (degrees)	ER (mm)	Field stop (mm)	Weight (oz)	Design (Elem/Grp)	Coatings	Barrel (inches)
6.3	52	4.1	5.0	2.0	4/2	MC	1.25
7.5	52	4.9	5.8	2.1	4/2	MC	1.25
10	52	6.5	8.0	2.3	4/2	MC	1.25
12.5	52	8.1	10.2	2.6	4/2	MC	1.25
17	52	11.0	14.5[E]	3.2	4/2	MC	1.25
20	52	13.0	17.1	3.4	4/2	MC	1.25
25	52	16.9	24.1	3.9	4/2	MC	1.25
32	52	20.0	28.6	5.2	4/2	MC	1.25
40	43	20.0	28.5	6.5	4/2	MC	1.25

Source: www.telescope.com

This eyepiece line is within the following competition class of eyepieces with similar specifications and performance: Antares Plössl, Astro-Professional Plössl, Astro-Tech High Grade Plössl, Astro-Tech Value Line Plössl, Bresser 52° Super Plössl, Carton Plössl, Celestron Omni, Celestron Silvertop Plössl, Clavé Plössl, Coronado CeMax, Edmund Scientific Plössl, Garrett Optical Plössl, GSO Plössl, GTO Plössl, Long Perng Plössl, Meade Series 3000 Plössl, Meade Series 4000 Super Plössl, Olivon Plössl, Opt Plössl, Orion HighLight Plössl, Owl Black Night Plössl, Parks Silver Series, Sky-Watcher SP-Series Super Plössl, Smart Astronomy Sterling Plössl, TAL—Symmetrical Super Plössl, Telescope Service Plössl, Telescope Service Super Plössl, Tele Vue Plössl, Vixen NPL.

Orion: Stratus Wide-Field

Fig. 9.13 The complete line of Orion Stratus eyepieces (Image courtesy of Mike Wooldridge, Hamilton, Ontario, Canada)

Focal length (mm)	AFOV (degrees)	ER (mm)	Field stop (mm)	Weight (oz)	Design (Elem/Grp)	Coatings	Barrel (inches)
3.5	68	20	5.1	15	8/5	FMC	1.25/2
5	68	20	7.3	14	8/5	FMC	1.25/2
8	68	20	10.7	14	8/5	FMC	1.25/2
13	68	20	19.0	14	8/5	FMC	1.25/2
17	68	20	23.0	14	8/5	FMC	1.25/2
21	68	20	24.3	14	8/5	FMC	1.25/2
24	68	15	28.4	12	6^E	FMC	1.25/2

Source: www.telescope.com

This eyepiece line is within the following competition class of eyepieces with similar specifications and performance: Baader Hyperion, Orion Lanthanum Superwide, Vixen Lanthanum Superwide (LVW).

Orion: Ultrascopic (*Discontinued*)

Fig. 9.14 The Orion Ultrascopic eyepieces (© 2013 Orion Telescopes. Reprinted with permission from Orion Telescopes & Binoculars, www.telescope.com)

Focal length (mm)	AFOV (degrees)	ER (mm)	Field stop (mm)	Weight (oz)	Design (Elem/Grp)	Coatings	Barrel (inches)
3.8	52	5	3.3^{E}	4	7/4	FMC	1.25
5	52	6	4.4^{E}	4	7/4	FMC	1.25
7.5	52	5	6.5	3	5/3	FMC	1.25
10	52	6	8.3	3	5/3	FMC	1.25
15	52	10	12.4	3	5/3	FMC	1.25
20	52	13	17.4	4	5/3	FMC	1.25
25	52	17	25.7	4	5/3	FMC	1.25
30	52	21	26.1	6	5/3	FMC	1.25
35	49	25	28.9	7	5/3	FMC	1.25

Source: www.telescope.com

The 35 mm focal length of this line offers a distinctive view as the image can sometimes seems to float somewhat in front of your eye (similar, but not as pronounced as the Edmund 28 mm RKE). Makes a great addition to any eyepiece collection. These eyepieces were invented in Japan by one of its most respected optical designers and are parfocal. This eyepiece line is within the following competition class of eyepieces with similar specifications and performance: Antares Elite, Baader Eudiascopic, Bresser 60° Plössl, Celestron Ultima, Kasai Astroplan, Meade Series 4000 Super Plössl (pre-1994, smooth sided, 5-elements), Meade Series 5000 Super Plössl, Parks Gold Series Plössl, Takahashi LE. Note—the Omcon Ultima and Tuthill Super Plössl lines are also competing brands, but these lines are not detailed.

Owl Astronomy: Advanced Wide Angle

Focal length (mm)	AFOV (degrees)	ER (mm)	Field stop (mm)	Weight (oz)	Design (Elem/Grp)	Coatings	Barrel (inches)
8	60	9	8.4^{E}	1.5^{E}	4/3	FMC	1.25
12	60	10	12.6^{E}	2^{E}	4/3	FMC	1.25
17	65	16	19.3^{E}	3^{E}	4/3	FMC	1.25
20	67	20	22.7^{E}	3^{E}	4/3	FMC	1.25

Source: www.owlastronomy.com

This eyepiece line is within the following competition class of eyepieces with similar specifications and performance: Agena Wide Angle, Astro-Tech Series 6 Economy Wide Field, Burgess Wide Angle.

Owl Astronomy: Black Knight Super Plössl

Focal length (mm)	AFOV (degrees)	ER (mm)	Field stop (mm)	Weight (oz)	Design (Elem/Grp)	Coatings	Barrel (inches)
4	52	3^{E}	3.5^{E}	2^{E}	4/2	FMC	1.25
6.5	52	4^{E}	5.7^{E}	2^{E}	4/2	FMC	1.25
10	52	7^{E}	8.8^{E}	3^{E}	4/2	FMC	1.25
12.5	52	9^{E}	10.9^{E}	3^{E}	4/2	FMC	1.25
15	52	10^{E}	13.1^{E}	3^{E}	4/2	FMC	1.25
20	52	14^{E}	17.5^{E}	3^{E}	4/2	FMC	1.25
25	52	17^{E}	21.9^{E}	4^{E}	4/2	FMC	1.25
30	52	20^{E}	26.3^{E}	4^{E}	4/2	FMC	1.25
40	44	27^{E}	27.0^{E}	5^{E}	4/2	FMC	1.25

Source: www.owlastronomy.com

This eyepiece line is within the following competition class of eyepieces with similar specifications and performance: Antares Plössl, Astro-Professional Plössl, Astro-Tech High Grade Plössl, Astro-Tech Value Line Plössl, Bresser 52° Super Plössl, Carton Plössl, Celestron Omni, Celestron Silvertop Plössl, Clavé Plössl, Coronado CeMax, Edmund Scientific Plössl, Garrett Optical Plössl, GSO Plössl, GTO Plössl, Long Perng Plössl, Meade Series 3000 Plössl, Meade Series 4000 Super Plössl, Olivon Plössl, Opt Plössl, Orion HighLight Plössl, Orion Sirius Plössl, Parks Silver Series, Sky-Watcher SP-Series Super Plössl, Smart Astronomy Sterling Plössl, TAL—Symmetrical Super Plössl, Telescope Service Plössl, Telescope Service Super Plössl, Tele Vue Plössl, Vixen NPL.

Owl Astronomy: Enhanced Superwide

Fig. 9.15 Owl Enhanced Superwide eyepieces in 9 and 6 mm (Image by the author)

Focal length (mm)	AFOV (degrees)	ER (mm)	Field stop (mm)	Weight (oz)	Design (Elem/Grp)	Coatings	Barrel (inches)
6	66	17	6.6[E]	4.1	6/4	FMC	1.25
9	66	17	9.9[E]	4.0	7/4	FMC	1.25
15	66	13	16.5[E]	2.8	4/3	FMC	1.25
20	66	16	22.0[E]	3.8	5/3	FMC	1.25

Source: www.owlastronomy.com

The 6 and 9 mm focal lengths of the series have a reputation of being excellent performers even in faster focal ratio telescopes. These focal lengths of the line are some of the few budget wide fields that perform exceptionally well. Note the exposed optical assembly of the 9 mm shows six elements in a 1-2-1-2 configuration and not seven elements as specified by many resellers.

This eyepiece line is within the following competition class of eyepieces with similar specifications and performance: Agena Enhanced Wide Angle, Orion Expanse, Sky-Watcher Ultra Wide Angle, Telescope Service SWM Wide Angle, William Optics WA 66

Fig. 9.16 Exposed optical assembly of a 9 mm Owl Enhanced Superwide eyepieces (Image by the author)

Owl Astronomy: High Resolution Planetary

Focal length (mm)	AFOV (degrees)	ER (mm)	Field stop (mm)	Weight (oz)	Design (Elem/Grp)	Coatings	Barrel (inches)
2.5	58	14	2.4[E]	5	5/3	FMC	1.25
3.2	58	14	3.1[E]	5	5/3	FMC	1.25
4	58	14	3.9[E]	5	5/3	FMC	1.25
5	58	14	4.9[E]	5	5/3	FMC	1.25
6	58	14	5.8[E]	5	5/3	FMC	1.25
7.5	58	14	7.3[E]	5	5/3	FMC	1.25
9	58	14	8.7[E]	5	6/4	FMC	1.25

Source: www.owlastronomy.com

This eyepiece line is within the following competition class of eyepieces with similar specifications and performance: APM UWA Planetary, Burgess/TMB Planetary, Olivon Wide Angled Plössl, Telescope Service Planetary HR, TMB Planetary II.

Owl Astronomy: Knight Owl Ultrawide Angle

Focal length (mm)	AFOV (degrees)	ER (mm)	Field stop (mm)	Weight (oz)	Design (Elem/Grp)	Coatings	Barrel (inches)
11	80	14	14.5[E]	3[E]	5/3	FMC	1.25
15	80	20	19.8[E]	3[E]	6/4	FMC	2
16	80	20	21.1[E]	9[E]	6/4	FMC	1.25
20	80	24	26.4[E]	21[E]	7/5	FMC	2
30	80	28	39.6[E]	20[E]	5/3	FMC	2

Source: www.owlastronomy.com

This eyepiece line is within the following competition class of eyepieces with similar specifications and performance: Agena Ultra Wide Angle, Apogee Widescan (20 and 30 mm), Astrobuffet 1RPD (30 mm), BW Optik Ultrawide (30 mm), Kokusai Kohki Widescan I/II/III (20 and 30 mm), Moonfish Ultrawide (30 mm), Olivion 80° Ultra Wide Angle, Sky-Watcher Ultra Wide Angle, Surplus Shed Wollensak (30 mm), TAL Ultra Wide Angle, University Optics 80°, University Optics Widescan I/II/III (20 and 30 mm).

Parks: Gold Series

Fig. 9.17 The Astro-Tech Long Eye Relief eyepieces (Image courtesy of James Curry, St. Louis, MO, USA)

Focal length (mm)	AFOV (degrees)	ER (mm)	Field stop (mm)	Weight (oz)	Design (Elem/Grp)	Coatings	Barrel (inches)
3.8	52	6	3.3[E]	4	7/4	FMC	1.25
5	52	8	4.4[E]	4	7/4	FMC	1.25
7.5	52	6	6.6[E]	3	5/3	FMC	1.25
10	52	8	8.8[E]	3	5/3	FMC	1.25
15	52	12	13.1[E]	3	5/3	FMC	1.25
20	52	14	17.5[E]	4	5/3	FMC	1.25
25	52	16	21.9[E]	4	5/3	FMC	1.25
30	52	24	26.3[E]	6	5/3	FMC	1.25
35	49	25	28.9[E]	7	5/3	FMC	1.25
50	50	30	41.3[E]	—	5/3	FMC	2

Source: www.parksoptical.com

The 35 mm focal length of this line offers a distinctive view as the image can sometimes seems to float somewhat in front of your eye (similar, but not as pronounced as the Edmund 28 mm RKE). Makes a great addition to any eyepiece collection. This eyepiece line is within the following competition class of eyepieces with similar specifications and performance: Antares Elite, Baader Eudiascopic, Bresser 60° Plössl, Celestron Ultima, Kasai Astroplan, Meade Series 4000 Super Plössl (pre-1994, smooth sided, five-elements), Meade Series 5000 Super Plössl, Orion Ultrascopic, Takahashi LE. *Note*: The Omcon Ultima and Tuthill Super Plössl lines are also competing brands, but these lines are not detailed here.

Parks: Silver Series

Fig. 9.18 The complete Parks Silver Series Plössls (Image courtesy of AgenaAstro.com)

Focal length (mm)	AFOV (degrees)	ER (mm)	Field stop (mm)	Weight (oz)	Design (Elem/Grp)	Coatings	Barrel (inches)
6.3	50	4^{E}	5.3^{E}	2^{E}	4/2	FMC	1.25
7.5	50	5^{E}	6.3^{E}	2^{E}	4/2	FMC	1.25
10	50	7^{E}	8.4^{E}	2^{E}	4/2	FMC	1.25
12.5	50	9^{E}	10.5^{E}	3^{E}	4/2	FMC	1.25
17	50	12^{E}	14.3^{E}	3^{E}	4/2	FMC	1.25
20	50	14^{E}	16.9^{E}	3^{E}	4/2	FMC	1.25
25	50	18^{E}	21.1^{E}	4^{E}	4/2	FMC	1.25
32	50	22^{E}	27.0^{E}	5^{E}	4/2	FMC	1.25
40	44	28^{E}	29.8^{E}	6^{E}	4/2	FMC	1.25

Source: www.parksoptical.com

This eyepiece line is within the following competition class of eyepieces with similar specifications and performance: Antares Plössl, Astro-Professional Plössl, Astro-Tech High Grade Plössl, Astro-Tech Value Line Plössl, Bresser 52° Super Plössl, Carton Plössl, Celestron Omni, Celestron Silvertop Plössl, Clavé Plössl, Coronado CeMax, Edmund Scientific Plössl, Garrett Optical Plössl, GSO Plössl, GTO Plössl, Long Perng Plössl, Meade Series 3000 Plössl, Meade Series 4000 Super Plössl, Olivon Plössl, Opt Plössl, Orion HighLight Plössl, Orion Sirius Plössl, Owl Black Night Plössl, Sky-Watcher SP-Series Super Plössl, Smart Astronomy Sterling Plössl, TAL—Symmetrical Super Plössl, Telescope Service Plössl, Telescope Service Super Plössl, Tele Vue Plössl, Vixen NPL.

Pentax: SMC Ortho (*Discontinued*)

Fig. 9.19 A 7 mm Pentax SMC Ortho (Image by the author)

Focal length (mm)	AFOV (degrees)	ER (mm)	Field stop (mm)	Weight (oz)	Design (Elem/Grp)	Coatings	Barrel (inches)
5	42	4	3.7[E]	2[E]	4/2	FMC	.965
6	42	5	4.4[E]	2[E]	4/2	FMC	.965
7	42	6	5.1[E]	2[E]	4/2	FMC	.965
9	42	8	6.6[E]	2[E]	4/2	FMC	.965
12	42	10	8.8[E]	2[E]	4/2	FMC	.965
18	42	16	13.2[E]	3[E]	4/2	FMC	.965
25	45	13	19.6[E]	3[E]	3/2	FMC	.965
40	45	21	31.4[E]	—	3/2	FMC	1.25
60	42	30	44.0[E]	—	3/2	FMC	2

Source: www.pentaximaging.com

These are fully multi-coated with Pentax's proprietary Super Multi-Coatings (SMC); the design incorporates low-dispersion glass; and are parfocal. Considered a top planetary and double star eyepiece. During their production, quality control was not always effective, and some units on the used market show stains or internal debris between the elements. A thorough inspection is advised prior to purchasing from the used market. This eyepiece line is within the following competition class of eyepieces with similar specifications and performance: Astro-Physics Super Planetary AP-SPL, Pentax XO, Pentax XP, TMB Aspheric Ortho, TMB Supermonocentric, Zeiss CZJ Ortho, Zeiss ZAO I/ZAO II.

Pentax: XF

Fig. 9.20 Pentax 8.5 and 12 mm eyepieces (Image courtesy of Alexander Kupco, Ricany, Czech Republic)

Focal length (mm)	AFOV (degrees)	ER (mm)	Field stop (mm)	Weight (oz)	Design (Elem/Grp)	Coatings	Barrel (inches)
8.5	60	18	8.8	5.3	6/4	FMC	1.25
12	60	18	12.8	5.5	6/4	FMC	1.25

Source: www.pentaximaging.com

These are designed with high-refraction, low-dispersion lanthanum glass elements; use Pentax-original SMC full-surface multi-layer lens coatings; have a twist-up rubber eyeguard; and are built with JIS Class 4 weatherproof construction. This eyepiece line is within the following competition class of eyepieces with similar specifications and performance: Astro-Tech Paradigm Dual ED, BST Explorer ED, Olivon 60° ED Wide Angle, Orion Epic II ED, Telescope Service NED "ED" Flat Field.

Pentax: XL (*Discontinued*)

Fig. 9.21 Pentax 14 mm XL (Image courtesy of Stephen Bueltman, Newport News, VA, USA)

Focal length (mm)	AFOV (degrees)	ER (mm)	Field stop (mm)	Weight (oz)	Design (Elem/Grp)	Coatings	Barrel (inches)
5.2	65	20	5.6[E]	15.5	7/5	FMC	1.25
7	65	20	7.6[E]	15.0	7/5	FMC	1.25
10.5	65	20	11.4[E]	13.4	6/4	FMC	1.25
14	65	20	15.2[E]	12.7	6/4	FMC	1.25
21	65	20	22.8[E]	11.5	6/4	FMC	1.25
28	55	20	25.9[E]	12.0	5/4	FMC	1.25
40	65	20	43.4[E]	13.4	5/5	FMC	2

Source: www.pentaximaging.com

These are engineered using advanced computer simulation design technology to drastically reduce internal reflections; use high-refraction, low-dispersion lanthanum glass elements; and have JIS Class 4 weatherproof construction. This eyepiece line is within the following competition class of eyepieces with similar specifications and performance: Explore Scientific 68 Series, Meade 4000 SWA, Meade 5000 SWA, Nikon NAV-SW, Pentax XW, Tele Vue Delos, and Tele Vue Panoptic.

Pentax: XO (*Discontinued*)

Fig. 9.22 Pentax XO eyepieces (Image courtesy of Jim Barnett, Petaluma, CA, USA)

Focal length (mm)	AFOV (degrees)	ER (mm)	Field stop (mm)	Weight (oz)	Design (Elem/Grp)	Coatings	Barrel (inches)
2.58	44	3.9	1.9	4.9	6	FMC	1.25
5.1	44	3.6	3.9	4.2	5/3	FMC	1.25

Source: www.pentaximaging.com

These use high refraction, extra-low dispersion lanthanum glass elements; are designed for focal ratios as low as f/4; use Pentax-original SMC full-surface multi-layer lens coatings; use advanced computer simulation design technology incorporated to drastically reduce internal reflections; uses eco-friendly glass that does not contain substances harmful to the environment (lead, arsenic, etc.); and have JIS Class 4 weatherproof construction. This eyepiece line is within the following competition class of eyepieces with similar specifications and performance: Astro-Physics Super Planetary AP-SPL, Pentax SMC Ortho, Pentax XP, TMB Aspheric Ortho, TMB Supermonocentric, Zeiss CZJ Ortho, Zeiss ZAO I/ZAO II. The Pentax XOs are considered by many observers to be the best planetary eyepieces ever produced in their focal lengths.

Pentax: XP (*Discontinued*)

Focal length (mm)	AFOV (degrees)	ER (mm)	Field stop (mm)	Weight (oz)	Design (Elem/Grp)	Coatings	Barrel (inches)
3.8	42[E]	2.7	2.8[E]	1.4	5/3	FMC	.965
8	42[E]	2.1	5.9[E]	1.8	6/5	FMC	.965
14	—	—	—	3[E]	6/5	FMC	.965
24	—	8.6	—	3[E]	6/5	FMC	.965

Source: www.pentaximaging.com

The 3.8 mm focal length of this eyepiece line is within the following competition class of eyepieces with similar specifications and performance: Astro-Physics Super Planetary AP-SPL, Pentax SMC Ortho, Pentax XO, TMB Aspheric Ortho, TMB Supermonocentric, Zeiss CZJ Ortho, Zeiss ZAO I/ZAO II.

Pentax: XW

Fig. 9.23 Pentax XWs in 40 mm, 20 mm, 14 mm, 10 mm, 7 mm, and 5 mm (Image by the author)

Focal length (mm)	AFOV (degrees)	ER mm)	Field top (mm)	Weight (oz)	Design (Elem/Grp)	Coatings	Barrel (inches)
3.5	70	20	4.3	14.3	8/5	FMC	1.25
5	70	20	6.2	14.0	8/5	FMC	1.25
7	70	20	8.8	13.8	8/6	FMC	1.25
10	70	20	12.4	13.8	7/6	FMC	1.25
14	70	20	17.6	12.9	7/6	FMC	1.25
20	70	20	24.0	12.6	6/4	FMC	1.25
30	70	20	36.2	26.1	7	FMC	2
40	70	20	46.5	24.7	6	FMC	2

Source: www.pentaximaging.com

These use high refraction, extra-low dispersion lanthanum glass elements; are designed for focal ratios as low as f/4; use Pentax-original SMC full-surface multi-layer lens coatings; use advanced computer simulation design technology incorporated to drastically reduce internal reflections; uses eco-friendly glass that does not contain substances harmful to the environment (lead, arsenic, etc.); and have JIS Class 4 weatherproof construction.

Many observers consider the Pentax XW line the "perfect" wide-field with neutral tone, extremely sharp and bright views, and very comfortable to use. The 14 mm and longer focal lengths show varying degrees of field curvature, depending on the telescope design used by the observer. The 10 mm and shorter focal lengths show sharp-to-the-edge. Many also consider the XWs outstanding for planetary observing, even though they are not a classical design with a minimum of glass elements. This line commands a sizable loyal following of customers, and the line continues to compete well even with the most modern entries into the field. Considered by many to be a standard by which others are to be compared, offering the best in terms of eye relief, comfort, wide field, bright images, high contrast, and detailed views. The 10 mm, 7 mm, 5 mm, and 3.5 mm focal lengths are the best performing focal lengths of the line and used in the planetary role by many observers.

This eyepiece line is within the following competition class of eyepieces with similar specifications and performance: Explore Scientific 68 Series, Meade 4000 SWA, Meade 5000 SWA, Nikon NAV-SW, Pentax XL, Tele Vue Delos, and Tele Vue Panoptic.

Rini: Various Eyepieces

Fig. 9.24 Rini 16 mm, 22 mm, and 35 mm eyepieces (Image courtesy of Andy Howie, Paisley, Scotland)

Focal length (mm)	AFOV (degrees)	ER (mm)	Field stop (mm)	Weight (oz)	Design (Elem/Grp)	Coatings	Barrel (inches)
13	82E	5	18.6E	—	6	FC/FMC	1.25
14	60E	—	14.7E	—	5	FC/FMC	1.25
16	60E	—	16.8E	—	—	FC/FMC	1.25
22	60E	—	23.0E	—	—	FC/FMC	1.25
30	50E	15	26.2E	—	5	FC	1.25
35	—	—	—	—	5	FC/FMC	2
40	—	26	—	—	5	FC/FMC	1.25
52	30E	36	27.2E	—	3	FC/FMC	1.25
26	60E	10	18.6E	—	4	FC	2

Source: www.telescope-warehouse.com

The 30 mm Plössl is an excellent performer usually available at a very reasonable price point. The manufacturer's statistics on AFOV, eye relief, weight, and optical design for some of the focal lengths is no longer readily available.

Russell Optics: 1.25 in. Series

Fig. 9.25 Russell Optics Königs in 21 mm, 14 mm, 12 mm, and 10 mm (Image by the author)

Focal length (mm)	AFOV (degrees)	ER (mm)	Field stop (mm)	Weight (oz)	Design (Elem/Grp)	Coatings	Barrel (inches)
6.7	70	18	8.2^E	$<4^E$	—	—	1.25/2
10	50^M	8.9^M	8.7^E	$<4^E$	4	—	1.25
12	43^M	17.9^M	9.0^E	$<4^E$	4	—	1.25
14	69^M	11.7^M	16.9^E	$<4^E$	4	—	1.25
18	—	—	—	$<4^E$	4	—	1.25
19	70	—	23.2^E	$<4^E$	4	—	1.25
21	49^M	23.5^M	18.0^E	$<4^E$	4	—	1.25
50	—	—	—	—	4	—	1.25

Source: www.russell-optics.com

Russell Optics 1.25 in. eyepieces provide very satisfying views in telescopes with longer focal ratios. Lightweight with excellent optics. The 10 and 14 mm are research-grade where the optics are claimed to have a 1/20th wavefront (regular eyepiece lenses are usually between 1/4 and 1/8 wave). The 50 mm is a Plössl design. The manufacturer's statistics on AFOV, eye relief, weight, optical design, and coatings for some of the focal lengths is not always consistently available.

Russell Optics: 2 in. Series

Fig. 9.26 Russell Optics 2 in. König and RKE eyepieces (Image courtesy of John W., MA, USA)

Focal length (mm)	AFOV (degrees)	ER (mm)	Field stop (mm)	Weight (oz)	Design (Elem/Grp)	Coatings	Barrel (inches)
7	60	—	7.3^{E}	—	4-König	MC	2
10.5	65	20	11.9^{E}	—	6-König	MC	2
12	70	—	14.7^{E}	—	4^{E}-König	MC	2
13	70	—	15.9^{E}	—	4^{E}-König	FMC	2
18	70	—	22.0^{E}	—	4^{E}-König	MC	2
19	70	18	23.2^{E}	—	4^{E}-König	MC	2
50	50^{E}	30	43.6^{E}	—	5/3-S. Plössl	MC	2
56	50^{E}	30	46.0^{E}	—	5/3-S. Plössl	MC	2
52	55–60	—	46.0^{E}	—	3/2-RKE	—	2
60	55–60	—	46.0^{E}	—	3/2-RKE	—	2
65	35	—	39.7^{E}	—	5/3-S. Plössl	MC	2
72	35	—	44.0^{E}	—	5/3-S. Plössl	MC	2
85	35	—	46.0^{E}	—	5/3-S. Plössl	MC	2

Source: www.russell-optics.com

These are constructed with Delrin housings and barrels, most barrels threaded for standard filters. The manufacturer's statistics on some of the focal lengths is not always consistently available. The 7, 12, and 18 mm eyepieces are advertised as having research grade optics (e.g., having 1/20th wavefront, whereas regular optics are typically between 1/4 and 1/8 wave). The optics used for Russell Optics research grade eyepieces were purportedly made by a company called "Japanese Special Optics" sometime in the 1970s or 1980s. The 19, 65, 72, and 85 are advertised to have "XL" optics, which is not defined but implied to be of higher quality and produced in Japan.

The 52 and 60 mm focal lengths of this eyepiece line is within the following competition class of eyepieces with similar specifications and performance (Kellners and Reverse Kellners, and RKE—Rank-Kaspereit-Erfle): Celestron E-Lux (2 in. models only), Celestron Kellner, Criterion Kellner, Edmund Scientific RKE, GSO Kellner, Kokusai Kohki Kellner, Orion DeepView, Orion E-Series, Sky-Watcher Kellner, Sky-Watcher Super MA Series, Telescope Service RK, Unitron Kellner.

Siebert Optics: MonoCentricID

Fig. 9.27 Siebert Monocentric ID eyepieces and Siebert Barlows (Image courtesy of John W., MA, USA)

Focal length (mm)	AFOV (degrees)	ER (mm)	Field stop (mm)	Weight (oz)	Design (Elem/Grp)	Coatings	Barrel (inches)
10	30[E]	—	5.2[E]	<2[E]	3/1	—	1.25
12	30[E]	—	6.3[E]	<2[E]	3/1	—	1.25
14.5	30[E]	—	7.6[E]	<2[E]	3/1	—	1.25
17.5	30[E]	—	9.2[E]	<2[E]	3/1	—	1.25
20	30[E]	—	10.5[E]	<2[E]	3/1	—	1.25
24.5	30[E]	—	12.8[E]	<2[E]	3/1	—	1.25

Source: www.siebertoptics.com

According to the manufacturer these are best when used in telescopes with a focal ratio of f/4 and higher unless using a flattener/amplifier (except the 10 mm, which works best with at least f/7.5). All focal lengths use Delrin housings with aluminum barrels. The coatings used for this line are not specified by the manufacturer. Although not in the same class as top-tier planetary eyepieces such as the TMB Monocentric, they do approach the on-axis precision of such highly regarded planetary eyepieces such as the Astro-Physics SPL. Note that due to the Monocentric design, this eyepiece operates best with longer focal ratio telescopes for the off-axis to be sharp. Use with shorter focal ratio telescopes are expected to show moderate field curvature.

Siebert Optics: Observatory

Fig. 9.28 Siebert 2 in. Observatory eyepiece (Image courtesy of Charles Brault, Rindge, NH, USA)

Focal length (mm)	AFOV (degrees)	ER (mm)	Field stop (mm)	Weight (oz)	Design (Elem/Grp)	Coatings	Barrel (inches)
34	70	20	41.5[E]	<9	—	—	2
36	70	20	44.0[E]	<9	—	—	2

Source: www.siebertoptics.com

The design and coatings used for this line are not specified by the manufacturer. Many observers report that they perform well in f/6 and faster focal ratio telescopes. Some observers reports note minor ghosting on bright planets and that they often need significant out-travel of the focuser for many Dobsonian telescopes, for some needing an extension tube to reach focus.

Fig. 9.29 Older version of the Siebert 2 in. Observatory series (Image courtesy of D. Regan, RCP Observatory, Oswego, NY, USA)

Siebert Optics: Performance Series

Fig. 9.30 A 15 mm Siebert Standard (Image courtesy of Hernando Bautista, Manila, Philippines)

Focal length (mm)	AFOV (degrees)	ER (mm)	Field stop (mm)	Weight (oz)	Design (Elem/Grp)	Coatings	Barrel (inches)
7	60	12	7.3[E]	2[E]	3[E]	FMC	1.25
9	60	12	9.4[E]	2[E]	3	FMC	1.25
10	60	12	10.5[E]	2[E]	3	FMC	1.25
13	60	12	13.6[E]	2[E]	5	FMC	1.25
15	60	12	15.7[E]	2[E]	5[E]	FMC	1.25
21	60	12	22.0[E]	2[E]	5[E]	FMC	1.25

Source: www.siebertoptics.com

A fairly good performing eyepiece line. Observer reports indicate that the eye relief is much tighter than stated. The 21 mm requires a very long focal ratio telescope for the off-axis to show reduced aberrations. Not all focal lengths are currently available, and focal lengths other than those listed may be found on the used market due to the substantial customizations Siebert Optics performs for its customers.

Siebert Optics: Planisphere

Focal length (mm)	AFOV (degrees)	ER (mm)	Field stop (mm)	Weight (oz)	Design (Elem/Grp)	Coatings	Barrel (inches)
3	<20[E]	1.1[E]	1.0[E]	<2[E]	1/1	—	1.25
3.5	<20[E]	1.1[E]	1.2[E]	<2[E]	1/1	—	1.25
4	<20[E]	1.5[E]	1.4[E]	<2[E]	1/1	—	1.25
4.5	<20[E]	1.8[E]	1.6[E]	<2[E]	1/1	—	1.25
5	<20[E]	1.9[E]	1.7[E]	<2[E]	1/1	—	1.25
6	<20[E]	1.9[E]	2.1[E]	<2[E]	1/1	—	1.25
6.5	<20[E]	2.4[E]	2.3[E]	<2[E]	1/1	—	1.25
7.5	<20[E]	2.4[E]	2.6[E]	<2[E]	1/1	—	1.25

Source: www.siebertoptics.com

This is a highly specialized eyepiece that is constructed from a single spherical glass element. Best when used for critical on-axis planetary observing due to its very small AFOV. While sphere singlet eyepieces may show a larger AFOV, the portion of the AFOV that remains aberration-free for planetary observing is generally between 10° and 20°, depending on the focal ratio of the telescope used. Although the eye relief for these eyepieces is short, due to their narrow AFOV and the typical practice of mounting the ball lens so that a portion of it protrudes above the housing, their eye relief can still feel no tighter than a typical Abbe Ortho of similar focal length. The coatings used for this line are not specified by the manufacturer.

This eyepiece line is within the following competition class of eyepieces with similar specifications and performance: the Couture Ball Singlet.

Siebert Optics: Star Splitter/Super Star Splitter

Fig. 9.31 Siebert's 4.9 mm Star Splitter (Image courtesy of Andy Howie, Paisley, Scotland)

Focal length (mm)	AFOV (degrees)	ER (mm)	Field stop (mm)	Weight (oz)	Design Elem/Grp	Coatings	Barrel (inches)
2.9	60	12	3.0^{E}	$<4^{E}$	5/3	—	1.25
3.4	60	12	3.6^{E}	$<4^{E}$	5/3	—	1.25
3.9	60	12	4.1^{E}	$<4^{E}$	5/3	—	1.25
4.4	60	12	4.6^{E}	$<4^{E}$	5/3	—	1.25
4.9	60	12	5.1^{E}	$<4^{E}$	5/3	—	1.25
5.4	60	12	5.7^{E}	$<4^{E}$	5/3	—	1.25
5.9	60	12	6.2^{E}	$<4^{E}$	5/3	—	1.25
6.4	60	12	6.7^{E}	$<4^{E}$	5/3	—	1.25
6.9	60	12	7.2^{E}	$<4^{E}$	5/3	—	1.25
7.4	60	12	7.7^{E}	$<4^{E}$	5/3	—	1.25
7.9	60	12	8.3^{E}	$<4^{E}$	5/3	—	1.25
8.9	60	12	9.3^{E}	$<4^{E}$	5/3	—	1.25
9.4	60	12	9.8^{E}	$<4^{E}$	5/3	—	1.25
9.9	60	12	10.4^{E}	$<4^{E}$	5/3	—	1.25

Source: www.siebertoptics.com

Many observers often consider these as competing with planetary eyepieces such as the Abbe Ortho, even including the much revered Baader Genuine Orthos and University Optics HD or Volcano Orthos. These are designed to work best in telescope with f/6 and longer focal ratios. The coatings used for this line are not specified by the manufacturer.

This eyepiece line is within the following competition class of eyepieces with similar specifications and performance: Antares Ortho, Apogee Super Abbe Orthoscopic, Baader Planetarium Classic Ortho, Baader Planetarium Genuine Abbe Ortho, Cave Orthostar Orthoscopic, Celestron Ortho, Edmund Scientific Ortho, Kokusai Kohki Abbe Ortho, Kson Super Ortho, Masuyama Orthoscopic, Meade Research Grade Ortho, Meade Series II Orthoscopic, Nikon Ortho, Takahashi Ortho, Telescope Service Ortho, Unitron Ortho, University Optics Abbe HD Orthoscopic, University Optics Abbe Volcano Orthoscopic, University Optics O.P.S. Orthoscopic Planetary Series, University Optics Super Abbe Orthoscopic, VERNONscope Brandon.

Siebert Optics: Ultra

Fig. 9.32 Siebert's 11 and 7 mm ultra eyepieces (Image by the author)

Focal length (mm)	AFOV (degrees)	ER (mm)	Field stop (mm)	Weight (oz)	Design (Elem/Grp)	Coatings	Barrel inches)
7	70	20	8.1^{E}	$<5^{E}$	6	FMC	1.25 or 2
9	70	20	10.5^{E}	$<5^{E}$	6	FMC	1.25 or 2
11	70	20	12.8^{E}	$<5^{E}$	6	FMC	1.25 or 2
13	70	20	15.1^{E}	$<5^{E}$	6	FMC	1.25 or 2
15	70	20	17.5^{E}	$<5^{E}$	6	FMC	1.25 or 2
17	70	20	19.8^{E}	$<5^{E}$	4	FMC	1.25 or 2
18	70	20	20.9^{E}	$<5^{E}$	4	FMC	1.25 or 2
19	70	20	22.1^{E}	$<4^{E}$	4	FMC	1.25 or 2
24	65	20	26.0^{E}	$<4^{E}$	3	FMC	1.25 or 2
27	60	20	27.1^{E}	$<4^{E}$	—	—	1.25 or 2
28	58	20	27.2^{E}	$<4^{E}$	—	—	1.25 or 2

Source: www.siebertoptics.com

A good performing wide-field eyepiece line having expansive AFOV and very comfortable eye relief. The 15 mm and shorter focal lengths perform excellently even in fast focal ratio telescopes. The coatings used for this line are not specified by the manufacturer.

Sky-Watcher: AERO

Focal length (mm)	AFOV (degrees)	ER (mm)	Field stop (mm)	Weight (oz)	Design (Elem/Grp)	Coatings	Barrel (inches)
30	68	16.7	35.6^{E}	11^{E}	6	FMC	2
35	68	17.5	41.5^{E}	12^{E}	6	FMC	2
40	68	20	46.0^{E}	18^{E}	6	FMC	2

Source: www.skywatcher.com

These use extra low-dispersion glass; have a twist-up eyeguard; and are parfocal. This eyepiece line is within the following competition class of eyepieces with similar specifications and performance: Astro-Tech Titan II ED, TMB Paragon.

Sky-Watcher: Extra Flat

Focal length (mm)	AFOV (degrees)	ER (mm)	Field stop (mm)	Weight (oz)	Design (Elem/Grp)	Coatings	Barrel (inches)
16	60	17	16.6^E	4^E	6	FMC	1.25
19	65	19	21.3^E	4^E	5	FMC	1.25
27	53	23	24.4^E	6^E	6	FMC	1.25

Source: www.skywatcher.com

These use extra low-dispersion glass; have a twist-up eyeguard; and are parfocal. This eyepiece line is within the following competition class of eyepieces with similar specifications and performance: Astro-Professional EF Flatfield, Astro-Tech Flat Field, BST Flat Field, Orion Edge-On Flat-Field, Smart Astronomy Extra Flat Field, Telescope-Service Edge-On Flat Field.

Sky-Watcher: Kellner

Focal length (mm)	AFOV (degrees)	ER (mm)	Field stop (mm)	Weight (oz)	Design (Elem/Grp)	Coatings	Barrel (inches)
6.3	50	5.5^E	5.5^E	$<4^E$	3/2	FC	1.25
10	50	8.7^E	8.7^E	$<4^E$	3/2	FC	1.25
12.5	50	10.9^E	10.9^E	$<4^E$	3/2	FC	1.25
17	50	14.8^E	14.8^E	$<4^E$	3/2	FC	1.25
25	50	21.8^E	21.8^E	$<4^E$	3/2	FC	1.25

Source: www.skywatcher.com

This eyepiece line is within the following competition class of eyepieces with similar specifications and performance: Kellners and Reverse Kellners, and RKE—Rank-Kaspereit-Erfle: Celestron E-Lux (2 in. models only), Celestron Kellner, Criterion Kellner, Edmund Scientific RKE, GSO Kellner, Kokusai Kohki Kellner, Orion DeepView, Orion E-Series, Russell Optics (2 in. 52 and 60 mm only), Sky-Watcher Super MA Series, Telescope Service RK, Unitron Kellner.

Sky-Watcher: LET/Long Eye Relief (LER)

Focal length (mm)	AFOV (degrees)	ER (mm)	Field stop (mm)	Weight (oz)	Design (Elem/Grp)	Coatings	Barrel (inches)
2	45	20	1.6[E]	—	—	MC	1.25
5	45	20	3.9[E]	—	—	MC	1.25
9	50	20	7.9[E]	—	—	MC	1.25
15	50	20	13.1[E]	—	—	MC	1.25
20	50	20	17.5[E]	—	—	MC	1.25
25	50	20	21.8[E]	—	—	MC	1.25
28	56	20	27.4[E]	—	—	MC	2
35	56	20	34.2[E]	—	—	MC	2
42	52	20	38.1[E]	—	—	MC	2

Source: www.skywatcher.com

Some observers consider these good for public outreach due to their low cost and comfortable eye relief. The 5 mm is considered by some to the best of the series. The manufacturer does not specify the optical design or weight of these eyepieces.

Sky-Watcher: Nirvana UWA

Focal length (mm)	AFOV (degrees)	ER mm)	Field stop (mm)	Weight oz)	Design (Elem/Grp)	Coatings	Barrel (inches)
4	82	12	5.7[E]	7.1	7/4	FMC	1.25
7	82	12	10.0[E]	7.1	7/4	FMC	1.25
16	82	12	22.9[E]	7.1	7/4	FMC	1.25
28	82	18	40.1[E]	35.3	6/4	FMC	2

Source: www.skywatcher.com

This eyepiece line is within the following competition class of eyepieces with similar specifications and performance: Antares Speers-WALER Series, Astro-Professional UWA, Celestron Axiom LX, Celestron Luminos, Docter UWA, Explore Scientific 82° Nitrogen Purged, Meade Series 4000 Ultra Wide Angle, Meade Series 5000 Ultra Wide Angle, Orion MegaView Ultra-Wide, Sky-Watcher Sky Panorama, Tele Vue Nagler, Williams Optics UWAN.

Sky-Watcher: PanaView

Focal ength (mm)	AFOV (degrees)	ER (mm)	Field stop (mm)	Weight (oz)	Design (Elem/Grp)	Coatings	Barrel (inches)
26	70	20	31.8[E]	10[E]	5/3[E]	FMC	2
32	70	24	39.1[E]	14[E]	5/4[E]	FMC	2
38	70	28	46.0[E]	21[E]	5/3[E]	FMC	2

Source: www.skywatcher.com

These have a twist-up eyeguard and are parfocal. This eyepiece line is within the following competition class of eyepieces with similar specifications and performance: Agena Super Wide Angle, Astro-Professional SWA, Astro-Tech Wide Field, Garrett SuperWide Angle, GSO Superview, Meade QX, Olivon 70° Wide Angle, Opt Super View, Orion Q70 Super Wide-Field, Telescope-Service WA, University Optics 70°, William Optics SWAN.

Sky-Watcher: Sky Panorama

Focal length (mm)	AFOV (degrees)	ER (mm)	Field stop (mm)	Weight (oz)	Design (Elem/Grp)	Coatings	Barrel (inches)
7	82	14	10.0[E]	12.3	7	FMC	1.25
15	82	14	21.5[E]	14.5	7	FMC	1.25
23	82	14	32.9[E]	29.3	7	FMC	2

Source: www.skywatcher.com

This eyepiece line is within the following competition class of eyepieces with similar specifications and performance: Antares Speers-WALER Series, Astro-Professional UWA, Celestron Axiom LX, Celestron Luminos, Docter UWA, Explore Scientific 82° Nitrogen Purged, Meade Series 4000 Ultra Wide Angle, Meade Series 5000 Ultra Wide Angle, Orion MegaView Ultra-Wide, Sky-Watcher Nirvana UWA, Tele Vue Nagler, Williams Optics UWAN.

Sky-Watcher: SP-Series Super Plössl

Focal length (mm)	AFOV (degrees)	ER mm)	Field stop (mm)	Weight (oz)	Design (Elem/Grp)	Coatings	Barrel (inches)
6.3	52	4^{E}	5.7^{E}	$<2^{E}$	4/2	FMC	1.25
7.5	52	5^{E}	6.8^{E}	$<3^{E}$	4/2	FMC	1.25
10	52	7^{E}	9.1^{E}	$<3^{E}$	4/2	FMC	1.25
12.5	52	9^{E}	11.3^{E}	$<3^{E}$	4/2	FMC	1.25
17	52	12^{E}	15.4^{E}	$<4^{E}$	4/2	FMC	1.25
20	52	14^{E}	18.2^{E}	$<4^{E}$	4/2	FMC	1.25
26	52	18^{E}	23.6^{E}	$<4^{E}$	4/2	FMC	1.25
32	52	22^{E}	27.0^{E}	$<5^{E}$	4/2	FMC	1.25
40	44	27^{E}	27.0^{E}	$<7^{E}$	4/2	FMC	1.25

Source: www.skywatcher.com

This eyepiece line is within the following competition class of eyepieces with similar specifications and performance: Antares Plössl, Astro-Professional Plössl, Astro-Tech High Grade Plössl, Astro-Tech Value Line Plössl, Bresser 52° Super Plössl, Carton Plössl, Celestron Omni, Celestron Silvertop Plössl, Clavé Plössl, Coronado CeMax, Edmund Scientific Plössl, Garrett Optical Plössl, GSO Plössl, GTO Plössl, Long Perng Plössl, Meade Series 3000 Plössl, Meade Series 4000 Super Plössl, Olivon Plössl, Opt Plössl, Orion HighLight Plössl, Orion Sirius Plössl, Owl Black Night Plössl, Parks Silver Series, Plössl, Smart Astronomy Sterling Plössl, TAL—Symmetrical Super Plössl, Telescope Service Plössl, Telescope Service Super Plössl, Tele Vue Plössl, Vixen NPL.

Sky-Watcher: Super-MA Series

Focal length (mm)	AFOV (degrees)	ER (mm)	Field stop (mm)	Weight (oz)	Design (Elem/Grp)	Coatings	Barrel (inches)
3.6	40	2^{E}	2.5^{E}	2^{E}	3/2	FC	1.25
4	50	3^{E}	3.5^{E}	2^{E}	3/2	FC	1.25
10	52	6^{E}	9.1^{E}	2^{E}	3/2	FC	1.25
20	52	12^{E}	18.2^{E}	3^{E}	3/2	FC	1.25
25	50	15^{E}	21.8^{E}	3^{E}	3/2	FC	1.25

Source: www.skywatcher.com

These are designed for telescopes with mid-range to long focal ratios. This eyepiece line is within the following competition class of eyepieces with similar specifications and performance: Kellners and Reverse Kellners, and RKE—Rank-Kaspereit-Erfle; Celestron E-Lux (2 in. models only), Celestron Kellner, Criterion Kellner, Edmund Scientific RKE, GSO Kellner, Kokusai Kohki Kellner, Orion DeepView, Orion E-Series, Russell Optics (2 in. 52 and 60 mm only), Sky-Watcher Kellner, Telescope Service RK, Unitron Kellner.

Sky-Watcher: Ultra Wide Angle

Focal length (mm)	AFOV (degrees)	ER (mm)	Field stop (mm)	Weight (oz)	Design (Elem/Grp)	Coatings	Barrel (inches)
6	66	14.8	8.0	4	5	MC	1.25
9	66	15.0	15.0	4	6	MC	1.25
15	66	13.0	17.0	3	4	MC	1.25
20	66	18.0	23.5	4	4	MC	1.25
15	80	20.0	18.6^E	9^E	7	MC	2
20	80	20.0	24.2^E	21^E	7	MC	2
30	80	20.0	39.6^E	20^E	5	MC	2

Source: www.skywatcher.com

The 6 and 9 mm focal lengths of the series have a reputation of being excellent performers even in faster focal ratio telescopes. These eyepieces generally compete with the following other eyepiece lines:

1.25 in.—Agena Enhanced Wide Angle, Orion Expanse, Owl Enhanced Superwide, Telescope Service SWM Wide Angle, William Optics WA 66.

2 in.—Agena Ultra Wide Angle, Apogee Widescan (20 and 30 mm), Astrobuffet 1RPD (30 mm), BW Optik Ultrawide (30 mm), Kokusai Kohki Widescan I/II/III (20 and 30 mm), Moonfish Ultrawide (30 mm), Olivion 80° Ultra Wide Angle, Owl Astronomy Knight Owl Ultrawide Angle, Surplus Shed Wollensak (30 mm), TAL Ultra Wide Angle, University Optics 80°, University Optics Widescan I/II/III (20 and 30 mm).

Smart Astronomy: Extra Flat Field (*Discontinued*)

Focal length (mm)	AFOV (degrees)	ER (mm)	Field stop (mm)	Weight (oz)	Design (Elem/Grp)	Coatings	Barrel (inches)
16	60	17	16.6^E	4^E	5/3	FMC	1.25
19	65	19	21.3^E	4^E	5/3	FMC	1.25
27	53	23	24.4^E	6^E	5/4	FMC	1.25

Source: www.smartastronomy.com

These have a twist-up eyeguard and are parfocal. This eyepiece line is within the following competition class of eyepieces with similar specifications and performance: Astro-Professional EF Flatfield, Astro-Tech Flat Field, BST Flat Field, Orion Edge-On Flat-Field, Sky-Watcher Extra Flat, Telescope-Service Edge-On Flat Field.

Smart Astronomy: SA's Solar System Long Eye Relief (*Discontinued*)

Focal length (mm)	AFOV (degrees)	ER (mm)	Field top (mm)	Weight (oz)	Design (Elem/Grp)	Coatings	Barrel (inches)
3	55	20	2.8[E]	10[E]	7/4	FMC	1.25
5	55	20	4.6[E]	10[E]	7/4	FMC	1.25
6	55	20	5.5[E]	10[E]	7/4	FMC	1.25
9	55	20	8.3[E]	9[E]	7/4	FMC	1.25
12.5	55	20	11.5[E]	8[E]	7/4	FMC	1.25
14.5	55	20	13.4[E]	7[E]	7/4	FMC	1.25
18	55	20	16.6[E]	6[E]	5/3	FMC	1.25

Source: www.smartastronomy.com

This eyepiece line is within the following competition class of eyepieces with similar specifications and performance: Astro-Professional Long Eye Relief Planetary, Astro-Tech Long Eye Relief, Long Perng Long Eye Relief, Orion Edge-On Planetary, Stellarvue Planetary, William Optics SPL, Zhumell Z Series.

Smart Astronomy: Sterling Plössl

Fig. 9.33 Smart Astronomy's Sterling Plössls in 25 mm, 20 mm, 17 mm, 12.5 mm, and 6 mm (Image by the author)

Focal length (mm)	AFOV (degrees)	ER mm)	Field stop (mm)	Weight (oz)	Design (Elem/Grp)	Coatings	Barrel (inches)
4	55	2.4	3.7[E]	2.1	4/2	FMC	1.25
6	55	3.6	5.5[E]	2.2	4/2	FMC	1.25
12.5	55	7.5	11.5[E]	3.0	4/2	FMC	1.25
17	55	10.2	15.7[E]	3.4	4/2	FMC	1.25
20	55	12	18.5[E]	3.4	4/2	FMC	1.25
25	55	15	23.1[E]	4.7	4/2	FMC	1.25
30	55	18	27.7[E]	13.8	4/2	FMC	2
40	55	24	36.9[E]	13.3	4/2	FMC	2

Source: www.smartastronomy.com

These have 40 layer multi-coated optics with 98 % transmission and a unique distinction over most other Plössl lines as having an extended AFOV. Almost all other Plössl lines have an AFOV of 50–52°, whereas the Astro-Tech High Grade Plössl, Smart Astronomy Sterling Plössl, and Long Perng Plössl have extended this to an advertised 55° (closer to 57° when measured). This larger AFOV, although only a few degrees more than a standard Plössl, is visually more impressive, conveying the impression of an eyepiece in the next larger AFOV class. For improved off-axis performance compared to typical four-element Plössls, this line also uses unconventional concave lens surfaces instead of the standard flat surfaces on the outward-facing eye lens and field lens. Other four-element Plössl lines that use concave lens surfaces include the Astro-Tech High Grade Plössl, Meade 4000 (current four-element version), and Tele Vue Plössl.

Overall an excellent Plössl both in terms of optical performance as well as in price point, being very affordable. Solidly constructed, performs on-par with the standard for this class, the Tele Vue Plössl. Exceeds the Tele Vue in some areas, and runs very slightly behind in others. Off-axis performance slightly improved over the Tele Vue. As in a 2008 observation report by the author:

> *Observing Messier 57, the Ring Nebula, both the 12.5 mm Sterling and the 11 mm Tele Vue showed it very well. The dim star outside the Ring Nebula was visible with adverted vision using both. Again perhaps a little better in the Tele Vue. However, I much preferred the Ring Nebula itself in the Sterling, looking more pronounced with more definition to its shape. Also, the wider AFOV of the Sterling made the whole picture more interesting. A quick peek in the 12 mm Brandon and again, like with the Moon, the contrast improvement was obvious—Brandon's are simply very good performers if you don't mind the narrower 40° + AFOV.*

This eyepiece line is within the following competition class of eyepieces with similar specifications and performance: Antares Plössl, Astro-Professional Plössl, Astro-Tech High Grade Plössl, Astro-Tech Value Line Plössl, Bresser 52° Super Plössl, Carton Plössl, Celestron Omni, Celestron Silvertop Plössl, Clavé Plössl, Coronado CeMax, Edmund Scientific Plössl, Garrett Optical Plössl, GSO Plössl, GTO Plössl, Long Perng Plössl, Meade Series 3000 Plössl, Meade Series 4000 Super Plössl, Olivon Plössl, Opt Plössl, Orion HighLight Plössl, Orion Sirius Plössl, Owl Black Night Plössl, Parks Silver Series, Sky-Watcher SP-Series Super Plössl, TAL—Symmetrical Super Plössl, Telescope Service Plössl, Telescope Service Super Plössl, Tele Vue Plössl, Vixen NPL.

Stellarvue: Planetary

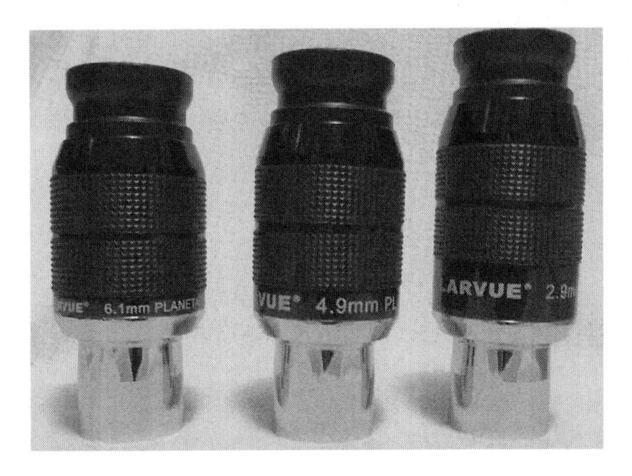

Fig. 9.34 The Stellarvue Planetary eyepieces (Image courtesy of David Elosser, NC, USA)

Focal length (mm)	AFOV (degrees)	ER (mm)	Field stop (mm)	Weight (oz)	Design (Elem/Grp)	Coatings	Barrel (inches)
2.9	53	20	2.7[E]	10[E]	7/4[E]	FMC	1.25
4.9	53	20	4.5[E]	10[E]	7/4[E]	FMC	1.25
6.1	53	20	5.3[E]	10[E]	7/4[E]	FMC	1.25

Source: www.stellarvue.com

This eyepiece line is within the following competition class of eyepieces with similar specifications and performance: Astro-Professional Long Eye Relief Planetary, Astro-Tech Long Eye Relief, Long Perng Long Eye Relief, Smart Astronomy SA Solar System Long Eye Relief, William Optics SPL, Zhumell Z Series.

Surplus Shed: Erfles

Fig. 9.35 Surplus Shed 26 and 38 mm Erfles (2 in.) (Image courtesy of www.SurplusShed.com. Used with permission)

Focal length (mm)	AFOV (degrees)	ER (mm)	Field stop (mm)	Weight (oz)	Design (Elem/Grp)	Coatings	Barrel (inches)
26	60	20	26.1[E]	—	5/3	FC	2
38	60	16	38.2[E]	—	5/3	FC	2

Source: www.surplusshed.com

These should be used with longer focal ratio telescopes for a best performing off-axis. Excellent public outreach eyepieces.

Surplus Shed: Wollensak

Fig. 9.36 Surplus Shed's 10 mm, 12 mm, and 48 mm Wollensak eyepieces (2 in.) (Image courtesy of www.SurplusShed.com. Used with permission.)

Focal length (mm)	AFOV (degrees)	ER (mm)	Field stop (mm)	Weight (oz)	Design (Elem/Grp)	Coatings	Barrel (inches)
10	60	11	10.0^{E}	—	8/7	FC	2
12.5	55	10	11.5^{E}	9	9	FC	2
30	80	22	46.2	19.2	5	FMC	2
48	55	20	44.3^{E}	9	6	FC	2
51	60	35	51.2^{E}	7.5	4/2	FC	2

Source: www.surplusshed.com

These carry the Wollensak 90-day warrantee. This eyepiece line is within the following competition class of eyepieces with similar specifications and performance: (30 mm only): Agena Ultra Wide Angle, Apogee Widescan (20 and 30 mm), Astrobuffet 1RPD (30 mm), BW Optik Ultrawide (30 mm), Kokusai Kohki Widescan I/II/III (20 and 30 mm), Moonfish Ultrawide (30 mm), Olivion 80° Ultra Wide Angle, Owl Astronomy Knight Owl Ultrawide Angle, Sky-Watcher Ultra Wide Angle, TAL Ultra Wide Angle, University Optics 80°, University Optics Widescan I/II/III (20 and 30 mm).

Fig. 9.37 Surplus Shed's 30 mm Wollensak (2 in.) (Image courtesy of www.SurplusShed.com. Used with permission.)

Chapter 10

Takahashi to Zooming Eyepieces

Takahashi: LE

Fig. 10.1 Takahashi LE eyepieces (Image courtesy of Tamiji Homma, Newbury Park, California, USA)

W. Paolini, *Choosing and Using Astronomical Eyepieces*, The Patrick Moore Practical Astronomy Series, DOI 10.1007/978-1-4614-7723-5_10,
© Springer Science+Business Media New York 2013

Focal length (mm)	AFOV (degrees)	ER (mm)	Field stop (mm)	Weight (oz)	Design (Elem/Grp)	Coatings	Barrel (inches)
2.8	42	5.8	2.1[E]	4.5	6/3	FMC	1.25
3.6	40	7.5	2.5[E]	4.5	6/3	FMC	1.25
5	52	10	4.5[E]	4.5	5/3	FMC	1.25
7.5	52	10	6.8[E]	4.0	5/3	FMC	1.25
10	52	6.2	9.1[E]	3.5	5/3	FMC	1.25
12.5	52	9	11	3.5	5/3	FMC	1.25
18	52	13	16	4.4	5/3	FMC	1.25
24	52	17	22	5.3	5/3	FMC	1.25
30	52	20	27	6.0	5/3	FMC	1.25
50	50	40	43	20.0	5/3	FMC	2

Source: www.takahashijapan.com; www.takahashiamerica.com; www.takahashi-europe.com

These use extra low-dispersion glass, are parfocal, and have an outstanding build quality. This eyepiece line is within the following competition class of eyepieces with similar specifications and performance: Antares Elite, Baader Eudiascopic, Bresser 60° Plössl, Celestron Ultima, Kasai Astroplan, Meade Series 4000 Super Plössl (pre-1994, smooth sided, five-elements), Meade Series 5000 Super Plössl, Orion Ultrascopic, Parks Gold Series Plössl. *Note*: The Omcon Ultima and Tuthill Super Plössl lines were also competing brands but these lines are not detailed.

Takahashi: Ortho (*Discontinued*)

Focal length (mm)	AFOV (degrees)	ER (mm)	Field stop (mm)	Weight (oz)	Design (Elem/Grp)	Coatings	Barrel (inches)
2.8	40	4	2.0[E]	<3[E]	4/2	–	.965
4	40	6	2.8[E]	<3[E]	4/2	–	.965
5	40	4	3.5[E]	<3[E]	4/2	–	.965
7	40	6	4.9[E]	<3[E]	4/2	–	.965
9	40	7	6.3[E]	<3[E]	4/2	–	.965
18	40	15	12.6[E]	<3[E]	4/2	–	.965
25	40	22	17.5[E]	<3[E]	4/2	–	.965

Source: www.takahashijapan.com; www.takahashiamerica.com; www.takahashi-europe.com

This eyepiece line is within the following competition class of eyepieces with similar specifications and performance: Antares Ortho, Apogee Super Abbe Orthoscopic, Baader Planetarium Classic Ortho, Baader Planetarium Genuine Abbe Ortho, Cave Orthostar Orthoscopic, Celestron Ortho, Edmund Scientific

Ortho, Kokusai Kohki Abbe Ortho, Kson Super Ortho, Masuyama Orthoscopic, Meade Research Grade Ortho, Meade Series II Orthoscopic, Nikon Ortho, Siebert Optics Star Splitter/Super Star Splitter, Telescope Service Ortho, Unitron Ortho, University Optics Abbe HD Orthoscopic, University Optics Abbe Volcano Orthoscopic, University Optics O.P.S. Orthoscopic Planetary Series, University Optics Super Abbe Orthoscopic, VERNONscope Brandon.

Takahashi: UW

Fig. 10.2 Takahashi 10 and 5.7 mm UW eyepieces (Image courtesy of Tamiji Homma, Newbury Park, CA, USA)

Focal length (mm)	AFOV (degrees)	ER (mm)	Field stop (mm)	Weight (oz)	Design (Elem/Grp)	Coatings	Barrel (inches)
3.3	90	12	5.2E	16.5	10/6	FMC	1.25
5.7	90	12	9.0E	15.8	10/6	FMC	1.25
7	90	12	11.0E	15.1	8/5	FMC	1.25
10	90	12	15.7E	14.8	8/5	FMC	1.25

Source: www.takahashijapan.com; www.takahashiamerica.com; www.takahashi-europe.com

These are aplanatic; free of coma and spherical aberration; are designed to be used with the flat field telescopes; and have an extendable rubber eyeguard. Due to their AFOV size they can compete well with either the 80° or 100° AFOV class eyepieces.

TAL: Super Wide Angle

Fig. 10.3 The complete line of TAL Super Wide Angle eyepieces (© 2013 TALteleoptics—www.talteleoptics.com)

Focal length (mm)	AFOV (degrees)	ER (mm)	Field stop (mm)	Weight (oz)	Design (Elem/Grp)	Coatings	Barrel (inches)
10	65	6.5	11.3[E]	3.2	5/3	FMC	1.25
15	65	9.8	17.0[E]	4.2	5/3	FMC	1.25
20	65	13	22.7[E]	4.2	5/3	FMC	1.25

Source: www.talteleoptics.com

All optical elements are multi-coated, reducing reflection to 0.8 % within 440–680 nm light spectrum.

TAL: Symmetrical (Super Plössl)

Fig. 10.4 The complete line of TAL Symmetrical Super Plössls (© 2013 TALteleoptics—www.talteleoptics.com)

Focal length (mm)	AFOV (degrees)	ER (mm)	Field stop (mm)	Weight (oz)	Design (Elem/Grp)	Coatings	Barrel (inches)
6.3	45	6	4.9^E	2.1	4/2	FMC	1.25
7.5	45	6	5.9^E	2.3	4/2	FMC	1.25
10	45	7.5	7.9^E	2.5	4/2	FMC	1.25
12.5	45	10	9.8^E	2.8	4/2	FMC	1.25
17	45	11	13.4^E	3.2	4/2	FMC	1.25
20	45	14	15.7^E	3.5	4/2	FMC	1.25
25	45	18.7	19.6^E	3.9	4/2	FMC	1.25
32	45	22	25.1^E	6.0	4/2	FMC	1.25
40	45	32	27.0^E	7.4	4/2	FMC	1.25

Source: www.talteleoptics.com

All optical elements are multicoated reducing reflection to 0.8 % within 440–680 nm light spectrum; parfocal. This eyepiece line is within the following competition class of eyepieces with similar specifications and performance: Antares Plössl, Astro-Professional Plössl, Astro-Tech High Grade Plössl, Astro-Tech Value Line Plössl, Bresser 52° Super Plössl, Carton Plössl, Celestron Omni, Celestron Silvertop Plössl, Clavé Plössl, Coronado CeMax, Edmund Scientific Plössl, Garrett Optical Plössl, GSO Plössl, GTO Plössl, Long Perng Plössl, Meade Series 3000

Plössl, Meade Series 4000 Super Plössl, Olivon Plössl, Opt Plössl, Orion HighLight Plössl, Orion Sirius Plössl, Owl Black Night Plössl, Parks Silver Series, Sky-Watcher SP-Series Super Plössl, Smart Astronomy Sterling Plössl, Telescope Service Plössl, Telescope Service Super Plössl, Tele Vue Plössl, Vixen NPL.

TAL: Ultra Wide Angle

Fig. 10.5 The complete line of TAL Symmetrical Super Plössls (© 2013 TALteleoptics - www.talteleoptics.com)

Focal length (mm)	AFOV (degrees)	ER (mm)	Field stop (mm)	Weight (oz)	Design (Elem/Grp)	Coatings	Barrel (inches)
15	80	7.3	20.9[E]	4.6	6/4	FMC	1.25
20	80	9.8	27.9[E]	11.3	6/4	FMC	2
24	80	11.6	33.5[E]	15.9	8/5	FMC	2
25	80	12.2	34.9[E]	12.7	6/4	FMC	2

Source: www.talteleoptics.com

All optical elements are multicoated reducing reflection to 0.8 % within 440–680 nm light spectrum. This eyepiece line is within the following competition class of eyepieces with similar specifications and performance: Agena Ultra Wide Angle, Apogee Widescan (20 and 30 mm), Astrobuffet 1RPD (30 mm), BW Optik Ultrawide (30 mm), Kokusai Kohki Widescan I/II/III (20 and 30 mm), Moonfish Ultrawide (30 mm), Olivion 80° Ultra Wide Angle, Owl Astronomy Knight Owl Ultrawide Angle, Sky-Watcher Ultra Wide Angle, Surplus Shed Wollensak (30 mm), University Optics 80°, University Optics Widescan I/II/III (20 and 30 mm).

Telescope Service: Edge-On Flat Field

Focal length (mm)	AFOV (degrees)	ER (mm)	Field stop (mm)	Weight (oz)	Design (Elem/Grp)	Coatings	Barrel (inches)
8	60	9	8.4[E]	3[E]	6	FMC	1.25
12	60	12	12.6[E]	4	6	FMC	1.25
16	60	17	16.8[E]	4	6	FMC	1.25
19	65	17	21.6[E]	4	5	FMC	1.25
27	53	21	25.0[E]	6	6	FMC	1.25

Source: www.telescope-service.com

These are parfocal. This eyepiece line is within the following competition class of eyepieces with similar specifications and performance: Astro-Professional EF Flatfield, Astro-Tech Flat Field, BST Flat Field, Orion Edge-On Flat-Field, Sky-Watcher Extra Flat, Smart Astronomy Extra Flat Field.

Telescope Service: Expanse ED

Focal length (mm)	AFOV (degrees)	ER (mm)	Field stop (mm)	Weight (oz)	Design (Elem/Grp)	Coatings	Barrel (inches)
3.5	70	20	4.3[E]	17.3	8[E]	FMC	1.25 or 2
5	70	20	6.1[E]	16.9	8[E]	FMC	1.25 or 2
8	70	20	9.8[E]	16.6	8[E]	FMC	1.25 or 2
13	70	20	15.9[E]	16.6	8[E]	FMC	1.25 or 2
17	70	20	20.8[E]	15.9	8[E]	FMC	1.25 or 2
22	70	20	26.9[E]	17.3	8[E]	FMC	1.25 or 2

Source: www.telescope-service.com

These have an integrated ED element, use an optional Hyp-T2 connector to allow a camera to be directly screwed to the eyepiece, and have dual barrels. This eyepiece line is within the following competition class of eyepieces with similar specifications and performance: Astro-Tech AF Series 70, Olivon 70° Ultra Wide Angle.

Telescope Service: NED "ED" Flat Field

Focal length (mm)	AFOV (degrees)	ER (mm)	Field stop (mm)	Weight (oz)	Design (Elem/Grp)	Coatings	Barrel (inches)
5	60	16	5.2[E]	6.8	6	FMC	1.25
8	60	16	8.4[E]	6.2	6	FMC	1.25
12	60	16	12.6[E]	6.6	6	FMC	1.25
15	60	16	15.7[E]	6.0	6	FMC	1.25
18	60	16	18.8[E]	5.6	6	FMC	1.25
25	60	16	26.2[E]	6.4	6	FMC	1.25

Source: www.telescope-service.com

These use an extra low-dispersion element and have a twist-up eyeguard. This eyepiece line is within the following competition class of eyepieces with similar specifications and performance: Astro-Tech Paradigm Dual ED, BST Explorer ED, Olivon 60° ED Wide Angle, Orion Epic II ED, Pentax XF.

Telescope Service: Ortho

Focal length (mm)	AFOV (degrees)	ER (mm)	Field stop (mm)	Weight (oz)	Design (Elem/Grp)	Coatings	Barrel (inches)
4.8	48	4[E]	3.9[E]	<4[E]	4/2	FMC	1.25
7.7	48	6[E]	6.5	<4[E]	4/2	FMC	1.25
10.5	48	9[E]	8.4[E]	<4[E]	4/2	FMC	1.25
16.8	48	14[E]	13.5[E]	<4[E]	4/2	FMC	1.25
24	48	20[E]	20.5	<4[E]	4/2	FMC	1.25

Source: www.telescope-service.com

This eyepiece line is within the following competition class of eyepieces with similar specifications and performance: Antares Ortho, Apogee Super Abbe Ortho, Baader Planetarium Classic Ortho, Baader Planetarium Genuine Abbe Ortho, Cave Orthostar Orthoscopic, Celestron Ortho, Edmund Scientific Ortho, Kokusai Kohki Abbe Ortho, Kson Super Ortho, Masuyama Orthoscopic, Meade Research Grade Ortho, Meade Series II Orthoscopic, Nikon Ortho, Siebert Optics Star Splitter/Super Star Splitter, Takahashi Ortho, Telescope Service Ortho, Unitron Ortho, University Optics Abbe HD Orthoscopic, University Optics Abbe Volcano Orthoscopic, University Optics O.P.S. Orthoscopic Planetary Series, University Optics Super Abbe Orthoscopic, VERNONscope Brandon.

Telescope Service: Paragon ED

Focal length (mm)	AFOV (degrees)	ER (mm)	Field stop (mm)	Weight (oz)	Design (Elem/Grp)	Coatings	Barrel (inches)
3.8	50	20	3.3^E	7.5^E	6/4	FMC	1.25
5.2	50	20	4.5^E	7.2^E	6/4	FMC	1.25
7.5	50	20	6.5^E	6.7^E	6/4	FMC	1.25
9.5	50	20	8.3^E	6.4^E	6/4	FMC	1.25
12.5	50	20	10.9^E	6.0^E	6/4	FMC	1.25
14	50	20	12.2^E	5.8^E	6/4	FMC	1.25
18	50	20	15.7^E	7.0^E	6/4	FMC	1.25
21	50	20	18.3^E	6.9^E	6/4	FMC	1.25
25	50	20	21.8^E	7.0^E	6/4	FMC	1.25

Source: www.telescope-service.com

These are designed with extra low-dispersion glass. This eyepiece line is within the following competition class of eyepieces with similar specifications and performance: Agena Astro ED, Vixen LV.

Telescope Service: Planetary HR

Fig. 10.6 Telescope-Service HR Planetary eyepieces in 2.5 mm through 9 mm (Image courtesy of William Rose, Larkspur, CO, USA)

Focal length (mm)	AFOV (degrees)	ER (mm)	Field stop (mm)	Weight (oz)	Design (Elem/Grp)	Coatings	Barrel (inches)
2.5	60	16	2.5[E]	5	6	FMC	1.25
3.2	60	16	3.4[E]	5	6	FMC	1.25
4	60	16	4.0[E]	5	6	FMC	1.25
5	60	16	5.1[E]	5	6	FMC	1.25
6	60	16	6.1[E]	5	6	FMC	1.25
7	60	16	7.1[E]	5	6	FMC	1.25
8	60	16	8.1[E]	5	6	FMC	1.25
9	60	16	9.1[E]	5	6	FMC	1.25
15	60	16	15.7[E]	5	6	FMC	1.25
20	60	16	20.9[E]	5	6	FMC	1.25
25	60	16	26.2[E]	5	6	FMC	1.25

Source: www.telescope-service.com

A digital camera can be attached to the Planetary HR eyepiece using a HR-T2 adaptor. This eyepiece line is within the following competition class of eyepieces with similar specifications and performance: APM UWA Planetary, Burgess/TMB Planetary, Olivon Wide Angled Plössl, Owl Astronomy High Resolution Planetary, TMB Planetary II.

Telescope Service: RK

Focal length (mm)	AFOV (degrees)	ER (mm)	Field stop (mm)	Weight (oz)	Design (Elem/Grp)	Coatings	Barrel (inches)
26	65[E]	18	29.5[E]	9.3[E]	3/2[E]	FMC	2
32	65[E]	30	36.3[E]	10.5[E]	3/2[E]	FMC	2
40	56[E]	30	39.1[E]	10.5[E]	3/2[E]	FMC	2

Source: www.telescope-service.com

Best for telescopes with focal ratios of f/6-f/8 and longer. This eyepiece line is within the following competition class of eyepieces with similar specifications and performance (Kellners and Reverse Kellners, and RKE—Rank-Kaspereit-Erfle): Celestron E-Lux (2 in. models only), Celestron Kellner, Criterion Kellner, Edmund Scientific RKE, GSO Kellner, Kokusai Kohki Kellner, Orion DeepView, Orion E-Series, Russell Optics (2 in. 52 and 60 mm only), Sky-Watcher Kellner, Sky-Watcher Super MA Series, Unitron Kellner.

Telescope Service: Plössl

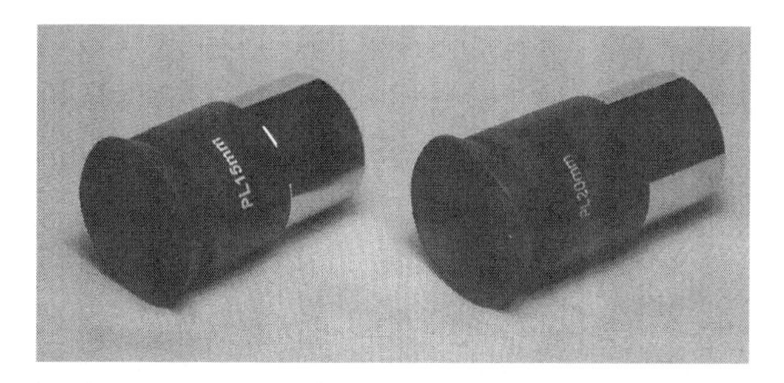

Fig. 10.7 TS Plössls in 15 and 20 mm focal lengths (Image courtesy of William Rose, Larkspur, CO, USA)

Focal length (mm)	AFOV (degrees)	ER (mm)	Field stop (mm)	Weight (oz)	Design (Elem/Grp)	Coatings	Barrel (inches)
4	50	2.7[E]	3.5[E]	<3[E]	4/2	FC	1.25
6.5	50	4.4[E]	5.7[E]	<3[E]	4/2	FC	1.25
10	50	6.8[E]	8.7[E]	<3[E]	4/2	FC	1.25
12.5	50	8.5[E]	10.9[E]	<3[E]	4/2	FC	1.25
17	50	11.6[E]	14.8[E]	<4[E]	4/2	FC	1.25
20	50	13.6[E]	17.5[E]	<4[E]	4/2	FC	1.25
25	50	17.0[E]	21.8[E]	<4[E]	4/2	FC	1.25
30	50	20.4[E]	26.2[E]	<5[E]	4/2	FC	1.25
32	50	21.8[E]	27.0[E]	<5[E]	4/2	FC	1.25
40	46	27.2[E]	27.0[E]	<7[E]	4/2	FC	1.25

Source: www.telescope-service.com

This eyepiece line is within the following competition class of eyepieces with similar specifications and performance: Antares Plössl, Astro-Professional Plössl, Astro-Tech High Grade Plössl, Astro-Tech Value Line Plössl, Bresser 52° Super Plössl, Carton Plössl, Celestron Omni, Celestron Silvertop Plössl, Clavé Plössl, Coronado CeMax, Edmund Scientific Plössl, Garrett Optical Plössl, GSO Plössl, GTO Plössl, Long Perng Plössl, Meade Series 3000 Plössl, Meade Series 4000 Super Plössl, Olivon Plössl, Opt Plössl, Orion HighLight Plössl, Orion Sirius Plössl, Owl Black Night Plössl, Parks Silver Series, Sky-Watcher SP-Series Super Plössl, Smart Astronomy Sterling Plössl, TAL—Symmetrical Super Plössl, Telescope Service Super Plössl, Tele Vue Plössl, Vixen NPL.

Telescope Service: Super Plössl

Focal length (mm)	AFOV (degrees)	ER (mm)	Field stop (mm)	Weight (oz)	Design (Elem/Grp)	Coatings	Barrel (inches)
4	52	2.7[E]	3.6[E]	<3[E]	4/2	FMC	1.25
6	52	4.1[E]	5.4[E]	<3[E]	4/2	FMC	1.25
9	52	6.1[E]	8.2[E]	<3[E]	4/2	FMC	1.25
12	52	8.2[E]	10.9[E]	<3[E]	4/2	FMC	1.25
15	52	10.2[E]	13.6[E]	<4[E]	4/2	FMC	1.25
20	52	13.6[E]	18.2[E]	<4[E]	4/2	FMC	1.25
25	52	17.0[E]	22.7[E]	<4[E]	4/2	FMC	1.25
32	52	21.8[E]	27.0[E]	<5[E]	4/2	FMC	1.25
40	46	27.2[E]	27.0[E]	<7[E]	4/2	FMC	1.25

Source: www.telescope-service.com

This eyepiece line is within the following competition class of eyepieces with similar specifications and performance: Antares Plössl, Astro-Professional Plössl, Astro-Tech High Grade Plössl, Astro-Tech Value Line Plössl, Bresser 52° Super Plössl, Carton Plössl, Celestron Omni, Celestron Silvertop Plössl, Clavé Plössl, Coronado CeMax, Edmund Scientific Plössl, Garrett Optical Plössl, GSO Plössl, GTO Plössl, Long Perng Plössl, Meade Series 3000 Plössl, Meade Series 4000 Super Plössl, Olivon Plössl, Opt Plössl, Orion HighLight Plössl, Orion Sirius Plössl, Owl Black Night Plössl, Parks Silver Series, Sky-Watcher SP-Series Super Plössl, Smart Astronomy Sterling Plössl, TAL—Symmetrical Super Plössl, Telescope Service Plössl, Tele Vue Plössl, Vixen NPL.

Telescope Service: SWM Wide Angle Eyepieces

Focal length (mm)	AFOV (degrees)	ER (mm)	Field stop (mm)	Weight (oz)	Design (Elem/Grp)	Coatings	Barrel (inches)
6	66	17[E]	6.6[E]	4.1	6/4[E]	FMC	1.25
9	66	17[E]	9.9[E]	4.0	7/4[E]	FMC	1.25
15	66	13[E]	16.5[E]	2.8	4/3[E]	FMC	1.25
20	66	16[E]	22.0[E]	3.8	5/3[E]	FMC	1.25

Source: www.telescope-service.com

The 6 and 9 mm focal lengths of the series have a reputation of being excellent performers even in faster focal ratio telescopes. This eyepiece line is within the following competition class of eyepieces with similar specifications and performance: Agena Enhanced Wide Angle, Orion Expanse, Owl Enhanced Superwide, Sky-Watcher Ultra Wide Angle, William Optics WA 66.

Telescope-Service: Wide Angle (WA)

Fig. 10.8 An 8 mm TS WA eyepiece (Image courtesy of Uwe Pilz, Leipzig, Germany)

Focal length (mm)	AFOV (degrees)	ER (mm)	Field stop (mm)	Weight (oz)	Design (Elem/Grp)	Coatings	Barrel (inches)
26	70	13[E]	32.5	10[E]	5/3	FMC	2
30	68	15	35.6[E]	13[E]	5/3	FMC	2
32	67	16	39.1[E]	14[E]	6/3	FMC	2
38	70	18[E]	46.0	21[E]	5/3	FMC	2
42	60	20	46.0[E]	13[E]	5/3	FMC	2
50	55	22	52.4[E]	14[E]	5/3	FMC	2

Source: www.telescope-service.com

This eyepiece line is within the following competition class of eyepieces with similar specifications and performance: Agena Super Wide Angle, Astro-Professional SWA, Astro-Tech Wide Field, Garrett SuperWide Angle, GSO Superview, Meade QX, Olivon 70° Wide Angle, Opt Super View, Orion Q70 Super Wide-Field, Sky-Watcher PanaView, University Optics 70°, William Optics SWAN.

Tele Vue: Delos

Fig. 10.9 The Tele Vue Delos in 6 and 10 mm (Image courtesy of Tamiji Homma, Newbury Park, CA, USA)

Focal length (mm)	AFOV (degrees)	ER (mm)	Field stop (mm)	Weight (oz)	Design (Elem/Grp)	Coatings	Barrel (inches)
3.5	72	20	4.4	17.6	–	FMC	1.25
4.5	72	20	5.6	17.6	–	FMC	1.25
6	72	20	7.6	16	–	FMC	1.25
8	72	20	9.9	16	–	FMC	1.25
10	72	20	12.7	14.4	–	FMC	1.25
12	72	20	15	14.4	–	FMC	1.25
14	72	20	17.3	14.4	–	FMC	1.25
17.3	72	20	21.2	14.4	–	FMC	1.25

Source: www.televue.com

These have full field sharpness with a virtually perfect f θ (theta) distortion mapping, large 35 mm diameter eye lens on all focal lengths, continuously adjustable height eyeguard system that can be locked in any position, and indicator marks on the body to reference eyeguard positioning. Tele Vue manufactures all their eyepiece lines with the following standard features: fine polished glass lenses; fully multi-coated glass matched multi-coatings; blackened lens edges; anti-reflection surfaces throughout; baffles to reduce grazing incidence reflections; barrels safety bevel-undercut and threaded for standard filters; and 100 % of the eyepieces are quality checked at the Tele Vue New York facility. The manufacturer does not provide information on the number of glass elements or groupings in this design. This eyepiece line is within the following competition class of eyepieces with similar specifications and performance: Explore Scientific 68 Series, Meade 4000 SWA, Meade Series 5000 SWA, Nikon NAV-SW, Pentax XL, and Pentax XW.

Tele Vue: Ethos

Fig. 10.10 The Tele Vue Ethos in 6 mm, 8 mm, 10 mm, 13 mm, 17 mm and 21 mm (Image courtesy of Neville Edwards, Norwich, Norfolk, U. K.)

Focal length (mm)	AFOV (degrees)	ER (mm)	Field stop (mm)	Weight (oz)	Design (Elem/Grp)	Coatings	Barrel (inches)
3.7	110	15	7.0	17.6	–	FMC	1.25/2
4.7	110	15	8.9	20.8	–	FMC	1.25/2
6	100	15	10.4	15.5	–	FMC	1.25/2
8	100	15	13.9	15.2	7/4[E]	FMC	1.25/2
10	100	15	17.7	17.6	–	FMC	1.25/2
13	100	15	22.3	20.8	9/5[E]	FMC	1.25/2
17	100	15	29.6	24.8	–	FMC	2
21	100	15	36.2	36.0	–	FMC	2

Source: www.televue.com

Although Tele Vue has never published the number of elements and groups in the Ethos design, amateur astronomers have both X-rayed the 13 mm Ethos (www.svenwienstein.de/HTML/es_14mm_ethos_13mm_English.html), as well as disassembled the 8 mm Ethos (http://www.cloudynights.com/ubbthreads/showflat.php/Cat/0/Number/2710707/page/0/view/collapsed/sb/5/o/all/fpart/all/vc/1). The X-rayed 13 mm Ethos appears to be approximately nine elements in five groups, while the disassembled 8 mm Ethos reveals approximately seven elements in four groups. Despite the large number of glass elements and air-to-glass interfaces, observers still report that the Ethos provides views as bright as even those eyepieces with many fewer elements, testifying to the effectiveness of the special "tuned" multi-coatings used by Tele Vue for their eyepieces.

These have full field sharpness with well-controlled astigmatism, field curvature, lateral color, angular magnification distortion correction, and low exit pupil sensitivity (e.g., kidney bean effect). Tele Vue manufactures all their eyepiece lines with the following standard features: fine polished glass lenses; fully multi-coated glass matched multi-coatings; blackened lens edges; anti-reflection surfaces throughout; baffles to reduce grazing incidence reflections; barrels safety bevel-undercut and threaded for standard filters; and 100 % of the eyepieces are quality checked at the Tele Vue New York facility. This eyepiece line is within the following competition class of eyepieces with similar specifications and performance: Explore Scientific 100 Series, Nikon NAV-HW.

Tele Vue: Nagler

Fig. 10.11 The Tele Vue Naglers in 31 mm, 26 mm, 17 mm, 13 mm, 11 mm, 9 mm, 7 mm, and 5 mm (Image courtesy of Chris Mohr, Raleigh, NC, USA)

Focal length (mm)	AFOV (degrees)	ER (mm)	Field stop (mm)	Weight (oz)	Design (Elem/Grp)	Coatings	Barrel (inches)
2.5 (T6)	82	12	3.4	8.7	7/4	FMC	1.25
3.5 (T6)	82	12	4.8	8.5	7/4	FMC	1.25
5 (T6)	82	12	7	7.9	7/4	FMC	1.25
7 (T6)	82	12	9.7	8	7/4	FMC	1.25
9 (T6)	82	12	12.4	6.7	7/4	FMC	1.25
11 (T6)	82	12	14.9	6.7	7/4	FMC	1.25
12 (T4)	82	17	17.1	16.2	6/4	FMC	1.25/2
13 (T6)	82	12	17.6	6.4	7/4	FMC	1.25
16 (T5)	82	10	22.1	7.1	6/4	FMC	1.25
17 (T4)	82	17	24.3	25.6	7/5	FMC	2
20 (T5)	82	12	27.4	16.6	6/4	FMC	2
22 (T4)	82	19	31.1	24	7/5	FMC	2
26 (T5)	82	16	35	25.6	6/4	FMC	2
31 (T5)	82	19	42	35.2	6/4	FMC	2

Source: www.televue.com

These have an 82° "spacewalk" experience for immersive views and a design that incorporates exotic glass types. Tele Vue manufactures all their eyepiece lines with the following standard features: fine polished glass lenses; fully multi-coated glass matched multi-coatings; blackened lens edges; anti-reflection surfaces throughout; baffles to reduce grazing incidence reflections; barrels safety bevel-undercut and threaded for standard filters; and 100 % of the eyepieces are quality checked at the Tele Vue New York facility. This eyepiece line is within the following competition class of eyepieces with similar specifications and performance: Antares Speers-WALER Series, Astro-Professional UWA, Celestron Axiom LX, Celestron Luminos, Docter UWA, Explore Scientific 82° Nitrogen Purged, Meade Series 4000 Ultra Wide Angle, Meade Series 5000 Ultra Wide Angle, Orion MegaView Ultra-Wide, Sky-Watcher Nirvana UWA, Sky-Watcher Sky Panorama, Williams Optics UWAN.

Tele Vue: Panoptic

Fig. 10.12 The Tele Vue Panoptics in 27 and 19 mm (Image courtesy of Alexander Kupco, Ricany, Czech Republic)

Focal length (mm)	AFOV (degrees)	ER (mm)	Field stop (mm)	Weight (oz)	Design (Elem/Grp)	Coatings	Barrel (inches)
15	68	10	17.1	4.8	6/4	FMC	1.25
19	68	13	21.3	6.6	6/4	FMC	1.25
24	68	15	27	8.2	6/4	FMC	1.25
22	68	15	25	16^{E}	6/4	FMC	1.25/2
27	68	19	30.5	16.4	6/4	FMC	2
35	68	24	38.7	25.6	6/4	FMC	2
41	68	27	46	33.6	6/4	FMC	2

Source: www.televue.com

These bring the Tele Vue Wide-Field eyepiece concept to Nagler-like performance. Tele Vue manufactures all their eyepiece lines with the following standard features: fine polished glass lenses; fully multi-coated glass matched multi-coatings; blackened lens edges; anti-reflection surfaces throughout; baffles to reduce grazing incidence reflections; barrels safety bevel-undercut and threaded for standard filters; and 100 % of the eyepieces are quality checked at the Tele Vue New York facility. This eyepiece line is within the following competition class of eyepieces with similar specifications and performance: Explore Scientific 68 Series, Meade 4000 SWA, Meade 5000 SWA, Pentax XL, Pentax XW.

Tele Vue: Plössl

Fig. 10.13 The Tele Vue Plössls in 8 mm, 11 mm, 15 mm, 20 mm, and 25 mm (© 2013 Tele Vue Optics, TeleVue.com)

Focal length (mm)	AFOV (degrees)	ER (mm)	Field stop (mm)	Weight (oz)	Design (Elem/Grp)	Coatings	Barrel (inches)
8	50	6	6.5	1.8	4/2	FMC	1.25
11	50	8	9.1	2.2	4/2	FMC	1.25
15	50	10	12.6	2.6	4/2	FMC	1.25
20	50	14	17.1	3	4/2	FMC	1.25
25	50	17	21.2	4.3	4/2	FMC	1.25
32	50	22	27	6.2	4/2	FMC	1.25
40	43	28	27	6.6	4/2	FMC	1.25
55	50	38	46	18.1	4/2	FMC	2

Source: www.televue.com

These deliver sharper images than any other Plössl or orthoscopic design; have a patented Plössl-type design that reduces astigmatism, lateral color, and coma; and are excellent in telescopes with focal ratios as fast as f/4. Tele Vue manufactures all their eyepiece lines with the following standard features: fine polished glass lenses;

fully multi-coated glass matched multi-coatings; blackened lens edges; anti-reflection surfaces throughout; baffles to reduce grazing incidence reflections; barrels safety bevel-undercut and threaded for standard filters; and 100 % of the eyepieces are quality checked at the Tele Vue New York facility.

Fig. 10.14 The original smooth-sided Tele Vue Plössls (© 2013 Tele Vue Optics, TeleVue.com)

For improved off-axis performance compared to typical Plössls, this line uses an unconventional patented design with high index glasses and concave lens surfaces instead of the standard flat surfaces. Other Plössl lines that use concave lens surfaces include: Astro-Tech High Grade Plössl, Meade 4000 Plössl (current four-element version), and the Smart Astronomy Sterling Plössl.

This eyepiece line is within the following competition class of eyepieces with similar specifications and performance: Antares Plössl, Astro-Professional Plössl, Astro-Tech High Grade Plössl, Astro-Tech Value Line Plössl, Bresser 52° Super Plössl, Carton Plössl, Celestron Omni, Celestron Silvertop Plössl, Clavé Plössl, Coronado CeMax, Edmund Scientific Plössl, Garrett Optical Plössl, GSO Plössl, GTO Plössl, Long Perng Plössl, Meade Series 3000 Plössl, Meade Series 4000 Super Plössl, Olivon Plössl, Opt Plössl, Orion HighLight Plössl, Orion Sirius Plössl, Owl Black Night Plössl, Sky-Watcher SP-Series Super Plössl, Smart Astronomy Sterling Plössl, TAL—Symmetrical Super Plössl, Telescope Service Plössl, Telescope Service Super Plössl, Tele Vue Plössl, Vixen NPL.

Tele Vue: Radian (*Partially Discontinued*)

Fig. 10.15 The Tele Vue Radians in 5 and 18 mm (Image courtesy of Tamiji Homma, Newbury Park, CA, USA)

Focal length (mm)	AFOV (degrees)	ER (mm)	Field stop (mm)	Weight (oz)	Design (Elem/Grp)	Coatings	Barrel (inches)
3	60	20	3.3	12.3	7/5	FMC	1.25
4	60	20	4.2	12.1	7/5	FMC	1.25
5	60	20	5.3	12.8	7/5	FMC	1.25
6	60	20	6.3	12.6	7/5	FMC	1.25
8	60	20	8.3	9.6	6/4	FMC	1.25
10	60	20	10.5	8.7	6/4	FMC	1.25
12	60	20	12.6	8.5	6/4	FMC	1.25
14	60	20	14.4	8.7	6/4	FMC	1.25
18	60	20	18.3	7.8	6/4	FMC	1.25

Source: www.televue.com

As of late 2012, all focal lengths were discontinued except for the 3 and 4 mm. These utilize exotic glasses; have a click-stop, Instadjust eyeguard for fast, accurate positioning of the eye to the exit-pupil with maximum suppression of environmental stray light; and have freedom from distortion for an orthoscopic field of view. Tele Vue manufactures all their eyepiece lines with the following standard features: fine polished glass lenses; fully multi-coated glass matched multi-coatings; blackened lens edges; anti-reflection surfaces throughout; baffles to reduce grazing incidence reflections; barrels safety bevel-undercut and threaded for standard filters; and 100 % of the eyepieces are quality checked at the Tele Vue New York facility. This eyepiece line is within the following competition class of eyepieces with similar specifications and performance: Astro-Tech Paradigm Dual ED, HR Planetary, Meade HD 60, Pentax XF, Telescope Service HR Planetary, TMB Planetary.

Tele Vue: Wide Field (*Discontinued*)

Fig. 10.16 The Tele Vue second generation Wide Fields with rubber grip panels (© 2013 Tele Vue Optics, TeleVue.com)

Focal length (mm)	AFOV (degrees)	ER (mm)	Field stop (mm)	Weight (oz)	Design (Elem/Grp)	Coatings	Barrel (inches)
15	65	9	15.4	4.8	6/4	FMC	1.25
19	65	11	19.6	4.8	6/4	FMC	1.25
24	65	14	24.7	8	6/4	FMC	1.25
32	65	19	33.0	17.6	6/4	FMC	2
40	65	24	41.2	32	6/4	FMC	2

Source: www.televue.com

This eyepiece line was the predecessor to the Tele Vue Panoptic and various focal lengths were introduced between 1982 and 1984. Observer reports indicate these provide an engaging view for the observer, but to expect levels of rectilinear and/or angular magnification distortions off-axis that will result in the Moon and planets going "out of round" into more of an egg or football shape when positioned close to the field stop. Tele Vue manufactures all their eyepiece lines with the following standard features: fine polished glass lenses; fully multi-coated glass matched multi-coatings; blackened lens edges; anti-reflection surfaces throughout; baffles to reduce grazing incidence reflections; barrels safety bevel-undercut and threaded for standard filters; and 100 % of the eyepieces are quality checked at the Tele Vue New York facility.

TMB: 100

Focal length (mm)	AFOV (degrees)	ER (mm)	Field stop (mm)	Weight (oz)	Design (Elem/Grp)	Coatings	Barrel (inches)
9	100	12	15.7^{E}	14.4	8/5	FMC	1.25 or 2
16	100	12	27.9^{E}	14.4	8/5	FMC	2

Source: www.tmboptical.com

These are lightweight for easy telescope balance; the 9 mm has a convertible 1.25 in.–2 in. barrel. Observers report the 9 mm focal length as having a fairly good off-axis; however reports for the 16 mm focal length indicate a longer focal ratio telescope is needed for best off-axis performance. This eyepiece line is within the following competition class of eyepieces with similar specifications and performance: Agena Mega Wide Angle, Orion GiantView, and Zhumell Z100.

TMB: Aspheric Ortho (*Discontinued*)

Fig. 10.17 The Tele Vue Radians in 5 and 18 mm (Image courtesy of Peter Sursock, Melbourne, Australia)

Focal length (mm)	AFOV (degrees)	ER (mm)	Field stop (mm)	Weight (oz)	Design (Elem/Grp)	Coatings	Barrel (inches)
25	54	20	24	–	3/1	FMC	1.25

Source: www.tmboptical.com

This eyepiece line is within the following competition class of eyepieces with similar specifications and performance: Astro-Physics Super Planetary AP-SPL, Pentax SMC Ortho, Pentax XO, Pentax XP, TMB Supermonocentric, Zeiss CZJ Ortho, Zeiss ZAO I/ZAO II.

TMB: Paragon (*Discontinued*)

Fig. 10.18 TMB 40 and 30 mm eyepieces (2 in.) (Image courtesy of Tamiji Homma, Newbury Park, CA, USA)

Focal length (mm)	AFOV (degrees)	ER (mm)	Field stop (mm)	Weight (oz)	Design (Elem/Grp)	Coatings	Barrel (inches)
30	69	16	36	10	6/4	FMC	2
40	69	16	46	17	6/4	FMC	2

Source: www.tmboptical.com

These were designed by Thomas Back of TMB Optical; the coatings used have 99.98–99.99 % transmission efficiency; and the lettering/graphics are red-light reflective. The 40 mm is the best performing member of the line, offering performance indicative of much more expensive wide fields. Offers a very good price-performance ratio. This eyepiece line is within the following competition class of eyepieces with similar specifications and performance: Astro-Tech Titan II ED, Sky-Watcher Aero.

TMB: Planetary II

Fig. 10.19 A 6 mm TMB Planetary II (Image by the author)

Focal length (mm)	AFOV (degrees)	ER (mm)	Field stop (mm)	Weight (oz)	Design (Elem/Grp)	Coatings	Barrel (inches)
2.5	58	14	2.5[E]	5	6	FMC	1.25
3.2	58	12	3.2[E]	5	6	FMC	1.25
4	58	10	4.0[E]	5	6	FMC	1.25
5	58	12	5.1[E]	5	6	FMC	1.25
6	58	12	6.1[E]	5	6	FMC	1.25
7	58	12	7.1[E]	5	6	FMC	1.25
8	58	14	8.1[E]	5	6	FMC	1.25
9	58	12	9.1[E]	5	6	FMC	1.25

Source: www.tmboptical.com

These have a twist-up eyeguard. This eyepiece line is within the following competition class of eyepieces with similar specifications and performance: APM UWA Planetary, Burgess/TMB Planetary, Olivon Wide Angled Plössl, Owl Astronomy High Resolution Planetary, Telescope Service Planetary HR.

TMB: Supermonocentric (*Partially Discontinued*)

Fig. 10.20 TMB Supermonocentrics in 4 mm (*top*), 5 mm, 6 mm, 7 mm, and 8 mm (*bottom*) focal lengths (Image courtesy of Jim Barnett, Petaluma, CA, USA)

Focal length (mm)	AFOV (degrees)	ER (mm)	Field stop (mm)	Weight (oz)	Design (Elem/Grp)	Coatings	Barrel (inches)
4	30	3.4	2.2	<3[E]	3/1	FMC	1.25
5	30	4.3	2.7	<3[E]	3/1	FMC	1.25
6	30	5.1	3.3	<3[E]	3/1	FMC	1.25
7	30	6.0	3.7[E]	<3[E]	3/1	FMC	1.25
8	30	6.8	4.3	<3[E]	3/1	FMC	1.25
9	30	7.7	4.7[E]	<3[E]	3/1	FMC	1.25
10	30	8.5	5.3	<3[E]	3/1	FMC	1.25
12	30	10.2	6.3[E]	<3[E]	3/1	FMC	1.25
14	30	11.9	7.3[E]	<3[E]	3/1	FMC	1.25
16	30	13.6	8.4[E]	<3[E]	3/1	FMC	1.25
25	55	20	24.0	<3[E]	3/1	FMC	1.25

Source: www.tmboptical.com; www.apm-telescopes.de

These are made with Schott water white, low iron, ultra clear glass and have 99 % transmission efficiency. Considered one of the very best eyepieces for planetary observing with very high contrast. Requires longer focal ratio telescopes for the off-axis to not show field curvature. Although these eyepieces are discontinued, infrequently APM Telescopes of Germany offers new production runs of some focal lengths.

This eyepiece line is within the following competition class of eyepieces with similar specifications and performance: Astro-Physics Super Planetary AP-SPL, Pentax SMC Ortho, Pentax XO, Pentax XP, TMB Aspheric Ortho, Zeiss CZJ Ortho, Zeiss ZAO I/ZAO II.

Unitron: Kellner/Ortho/Symmetrical Achromat (*Discontinued*)

Fig. 10.21 The Unitron eyepieces (Image courtesy of Steve G. McFarland, WI, USA)

Focal length (mm)	AFOV (degrees)	ER (mm)	Field stop (mm)	Weight (oz)	Design (Elem/Grp)	Coatings	Barrel (inches)
4	42^{E}	3^{E}	2.9^{E}	$<3^{E}$	4/2-Ortho	FC	.965
6	42^{E}	5^{E}	4.4^{E}	$<3^{E}$	4/2-Ortho	FC	.965
7	42^{E}	6^{E}	5.1^{E}	$<3^{E}$	4/2-Ortho	FC	.965
9	45^{E}	6^{E}	7.1^{E}	$<3^{E}$	4/2-Symm	FC	.965
12.5	45^{E}	6^{E}	9.8^{E}	$<3^{E}$	3/2-Kellner	FC	.965
18	45^{E}	8^{E}	14.1^{E}	$<3^{E}$	3/2-Kellner	FC	.965
25	40^{E}	4^{E}	17.5^{E}	$<3^{E}$	2/2-Huygen	FC	.965
40	45^{E}	26^{E}	31.4^{E}	$<4^{E}$	5/3-Monochro	FC	1.25

The Orthos of this eyepiece line are within the following competition class of eyepieces with similar specifications and performance: Antares Ortho, Apogee Super Abbe Orthoscopic, Baader Planetarium Classic Ortho, Baader Planetarium Genuine Abbe Ortho, Cave Orthostar Orthoscopic, Celestron Ortho, Edmund Scientific Ortho, Kokusai Kohki Abbe Ortho, Kson Super Ortho, Masuyama Orthoscopic, Meade Research Grade Ortho, Meade Series II Orthoscopic, Nikon Ortho, Siebert Optics Star Splitter/Super Star Splitter, Takahashi Ortho, Telescope Service Ortho, University Optics Abbe HD Orthoscopic, University Optics Abbe Volcano Orthoscopic, University Optics O.P.S. Orthoscopic Planetary Series, University Optics Super Abbe Orthoscopic, VERNONscope Brandon.

The Kellners of this eyepiece line are within the following competition class of eyepieces with similar specifications and performance (Kellners and Reverse Kellners, and RKE—Rank-Kaspereit-Erfle): Celestron E-Lux (2 in. models only), Celestron Kellner, Criterion Kellner, Edmund Scientific RKE, GSO Kellner, Kokusai Kohki Kellner, Orion DeepView, Orion E-Series, Russell Optics (2 in. 52 and 60 mm only), Sky-Watcher Kellner, Sky-Watcher Super MA Series, Telescope Service RK, Unitron Kellner.

University Optics: 70°

Focal length (mm)	AFOV (degrees)	ER (mm)	Field stop (mm)	Weight (oz)	Design (Elem/Grp)	Coatings	Barrel (inches)
10	70	10^E	12.2^E	2.9^E	5/4	FMC	1.25
15	70	13^E	18.3^E	3.6^E	5/4	FMC	1.25
20	70	16^E	24.4^E	4.2^E	5/4	FMC	1.25
26	70	20^E	31.8^E	10.2^E	$5/3^E$	FMC	2
32	70	24^E	39.1^E	14.4^E	$5/4^E$	FMC	2
38	70	28^E	46.0^E	21.3^E	$5/3^E$	FMC	2

Source: www.universityoptics.com

This eyepiece line is within the following competition class of eyepieces with similar specifications and performance: Agena Super Wide Angle, Astro-Professional SWA, Astro-Tech Wide Field, Garrett SuperWide Angle, GSO Superview, Meade QX, Olivon 70° Wide Angle, Opt Super View, Orion Q70 Super Wide-Field, Sky-Watcher PanaView, Telescope-Service WA, William Optics SWAN.

University Optics: 80°

Focal length (mm)	AFOV (degrees)	ER (mm)	Field stop (mm)	Weight (oz)	Design (Elem/Grp)	Coatings	Barrel (inches)
11	80	9	14.2[E]	3.4[E]	5/3	FMC	1.25
15	80	18	19.6[E]	9.2[E]	6/4[E]	FMC	2
20	80	20	25.5[E]	21.3[E]	7/4[E]	FMC	2
30	80	22	41.9[E]	20.1[E]	5/3[E]	FMC	2

Source: www.universityoptics.com

The 2 in. models of this line use high index glasses. This eyepiece line is within the following competition class of eyepieces with similar specifications and performance: Agena Ultra Wide Angle, Apogee Widescan (20 and 30 mm), Astrobuffet 1RPD (30 mm), BW Optik Ultrawide (30 mm), Kokusai Kohki Widescan I/II/III (20 and 30 mm), Moonfish Ultrawide (30 mm), Olivion 80° Ultra Wide Angle, Owl Astronomy Knight Owl Ultrawide Angle, Sky-Watcher Ultra Wide Angle, Surplus Shed Wollensak (30 mm), TAL Ultra Wide Angle, University Optics Widescan I/II/III (20 and 30 mm).

University Optics: Abbe HD Orthoscopic (*Discontinued*)

Fig. 10.22 The University Optic HD Orthoscopic eyepieces (Image courtesy of James Curry, St. Louis, MO, USA)

Focal length (mm)	AFOV (degrees)	ER (mm)	Field stop (mm)	Weight (oz)	Design (Elem/Grp)	Coatings	Barrel (inches)
5	43	4	3.8[E]	3.0	4/2	FMC	1.25
6	43	4.8	4.5[E]	3.2	4/2	FMC	1.25
7	42	5.6	5.1[E]	3.2	4/2	FMC	1.25
9	42	7.2	6.6[E]	3.2	4/2	FMC	1.25
12.5	44	10	9.6[E]	3.4	4/2	FMC	1.25
18	46	14.4	14.5[E]	3.4	4/2	FMC	1.25

Source: www.universityoptics.com

Along with the Baader Planetarium Genuine Orthos, these eyepieces have been considered to offer the very best performance for the price for planetary observing. Often referred to as the poor man's Zeiss Abbe Ortho. This line uses high index glass. This eyepiece line is within the following competition class of eyepieces with similar specifications and performance: Antares Ortho, Apogee Super Abbe Orthoscopic, Baader Planetarium Classic Ortho, Baader Planetarium Genuine Abbe Ortho, Cave Orthostar Orthoscopic, Celestron Ortho, Edmund Scientific Ortho, Kokusai Kohki Abbe Ortho, Kson Super Ortho, Masuyama Orthoscopic, Meade Research Grade Ortho, Meade Series II Orthoscopic, Nikon Ortho, Siebert Optics Star Splitter/Super Star Splitter, Takahashi Ortho, Telescope Service Ortho, Unitron Ortho, University Optics Abbe Volcano Orthoscopic, University Optics O.P.S. Orthoscopic Planetary Series, University Optics Super Abbe Orthoscopic, VERNONscope Brandon.

University Optics: Abbe Volcano Orthoscopic (*Discontinued*)

Fig. 10.23 University Abbe "Volcano" Orthos in 4–25 mm focal lengths (Image courtesy of Doug Bailey, Franklinton, LA, USA)

Focal length (mm)	AFOV (degrees)	ER (mm)	Field stop (mm)	Weight (oz)	Design (Elem/Grp)	Coatings	Barrel (inches)
4	41	3.5	2.9^{E}	2.8	4/2	FC	1.25
5	43	4	3.8^{E}	3.0	4/2	FC	1.25
6	43	4.8	4.5^{E}	3.2	4/2	FC	1.25
7	42	5.6	5.1^{E}	3.2	4/2	FC	1.25
9	42	7.2	6.6^{E}	3.2	4/2	FC	1.25
12.5	44	10	9.6^{E}	3.4	4/2	FC	1.25
18	46	14.4	14.5^{E}	3.4	4/2	FC	1.25
25	47	20	20.5^{E}	3.5	4/2	FC	1.25

Source: www.universityoptics.com

These "volcano" Orthos have a long-standing reputation as being the standard for a workhorse planetary eyepiece. The volcano design of the housing greatly aids in making the eye relief feel less tight for the shorter focal length eyepieces. Highly recommended by almost all observers.

This eyepiece line is within the following competition class of eyepieces with similar specifications and performance: Antares Ortho, Apogee Super Abbe Orthoscopic, Baader Planetarium Classic Ortho, Baader Planetarium Genuine Abbe Ortho, Cave Orthostar Orthoscopic, Celestron Ortho, Edmund Scientific Ortho, Kokusai Kohki Abbe Ortho, Kson Super Ortho, Masuyama Orthoscopic, Meade Research Grade Ortho, Meade Series II Orthoscopic, Nikon Ortho, Siebert Optics Star Splitter/Super Star Splitter, Takahashi Ortho, Telescope Service Ortho, Unitron Ortho, University Optics Abbe HD Orthoscopic, University Optics O.P.S. Orthoscopic Planetary Series, University Optics Super Abbe Orthoscopic, VERNONscope Brandon.

University Optics: König/König II/MK-70/MK-80 (*Discontinued*)

Fig. 10.24 University Optics Königs in 32 mm, 24 mm, 16 mm, and 12 mm focal lengths (Image courtesy of Jamie Crona, Fanwood, NJ, USA)

Focal length (mm)	AFOV (degrees)	ER (mm)	Field stop (mm)	Weight (oz)	Design (Elem/Grp)	Coatings	Barrel (inches)
6.5	60	8	6.8[E]	–	4	MC	1.25
8	60	–	8.4[E]	–	4	MC	1.25
12	60	7[E]	9.9	–	4	MC	1.25
16	68	6[E]	19.0[E]	–	4	MC	1.25
24	60	–	25.1[E]	–	5	MC	1.25
32	52	–	29.0[E]	–	4	MC	1.25
32	60	–	33.5[E]	–	4	MC/FMC	2
40	60	–	41.9[E]	–	5	MC	2
25-MK70	70	–	29	13.2	7	MC	2
32-MK80	80	–	46	16.4	6	FMC	2
40-MK70	70	–	45	18.4	7	FMC	2

Source: www.universityoptics.com

The manufacturer's statistics on eye relief and weight are no longer available. The shorter focal lengths have a reputation as being very good for planetary observing, although the eye relief is reported to feel shorter than the stated eye relief. The 32 mm in the 1.25 in. barrel provides the widest TFOV of any 1.25 in. barreled eyepiece.

Fig. 10.25 Vintage University Optics "Flat Top" Königs (Image courtesy of William Rose, Larkspur, CO, USA)

Fig. 10.26 The University Optics 32 mm König (Image courtesy of Andy Howie, Paisley, Scotland)

University Optics: O.P.S. Orthoscopic Planetary Series (*Discontinued*)

Focal length (mm)	AFOV (degrees)	ER (mm)	Field stop (mm)	Weight (oz)	Design (Elem/Grp)	Coatings	Barrel (inches)
9	42	7.2	6.6[E]	3.2	4/2	Special	1.25
12.5	44	10	9.6[E]	3.4	4/2	Special	1.25
18	46	14.4	14.5[E]	3.4	4/2	Special	1.25

Source: www.universityoptics.com

This eyepiece line was a joint venture between UO, ITE, and Sirius Optics; uses a special 20-layer band-pass filter to further enhance the subtle planetary details (e.g., the filter has been incorporated into the ocular's lenses). Although an excellent idea the line was short-lived, apparently because it limited the usefulness of the eyepieces for other observing. This eyepiece line is within the following competition class of eyepieces with similar specifications and performance: Antares Ortho, Apogee Super Abbe Orthoscopic, Baader Planetarium Classic Ortho, Baader Planetarium Genuine Abbe Ortho, Cave Orthostar Orthoscopic, Celestron Ortho, Edmund Scientific Ortho, Kokusai Kohki Abbe Ortho, Kson Super Ortho, Masuyama Orthoscopic, Meade

Research Grade Ortho, Meade Series II Orthoscopic, Nikon Ortho, Siebert Optics Star Splitter/Super Star Splitter, Takahashi Ortho, Telescope Service Ortho, Unitron Ortho, University Optics Abbe HD Orthoscopic, University Optics Abbe Volcano Orthoscopic, University Optics Super Abbe Orthoscopic, VERNONscope Brandon.

University Optics: Super Abbe Orthoscopic

Focal length (mm)	AFOV (degrees)	ER (mm)	Field stop (mm)	Weight (oz)	Design (Elem/Grp)	Coatings	Barrel (inches)
4.8	46	4[E]	3.9[E]	<3[E]	4/2	FMC	1.25
7.7	46	6[E]	6.5	<3[E]	4/2	FMC	1.25
10.5	46	9[E]	8.4[E]	<3[E]	4/2	FMC	1.25
16.8	46	14[E]	13.5[E]	<4[E]	4/2	FMC	1.25
24	46	20[E]	20.5	<4[E]	4/2	FMC	1.25

Source: www.universityoptics.com

This eyepiece line is within the following competition class of eyepieces with similar specifications and performance: Antares Ortho, Apogee Super Abbe Orthoscopic, Baader Planetarium Classic Ortho, Baader Planetarium Genuine Abbe Ortho, Cave Orthostar Orthoscopic, Celestron Ortho, Edmund Scientific Ortho, Kokusai Kohki Abbe Ortho, Kson Super Ortho, Masuyama Orthoscopic, Meade Research Grade Ortho, Meade Series II Orthoscopic, Nikon Ortho, Siebert Optics Star Splitter/Super Star Splitter, Takahashi Ortho, Telescope Service Ortho, Unitron Ortho, University Optics Abbe HD Orthoscopic, University Optics Abbe Volcano Orthoscopic, University Optics O.P.S. Orthoscopic Planetary Series, VERNONscope Brandon.

University Optics: Super Erfle (*Discontinued*)

Fig. 10.27 University Optics 16 mm Super Erfle (Image courtesy of Malcolm Neo, Singapore)

Focal length (mm)	AFOV (degrees)	ER (mm)	Field stop (mm)	Weight (oz)	Design (Elem/Grp)	Coatings	Barrel (inches)
16	65	14–17	17[E]	6.2	5/3	MC	1.25
20	65	14–17	22[E]	6.2	5/3	MC	1.25
25	60	14–17	26[E]	6.2	5/3	MC	1.25

Source: www.universityoptics.com

This eyepiece line is within the following competition class of eyepieces with similar specifications and performance: Celestron Erfle, Kokusai Kohki Erfle.

University Optics: Widescan II/III (*Discontinued*)

Fig. 10.28 University Widescan eyepieces in 8 mm, 10 mm, 13 mm, 16 mm, 20 mm, 25 mm, and 32 mm (Image courtesy of Doug Richter, Fort Atkinson, WI, USA)

Focal length (mm)	AFOV (degrees)	ER (mm)	Field stop (mm)	Weight (oz)	Design (Elem/Grp)	Coatings	Barrel (inches)
20	80	20	25.5^E	21.3^E	$7/4^E$	FMC	2
30	80	22	41.9^E	20.1^E	$5/3^E$	FMC	2

Source: www.universityoptics.com

The 2 in. models of this line use high index glass. This eyepiece line is within the following competition class of eyepieces with similar specifications and performance: Agena Ultra Wide Angle, Apogee Widescan (20 and 30 mm), Astrobuffet 1RPD (30 mm), BW Optik Ultrawide (30 mm), Kokusai Kohki Widescan I/II/III (20 and 30 mm), Moonfish Ultrawide (30 mm), Olivion 80° Ultra Wide Angle, Owl Astronomy Knight Owl Ultrawide Angle, Sky-Watcher Ultra Wide Angle, Surplus Shed Wollensak (30 mm), TAL Ultra Wide Angle, University Optics 80°.

VERNONscope: Brandon

Fig. 10.29 VERNONscope Brandon eyepieces (Image courtesy of John W., MA, USA)

Focal length (mm)	AFOV (degrees)	ER (mm)	Field stop (mm)	Weight (oz)	Design (Elem/Grp)	Coatings	Barrel (inches)
4	40^E	3^E	2.8^E	1.5^E	4/2	FC	1.25
6	40	5^E	4.1^E	1.5	4/2	FC	1.25
8	42	6^E	5.7^E	2	4/2	FC	1.25
12	42	9^E	8.5^E	2	4/2	FC	1.25
16	42	12^E	11.4^E	3	4/2	FC	1.25
24	53	18^E	21.4^E	4	4/2	FC	1.25
32	48	24^E	25.9^E	4	4/2	FC	1.25
48	46	36^E	37.3^E	10.4	4/2	FC	2

Source: www.vernonscope.com

Brandon oculars incorporate the original optical designs of Chester Brandon; are recommended for telescope focal ratios of f/7 and longer; and are parfocal. The Brandons have an extremely loyal following of users. Their optics are rumored to be polished to a higher level than conventional commercial optics. When compared directly to other very finely polished optics, such as the Zeiss Abbe Orthos, some observers report their superior polish is also evident as showing exceedingly pure black backgrounds, minimal scatter, and excellent color rendition. A highly regarded eyepiece line for planetary observation. The 6 mm though 32 mm focal lengths consistently receive positive observer reports. One of the few eyepiece lines that advertise themselves as being "Made in the USA."

This eyepiece line generally competes with the following other eyepiece lines, and by many observers is considered the top performer of these other lines for planetary observing: Antares Ortho, Apogee Super Abbe Orthoscopic, Baader Planetarium Classic Ortho, Baader Planetarium Genuine Abbe Ortho, Cave Orthostar Orthoscopic, Celestron Ortho, Edmund Scientific Ortho, Kokusai Kohki Abbe Ortho, Kson Super Ortho, Masuyama Orthoscopic, Meade Research Grade Ortho, Meade Series II Orthoscopic, Nikon Ortho, Siebert Optics Star Splitter/ Super Star Splitter, Takahashi Ortho, Telescope Service Ortho, Unitron Ortho, University Optics Abbe HD Orthoscopic, University Optics Abbe Volcano Orthoscopic, University Optics O.P.S. Orthoscopic Planetary Series, University Optics Super Abbe Orthoscopic.

Fig. 10.30 An 8 mm Brandon with eyeguard removed (Image by the author)

Fig. 10.31 Example of older (*left*) vs. newer (*right*) coatings used on Brandons (Image by the author)

Vixen: Lanthanum (LV) (*Discontinued*)

Fig. 10.32 The complete Vixen LV series (© 2013 Vixen Optics—www.vixenoptics.com/eye.htm)

Focal length (mm)	AFOV (degrees)	ER (mm)	Field stop (mm)	Weight (oz)	Design (Elem/Grp)	Coatings	Barrel (inches)
2.5	45	20	2.0[E]	5.6	8/4	FMC	1.25
4	45	20	3.1[E]	5.4	7/4	FMC	1.25
5	45	20	3.9[E]	5.3	7/4	FMC	1.25
6	45	20	4.7[E]	5.3	7/4	FMC	1.25
7	45	20	5.5[E]	5.2	7/4	FMC	1.25
9	50	20	7.9[E]	5.6	7/4	FMC	1.25
10	50	20	8.7[E]	5.6	7/4	FMC	1.25
12	50	20	10.5[E]	5.3	7/4	FMC	1.25
15	50	20	13.1[E]	5.0	7/4	FMC	1.25
18	50	20	15.7[E]	4.9	7/4	FMC	1.25
20	50	20	17.5[E]	5.0	6/4	FMC	1.25
25	50	20	21.8[E]	5.3	5/3	FMC	1.25
40	42	32	27.0[E]	4.9	4/2	FMC	1.25
30	50–60	20	26–31[E]	15.5	5/3	FMC	2
50	45	22	39.3[E]	17.6	5/3	FMC	2

Source: www.vixenoptics.com

These incorporate a Lanthanum (rare-Earth glass) field lens to provide 20 mm eye relief; optical performance is purported to approach that of a Plössl; and the short focal length LVs use a built-in Barlow to achieve high power without reducing eye relief. This line was within the following competition class of eyepieces with similar specifications and performance: Agena ED, Celestron X-Cel, Orion Epic ED II (older version), Vixen NLV.

Vixen: Lanthanum Superwide (LVW)

Fig. 10.33 The Vixen LVW series of eyepieces (Eyepieces courtesy of www.handsonoptics.com. Image by the author.)

Focal length (mm)	AFOV (degrees)	ER (mm)	Field stop (mm)	Weight (oz)	Design (Elem/Grp)	Coatings	Barrel (inches)
3.5	65	20	4.0[E]	16.8	8/5	FMC	1.25/2
5	65	20	5.7[E]	16.4	8/5	FMC	1.25/2
8	65	20	9.1[E]	17.4	8/5	FMC	1.25/2
13	65	20	14.7[E]	16.2	8/5	FMC	1.25/2
17	65	20	19.3[E]	15.5	8/5	FMC	1.25/2
22	65	20	25.0[E]	15.2	8/5	FMC	1.25/2
30	65	20	34.0[E]	12.8	8/5	FMC	2
42	72	20	46.6	19.2	8/5	FMC	2

Source: www.vixenoptics.com

The design of this line uses several elements made from rare-Earth Lanthanum glass; and has integrated dual 1.25/2 in. barrels (except on 30 and 42 mm units). This eyepiece line is within the following competition class of eyepieces with similar specifications and performance: Baader Planetarium Hyperion, Orion Lanthanum Superwide, Orion Stratus.

Vixen: NLV

Fig. 10.34 The complete Vixen NLV series (© 2013 Vixen Optics—www.vixenoptics.com/eye.htm)

Focal length (mm)	AFOV (degrees)	ER (mm)	Field stop (mm)	Weight (oz)	Design (Elem/Grp)	Coatings	Barrel (inches)
2.5	45	20	2.0^E	5.5	8/4	FMC	1.25
4	45	20	3.1^E	5.4	7/4	FMC	1.25
5	45	20	3.9^E	5.3	7/4	FMC	1.25
6	45	20	4.7^E	5.3	7/4	FMC	1.25
9	50	20	7.9^E	5.6	7/4	FMC	1.25
10	50	20	8.7^E	5.4	7/4	FMC	1.25
12	50	20	10.5^E	5.4	7/4	FMC	1.25
15	50	20	13.1^E	4.9	7/4	FMC	1.25
20	50	20	17.5^E	4.6	6/4	FMC	1.25
25	50	20	21.8^E	4.5	5/3	FMC	1.25
40	42	32	29.3^E	4.9	4/2	FMC	1.25
50	42	38	36.7^E	14.8	–	FMC	2

Source: www.vixenoptics.com

This design uses several elements made from rare-Earth Lanthanum glass, has twist-up rubberized eyeguard, and is color-coded band to help distinguish focal lengths.

Vixen: NPL

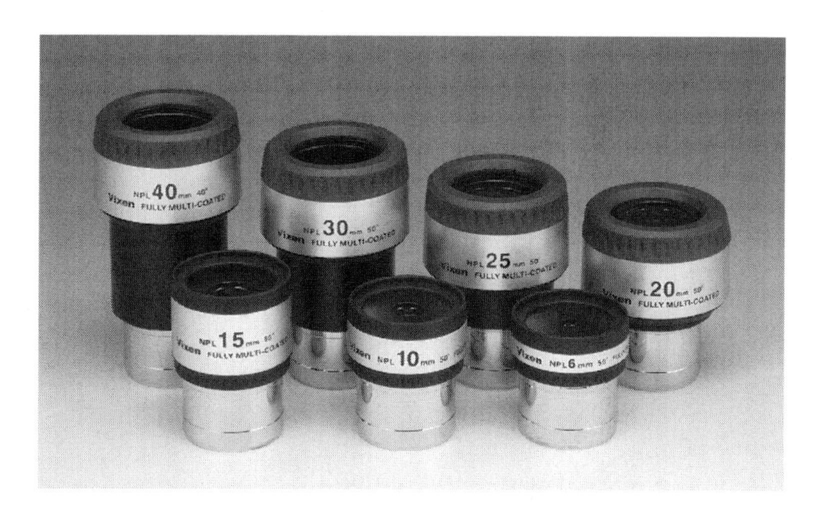

Fig. 10.35 The complete Vixen NPL series (© 2013 Vixen Optics—www.vixenoptics.com/eye.htm)

Focal length (mm)	AFOV (degrees)	ER (mm)	Field stop (mm)	Weight (oz)	Design (Elem/Grp)	Coatings	Barrel (inches)
6	50	3	5.2^E	2.5	4/2	FMC	1.25
10	50	6.5	8.7^E	2.8	4/2	FMC	1.25
15	50	11	13.1^E	3.5	4/2	FMC	1.25
20	50	15	17.5^E	3.9	4/2	FMC	1.25
25	50	19.5	21.8^E	4.6	4/2	FMC	1.25
30	50	24	26.2^E	4.2	4/2	FMC	1.25
40	40	36	27.0^E	4.3	4/2	FMC	1.25

Source: www.vixenoptics.com

This eyepiece line is within the following competition class of eyepieces with similar specifications and performance: Antares Plössl, Astro-Professional Plössl, Astro-Tech High Grade Plössl, Astro-Tech Value Line Plössl, Bresser 52° Super Plössl, Carton Plössl, Celestron Omni, Celestron Silvertop Plössl, Clavé Plössl, Coronado CeMax, Edmund Scientific Plössl, Garrett Optical Plössl, GSO Plössl, GTO Plössl, Long Perng Plössl, Meade Series 3000 Plössl, Meade Series 4000

Super Plössl, Olivon Plössl, Opt Plössl, Orion HighLight Plössl, Orion Sirius Plössl, Owl Black Night Plössl, Parks Silver Series, Sky-Watcher SP-Series Super Plössl, Smart Astronomy Sterling Plössl, TAL—Symmetrical Super Plössl, Telescope Service Super Plössl, Tele Vue Plössl, Vixen NPL.

William Optics: SPL (Super Planetary Long Eye Relief)

Focal length (mm)	AFOV (degrees)	ER (mm)	Field stop (mm)	Weight (oz)	Design (Elem/Grp)	Coatings	Barrel (inches)
3	55	20	2.9[E]	7.1	7/4	FMC	1.25
6	55	20	5.8[E]	5.3	7/4	FMC	1.25
12.5	55	20	12.0[E]	5.3	7/4	FMC	1.25

Source: www.williamoptics.com

This eyepiece line is within the following competition class of eyepieces with similar specifications and performance: Astro-Professional Long Eye Relief Planetary, Astro-Tech Long Eye Relief, Long Perng Long Eye Relief, Orion Edge-On Planetary, Smart Astronomy SA Solar System Long Eye Relief, Zhumell Z Series.

William Optics: SWAN

Fig. 10.36 The 40 mm SWAN eyepiece (Image courtesy of Eric Sheperd, USA)

Focal length (mm)	AFOV (degrees)	ER (mm)	Field stop (mm)	Weight (oz)	Design (Elem/Grp)	Coatings	Barrel (inches)
9	72	12	11.3[E]	70	5/4	FMC	1.25
15	72	14	18.8[E]	90	5/4	FMC	1.25
20	72	17	25.1[E]	100	5/4	FMC	1.25
25	72	21	31.4[E]	330	5/3	FMC	2
33	72	25	41.5[E]	410	5/4	FMC	2
40	72	28	50.3[E]	570	5/3	FMC	2

Source: www.williamoptics.com

These are parfocal. This eyepiece line is within the following competition class of eyepieces with similar specifications and performance: Agena Super Wide Angle, Astro-Professional SWA, Astro-Tech Wide Field, Garrett SuperWide Angle, GSO Superview, Meade QX, Olivon 70° Wide Angle, Opt Super View, Orion Q70 Super Wide-Field, Sky-Watcher PanaView, Telescope-Service WA, University Optics 70°.

William Optics: WA 66°

Fig. 10.37 Binoviewer pair of 20 mm Williams Optics wide angle eyepieces (Image by the author)

Focal length (mm)	AFOV (degrees)	ER (mm)	Field stop (mm)	Weight (oz)	Design (Elem/Grp)	Coatings	Barrel (inches)
20	66	–	–	–	–	FMC	1.25

Source: www.williamoptics.com

Sold bundled with the William Optics binoviewer package. Although similar in optical design to the Orion Expanse and Owl Enhanced Superwide, performs better off-axis and provides a brighter image.

This eyepiece line is within the following competition class of eyepieces with similar specifications and performance: Agena Enhanced Wide Angle, Orion Expanse, Owl Enhanced Superwide, Sky-Watcher Ultra Wide Angle, and Telescope Service SWM Wide Angle.

Williams Optics: UWAN

Fig. 10.38 William Optics 28 mm UWAN (Image courtesy of Andy Howie, Paisley, Scotland)

Focal length (mm)	AFOV (degrees)	ER (mm)	Field stop (mm)	Weight (oz)	Design (Elem/Grp)	Coatings	Barrel (inches)
4	82	12	6.0	200	7/4	FMC	1.25
7	82	12	12.0	200	7/4	FMC	1.25
16	82	12	28.6	200	7/4	FMC	1.25
28	82	18	43.5	1000	6/4	FMC	2

Source: www.williamoptics.com

These have twist-up eyeguard and are parfocal. Overall an excellent performing 82 ultra wide-field eyepiece line. Often compared to the standard-setting Tele Vue Naglers. Some observers feel the 16 mm UWAN is even more engaging than the Nagler 16 mm T5. The 28 mm UWAN often competes with observers as a candidate for a 2 in. eyepiece that can produce the maximum TFOV possible with the smallest exit pupil for richest dark background fields of view.

This eyepiece line is within the following competition class of eyepieces with similar specifications and performance: Antares Speers-WALER Series, Astro-Professional UWA, Celestron Axiom LX, Celestron Luminos, Docter UWA, Explore Scientific 82° Nitrogen Purged, Meade Series 4000 Ultra Wide Angle, Meade Series 5000 Ultra Wide Angle, Orion MegaView Ultra-Wide, Sky-Watcher Nirvana UWA, Sky-Watcher Sky Panorama, and Tele Vue Nagler.

Zeiss: CZJ Ortho (*Discontinued*)

Fig. 10.39 Carl Zeiss Jena eyepiece collection (.965 in. barrels) (Image courtesy of Alexander Kupco, Ricany, Czech Republic)

Focal length (mm)	AFOV (degrees)	ER (mm)	Field stop (mm)	Weight (oz)	Design (Elem/Grp)	Coatings	Barrel (inches)
4	40[E]	3[E]	2.7[E]	<3[E]	4/2	FC[E]	.965
6	40[E]	5[E]	4.1[E]	<3[E]	4/2	FC[E]	.965
8	40[E]	6[E]	5.4[E]	<3[E]	4/2	FC[E]	.965
10	40[E]	8[E]	6.8[E]	<3[E]	4/2	FC[E]	.965
12.5	40[E]	10[E]	8.5[E]	<3[E]	4/2	FC[E]	.965
16	40[E]	13[E]	10.9[E]	<4[E]	4/2	FC[E]	.965
25	40[E]	20[E]	17.0[E]	<4[E]	4/2	FC[E]	.965
40	40[E]	32[E]	27.1[E]	–	4/2	FC[E]	.965

Source: http://sportsoptics.zeiss.com/worldwide/en_us/home.html

Excellent planetary eyepieces. The CZJ annotation refers to Carl Zeiss Jena and the Zeiss production facility in Jena, Germany. This eyepiece line is within the following competition class of eyepieces with similar specifications and performance: Astro-Physics Super Planetary AP-SPL, Pentax SMC Ortho, Pentax XO, Pentax XP, TMB Aspheric Ortho, TMB Supermonocentric, Zeiss ZAO I/ZAO II.

Zeiss: ZAO I/ZAO II (*Discontinued*)

Fig. 10.40 The ZAO-I collection with walnut presentation case (Image courtesy of William Rose, Larkspur, CO, USA)

Focal length (mm)	AFOV (degrees)	ER (mm)	Field stop (mm)	Weight (oz)	Design (Elem/Grp)	Coatings	Barrel (inches)
4 (I)	45	2[a]	3.3	2.1[M]	4/2	FMC	1.25
6 (I)	45	3.7[a]	5.0	2.2[M]	4/2	FMC	1.25
10 (I)	45	5[a]	8.3	2.5[M]	4/2	FMC	1.25
16 (I)	45	4[a]	13.2	3.2[M]	4/2	FMC	1.25
25 (I)	45	5[a]	20.7	3.8[M]	4/2	FMC	1.25
34 (I)	40	5[a]	24.3	3.6[M]	4/2	FMC	1.25
4 (II)	43	3.1[E]	3.0[E]	2[E]	4/2	FMC	1.25
6 (II)	43	4.7[E]	4.5[E]	2[E]	4/2	FMC	1.25
10 (II)	43	7.8[E]	7.5[E]	3[E]	4/2	FMC	1.25
16 (II)	43	12.5[E]	12.0[E]	3[E]	4/2	FMC	1.25

Source: www.baader-planetarium.com
[a]Indicates "usable" eye relief from top of eyepiece housing as reported by Zeiss

ZAO-I

Uses high-index Lanthanum glass, well corrected on-axis to f/8, very high transmittance glass types, and is multi-coated with the Zeiss patented "T" multi-layer coatings for >97 % peak transmission and color purity.

ZAO-II

Well corrected on-axis to f/4 for improved performance beyond the original ZAO-I; uses a new lens design and improved glasses; uses the most transparent and colorless glasses that can be produced; flat field free of distortion (orthoscopic); produced entirely at Zeiss, in Jena; field stop contains innovative micrometric field marks of four opposing see-through triangles located outside the field of view edge for more accurate centering and field size drift measurements.

Although both these eyepieces were manufactured by Zeiss, they were commissioned by and made available to the market through Baader Planetarium. Considered the ultimate planetary eyepiece by most observers and is considered the reference standard, providing the classic "diamonds on velvet" views when used for star field and star cluster observing. The 34 mm focal length is extremely rare and commands a very high price on the used market. Both of these eyepiece lines are within the following competition class of eyepieces with similar specifications and performance: Astro-Physics Super Planetary AP-SPL, Pentax SMC Ortho, Pentax XO, Pentax XP, TMB Aspheric Ortho, TMB Supermonocentric, Zeiss CZJ Ortho.

Zhumell: Z100

Focal length (mm)	AFOV (degrees)	ER (mm)	Field stop (mm)	Weight (oz)	Design (Elem/Grp)	Coatings	Barrel (inches)
9	100	16	15.7^{E}	14^{E}	$8/5^{E}$	FMC	1.25 or 2
16	100	16	27.0^{E}	14^{E}	$8/5^{E}$	FMC	2

Source: www.zhumell.com

These come included with 1.25 in. and 2 in. adaptors (the 2 in. barrel is threaded for standard filters). Observers report the 9 mm focal length as having a fairly good off-axis; however reports for the 16 mm focal length indicate a longer focal ratio telescope is needed for best off-axis performance. This eyepiece line is within the following competition class of eyepieces with similar specifications and performance: Agena Mega Wide Angle, Orion GiantView, TMB 100.

Zhumell: Z Series Planetary

Fig. 10.41 Zhumell Z-Series eyepiece (Eyepieces courtesy of David Ittner, Newark, CA, USA)

Focal length (mm)	AFOV (degrees)	ER (mm)	Field stop (mm)	Weight (oz)	Design (Elem/Grp)	Coatings	Barrel (inches)
3	55	20	2.9^{E}	10	7/4	FMC	1.25
5	55	20	4.8^{E}	10	7/4	FMC	1.25
6	55	20	5.8^{E}	10	7/4	FMC	1.25
9	55	20	8.6^{E}	9	7/4	FMC	1.25
12.5	55	20	12.0^{E}	8	7/4	FMC	1.25
14.5	55	20	13.9^{E}	7	7/4	FMC	1.25
18	55	20	17.3^{E}	6	5/3	FMC	1.25

Source: www.zhumell.com

This eyepiece line is within the following competition class of eyepieces with similar specifications and performance: Astro-Professional Long Eye Relief Planetary, Astro-Tech Long Eye Relief, Long Perng Long Eye Relief, Orion Edge-On Planetary, Smart Astronomy SA Solar System Long Eye Relief, Stellarvue Planetary, William Optics SPL.

Zooming Eyepieces

Fig. 10.42 A variety of zoom eyepieces (Image courtesy of William Rose, Larkspur, CO, USA)

Zoom eyepieces are a highly specialized and unique type of astronomical eyepiece. These eyepieces offer the observer an extreme amount of flexibility, as a single zoom can perform the function three, four, or even five fixed length eyepieces. And when a zoom eyepiece is coupled with a Barlow, then these two items can serve as all that is needed for a complete range of eyepiece focal lengths.

The primary limitation of zoom eyepieces has been their inability to show a constant AFOV at their different focal length settings. As can be seen from the chart below, there are only two zoom eyepieces that have managed to maintain a constant AFOV while zooming, the 2–4 mm and 3–6 mm Tele Vue Nagler Zooms (the first production eyepiece to accomplish this), and the 7.9–15.8 mm Meopta Zoom available through APM Germany. The Nagler Zooms both have a very well established reputation, with many observers using them as planetary eyepieces, as they have only five optical elements and maintain a constant 10 mm of eye relief as well, even when they are set at the extremely short 2 mm focal length setting. The Meopta zoom is a relative newcomer, first being offered in late 2012. It is actually an eyepiece made for a spotting 'scope that has adapters attached to fit into standard 1.25 in. and 2 in. astronomical focusers.

At the premium end of the scale, there is one zoom which possesses a rather unparalleled reputation, the Leica Vario 25×–50× Aspheric zoom (e.g., 17.8–8.9 mm focal lengths). This zoom is also designed for spotting scopes and have conversion units available for use in standard 1.25 in. and 2 in. astronomical focusers. This zoom is unique, however, in that it has gained a reputation as being a very capable planetary eyepiece, even keeping pace with some of the top-tier fixed focal length planetary eyepieces available, such as the Astro-Physics SPLs and Pentax SMC Orthos.

Fig. 10.43 The Leica Vario Zoom pictured with the Baader Zeiss 2× Barlow and Pentax 10 mm XW (Leica Zoom courtesy of Larry Eastwood, Nashville, TN, USA. Image by the author)

As reported by the author in a 2012 review of this zoom for planetary observing:

> *... the Leica Vario Aspheric Zoom with the Baader Zeiss 2x Barlow provides a view almost as good as some of the more exotic classic planetary eyepieces. This testing demonstrated that it was a real battle-royal between the Leica Vario Aspheric Zoom and what I characterize as a Tier-2 premium classic planetary, the Astro-Physics Super Planetary eyepieces (AP-SPLs). Some evenings the Leica Vario Aspheric Zoom very slightly bested this top performing planetary, and on other evenings only very slightly underperforming the AP-SPL. However, when compared to the most highly renowned classic planetary eyepieces, the Zeiss Abbe Orthos (ZAO) and Pentax XOs, while the Leica Vario Aspheric Zoom could come close, it could not beat these best-in-class eyepieces.*
>
> *While it runs contrary to the conventional wisdom of trying to minimize the glass for the most critical planetary observations, this Leica Vario Aspheric Zoom challenges that wisdom and demonstrates that even though it is a complex multi-element design, it can still effectively compete with the "big boys" of classic planetary eyepieces, providing a level of comfort and flexibility that the classic minimum glass planetary eyepiece just cannot replicate....As the evenings wore on during my testing, I noted that when I finished my tests and turned to planetary observing for my own benefit, I always used just the Leica Vario Aspheric Zoom for the remainder of the evening, as it was a pure joy to use and the planetary imagery it provided was so detailed and precise....The Leica Vario Aspheric Zoom provided this planetary observer with a very satisfying experience, and it certainly demonstrated a capability for planetary far superior to other wide-fields I have tried in this role while it came close enough to the on-axis performance of best-in-class classic planetary eyepieces to make it a real planetary treat.*

Of all the available zooms on the market, though, the one that arguably has the largest loyal following of users is the Baader Hyperion Zoom, currently available in its latest "Mark II" version. This is an 8–24 mm Zoom that has a very generous 68° + AFOV at the 8 mm setting. Although this zoom does follow suit with most zooms and has an AFOV that varies as it zooms, it has seemingly managed to garner a large following, as its name always comes up when observers are discussing ways to reduce their eyepiece needs.

For some observers, the 8–24 mm Hyperion Zoom and a 3× Barlow can effectively serve as a one-eyepiece solution. With a 3× Barlow the Hyperion Zoom covers a range of focal lengths from 2.7to 24 mm, making this combo a highly effective choice for richer field telescopes that both need a very short focal length eyepiece capability to get to effective high magnifications, and whose fast focal ratio is more forgiving of smaller AFOVs at the longer focal lengths, as they will still show sizable TFOVs. And in addition to its generous AFOV and comfortable eye relief, the Hyperion Zoom also is highly adaptable for digital terrestrial or astro-photography, having M43 threads at eye lens and optional Hyperion digital t-rings for direct attachment to digital cameras, camcorders, and CCD cameras. Highly recommended by many observers, the Hyperion Zoom is a well liked and popular zoom eyepiece.

Fig. 10.44 The Baader Hyperion Mark III Zoom (Image courtesy of William Rose, Larkspur, CO, USA)

Given the number of zoom eyepieces available, should every observer have a zoom eyepiece in their arsenal of astronomy equipment? Although zooming eyepieces offer extreme flexibility, for some observers they can still be an "acquired taste." The varying AFOV that is common on most zoom eyepieces is probably the number one reason why zooms are not universally accepted as a best value and best approach. In addition to this, like Barlows, zoom eyepieces were plagued with an early reputation for not being "good" when compared to a fixed focal length eyepiece. They gained this reputation probably from the many "junk" eyepieces that were on the market in the earlier days of the hobby.

Today, however, the vast majority of zoom eyepieces are well made, with some of them of very high precision that can compete with the best fixed length eyepiece available. Before an observer ventures into the realm of the zoom eyepiece, he or she should carefully consider AFOV, TFOV, and eye relief needs from the various focal lengths the zoom will provide. When this is done, and the zoom is thoroughly researched, with all the pros and cons established, then a decision to choose a zooming eyepiece can be wise and cost effective. Additionally, once acquired many observers find that zooming eyepieces provide some thoroughly engaging observing experiences, providing the ability to go from distant to close up with the twist of the wrist, and allowing the observer to instantly tune their magnification to the precise level that shows the sharpest view with the highest degree of contrast. Although the zoom eyepiece is not for everyone, for many observers it has proven to be a best choice.

Brand	Focal length (mm)	AFOV (degrees)	ER (mm)	Design (elements)	Coatings	Barrel (inches)
Agena Astro	7–21	43–30	16–33	–	MC	1.25
Agena Astro	8–24	60–40	15–18	4/3	FMC	1.25
Antares Speer WALER	8.5–12	84	16	9	FMC	1.25
Antares Speer WALER	5–8	84	12	9	FMC	1.25
Apogee	7.4–22	–	–	–	–	1.25
Baader Hyperion	8–24	68–50	12–15	–	FMC	1.25/2
Celestron	8–24	60–40	15–18	4/3	FMC	1.25
Celestron Deluxe	8–24	66–43	15–20	–	FMC	1.25/2
Galileo	6.8–16	42	–	4	–	.965
GTO	7.4–22	40–22	12–14	–	–	1.25
Leica Vario ASPH	8.9–17.8	80–60	18	8/5	FMC	1.25/2
Leica	7.3–21	68–38	20	–	FMC	1.25
Meade 4000	8–24	55–40	15–20	7/4	FMC	1.25
Nikon	9–21	68–38	18	–	FMC	.965/1.25
Olivion ASPH	8–24	60–48	15–20	–	FMC	1.25/2
Olivion ASPH	9.5–19	60–48	15–20	–	FMC	1.25/2
Orion	7.2–21.5	60–40	15	7/4	FMC	1.25
Orion Explorer II	7–21	43–30	33	4	FC	1.25
Orion Pro	8–24	58–40	18.5	–	FMC	1.25
Pentax	8–24	60–49	12–18	6	FMC	1.25
Pentax XF	6.5–19	60–42	11–15	–	FMC	1.25
Sky-Watcher	7–21	43–30	33	4	FC	1.25
Sky-Watcher	8–24	60–40	15–18	4/3	FMC	1.25
Sky-Watcher	7.5–22.5	66–42	18–19.5	8/5	FMC	1.25
Swarovski	20–60×	65–40	17	9	FMC	1.25
Swarovski	25–50×	70–60	–	–	FMC	1.25
Tele Vue	8–24	55–40	15–20	7/4	FMC	1.25
Tele Vue	2–4	50	10	5/3	FMC	1.25
Tele Vue	3–6	50	10	5/3	FMC	1.25
Vixen LV	8–24	60–40	15–20	–	FMC	1.25
Vixen NLV	8–24	55–40	15–20	7/4	FMC	1.25
William Optics	8–24	60–40	20	8/5	FMC	1.25
William Optics	7.5–22.5	66–42	18–20	8/5	FMC	1.25
Zhumell	8–24	60–40	–	4/3	FMC	1.25

Appendix 1

Formulas and Optical Design Data

Calculating Eyepiece Magnification

$$\text{Magnification} = \frac{\text{FL[telescope]}}{\text{FL[eyepiece]}}$$

Notes: FL[telescope] means the focal length of the telescope in millimeters; FL[eyepiece] means the focal length of the eyepiece in millimeters.

Estimating Eyepiece AFOV (Accuracy Generally Within 10 % of Actual)

$$\text{AFOV} \approx \text{TFOV} \times \text{Magnification}$$

Notes: TFOV is true field of view in degrees; Magnification is the magnification produced by the eyepiece in the telescope.

W. Paolini, *Choosing and Using Astronomical Eyepieces*, The Patrick Moore Practical Astronomy Series, DOI 10.1007/978-1-4614-7723-5,
© Springer Science+Business Media New York 2013

Estimating Eyepiece Field Stop (Accuracy Generally Within 10 % of Actual)

$$\text{Field Stop} \approx \frac{\text{FL[telescope]} \times \text{TFOV}}{57.3}$$

Notes: The field stop results are in millimeters; FL[telescope] means the focal length of the telescope in millimeters; TFOV is true field of view in degrees; This method will only provide approximate results as it does not account for distortions present in the eyepiece, however it should be accurate within 10 % or less of the correct value.

Calculating TFOV (in Degrees) Based on Manufacturer Provided Data

$$\text{TFOV} \approx \frac{\text{AFOV}}{\textit{Magnification}}$$

Notes: AFOV is the apparent field of view of the eyepiece in degrees; magnification is the magnification produced by the eyepiece in the telescope.

$$\text{TFOV} \approx \frac{\text{AFOV}}{(\text{FL [telescope]} \div \text{FL [eyepiece]})}$$

Notes: AFOV means the apparent field of view of the eyepiece in degrees; FL[telescope] means the focal length of the telescope in millimeters; FL[eyepiece] means the focal length of the eyepiece in millimeters. This method is only approximate, as any distortions in the eyepiece's field of view will make the results inaccurate by as much as 10 %.

Calculating TFOV (in Degrees) Based on Drift Time Observations of a Star

$$\text{TFOV} = \frac{\text{DT[sec]}}{239}$$

Notes: DT[sec] means Drift Time in seconds. This method is only accurate when the star is near the celestial equator.

$$\text{TFOV} = \textit{ABS}(\text{DT[sec]} \times .0041781 \times \textit{COS}(\text{DEC[star]} \div 57.3))$$

Notes: ABS means absolute value; DT[sec] means drift time across the entire field of view in seconds; COS means cosine; DEC[star] means the declination of the star in degrees. This method is accurate for any star chosen to drift time since the declination of the star from the celestial equator is taken into account.

Calculating TFOV (in Degrees) Based on Field Measures

$$\text{TFOV} = \frac{\text{Eyepiece Field Stop Diameter}}{\text{FL[telescope]}} \times 57.3$$

Notes: Eyepiece field stop diameter is in millimeters; FL[telescope] means the focal length of the telescope in millimeters; COS means cosine; DEC[star] means the declination of the star in degrees.

$$\text{TFOV} = \frac{(\text{Tape Measure[observed]} \times 57.3)}{\text{Distance[telescope-to -tape measure]}}$$

Notes: Tape Measure[observed] means the number of inches (or millimeters) of the tape measure that are observed through the eyepiece in the telescope; Distance[telescope-to-tape measure] means the distance in inches (or millimeters) from surface of the objective of the telescope to the wall where the tape measure is mounted.

Calculating the Exit Pupil of the Eyepiece and Telescope Combination

$$\text{Exit Pupil} = \frac{\text{FL[eyepiece]}}{\text{FR[telescope]}}$$

Notes: FL[eyepiece] means the focal length of the eyepiece in millimeters; FR[telescope] means the focal ratio of the telescope, which is the focal length of the telescope in millimeters divided by the aperture of the telescope in millimeters.

$$\text{Exit Pupil} = \frac{\text{APERTURE[telescope]}}{\text{Magnification}}$$

Notes: APERTURE[telescope] means the diameter of the main objective of the telescope in millimeters; Magnification is the magnification produced by the eyepiece in the telescope, which is the focal length of the telescope divided by the focal length of the eyepiece.

Calculating Measurement Accuracy

$$\text{Results} = \frac{(Value1 + Value2 + ValueN)}{\text{N}}$$

Notes: When taking field measurements, the measurement should be repeated several times, then the average taken using the formula for averages above. Once the average measure is calculated, then the formula for Accuracy below can be used to express the accuracy of the average.

$$\text{Accuracy} = \pm\frac{(MaximumValue - MinimumValue)}{2}$$

Calculating Brightness Change

$$\text{Brightness} = \frac{1}{(\text{Magnification2} \div \text{Magnification1})^2}$$

Notes: Magnification1 is the magnification first used to observe an object; Magnification2 is the magnification used to observe an object the second time.

Estimating Eye Relief Based on Optical Design

Eyepiece design	Eye relief	AFOV	# Lens elements
Abbe (Ortho)	≈0.80 × Eyepiece focal length	45°	4
Brandon	≈0.80 × Eyepiece focal length	50°	4
Erfle	≈0.60 × Eyepiece focal length	60–70°	5
Explore Scientific 120°	Fixed 13 mm	120°	12
Huygens	≈0.10 × Eyepiece focal length	40°	2
Kaspereit	≈0.30 × Eyepiece focal length	68°	6
Kellner	≈0.45 × Eyepiece focal length	45°	3
Kepler	≈0.90 × Eyepiece focal length	10°	1
König	≈0.92 × Eyepiece focal length	55–65°	3–5
LE (Takahashi)	≈0.73 × Eyepiece focal length	52°	5
Lippershey/Galilean	–	10°	1
Meade 4000 SWA/UWA	≈0.70 × Eyepiece focal length	65°/82°	6–8

(continued)

(continued)

Eyepiece design	Eye relief	AFOV	# Lens elements
Monocentric	≈0.80 × Eyepiece focal length	25–30°	3
Nikon NAV-SW/HW	Fixed 16–19 mm	72°/102°	7–12
Pentax XL/XW	Fixed 20 mm	65°/70°	7
Plössl/Symmetrical	≈0.68 × Eyepiece focal length	50°	4
Ramsden	≈0.00 × Eyepiece focal length	35°	2
RKE	≈0.90 × Eyepiece focal length	45°	3
Takahashi-UW	Fixed 12 mm	90°	8–10
Tele Vue Delos	Fixed 20 mm	72°	–
Tele Vue Ethos	Fixed 15 mm	100–113°	≈7–9
Tele Vue Nagler	≈0.6–1.4 × Eyepiece focal length	82°	6–7
Tele Vue Panoptic	≈0.68 × Eyepiece focal length	68°	6
Tele Vue Wide Field	–	65°	6
Williams Optics UWAN	Fixed 12 mm	82°	7

Appendix 2

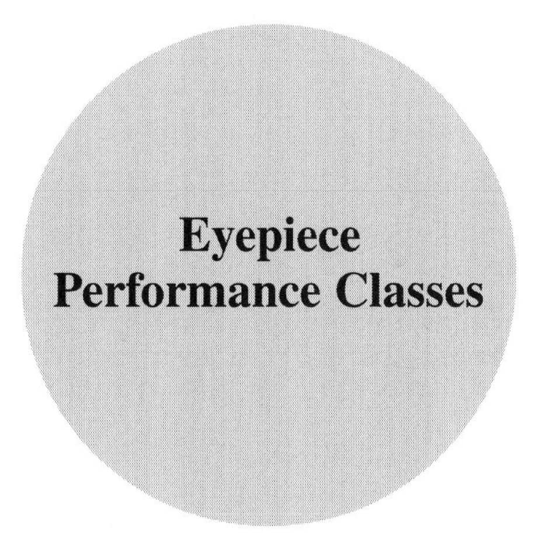

Although there are many brands and lines of eyepieces, many can be grouped into similar performing classes. As an example, although there are many brands of Plössl sold, they all perform relatively the same with the bigger distinction between most being their build quality and varying eye guard designs. Optically, however, with only a few exceptions most present the same AFOV of between 50° and 52° and all provide a similar level of off-axis correction.

The following table provides the brands and lines of eyepieces presented in this book, grouped into similar performing classes where their AFOVs and off-axis corrections are considered by other amateur astronomers to be very similar. When an observer is making a decision on choosing any eyepiece, researching the other eyepieces in these similar performance class listings will greatly assist the observer in making the best considered decision.

Performance class	Eyepieces
20°	Couture Ball Singlet, Siebert Planesphere
30°	Siebert Optics MonoCentricID
40°	Antares Ortho, Apogee Super Abbe Orthoscopic, Baader Planetarium Classic Ortho, Baader Planetarium Genuine Abbe Ortho, Cave Orthostar Orthoscopic, Celestron Ortho, Edmund Scientific Ortho, Kokusai Kohki Abbe Ortho, Kson Super Ortho, Masuyama Orthoscopic, Meade Research Grade Ortho, Meade Series II Orthoscopic, Nikon Ortho, Siebert Optics Star Splitter/Super Star Splitter, Takahashi Ortho, Telescope Service Ortho, Unitron Ortho, University Optics Abbe HD Orthoscopic, University Optics Abbe Volcano Orthoscopic, University Optics O.P.S. Orthoscopic Planetary Series, University Optics Super Abbe Orthoscopic, VERNONscope Brandon

(continued)

W. Paolini, *Choosing and Using Astronomical Eyepieces*, The Patrick Moore Practical Astronomy Series, DOI 10.1007/978-1-4614-7723-5,
© Springer Science+Business Media New York 2013

Performance class	Eyepieces
40°	Astro-Physics Super Planetary AP-SPL, Pentax XO, Pentax XP, TMB Aspheric Ortho, TMB Supermonocentric, Zeiss CZJ Ortho, Zeiss ZAO I/ZAO II
40°	Garrett Optical Orthoscopic
40°	Meade Series II Modified Achromatic MA
40–65°	Galland/Gailand/Galoc Ortho/Erfle/König
45–55°	Sky-Watcher LET/Long Eye Relief (LER)
50–52°	Antares Plössl, Astro-Professional Plössl, Astro-Tech High Grade Plössl, Astro-Tech Value Line Plössl, Bresser 52° Super Plössl, Carton Plössl, Celestron Omni, Celestron Silvertop Plössl, Clavé Plössl, Coronado CeMax, Edmund Scientific Plössl, Garrett Optical Plössl, GSO Plössl, GTO Plössl, Long Perng Plössl, Meade Series 3000 Plössl, Meade Series 4000 Super Plössl, Olivon Plössl, Opt Plössl, Orion HighLight Plössl, Orion Sirius Plössl, Owl Black Night Plössl, Parks Silver Series, Sky-Watcher SP-Series Super Plössl, Smart Astronomy Sterling Plössl, TAL – Symmetrical Super Plössl, Telescope Service Plössl, Telescope Service Super Plössl, Tele Vue Plössl, Vixen NPL
50–55°	Celestron E-Lux (2″ models only), Celestron Kellner, Criterion Kellner, Edmund Scientific RKE, GSO Kellner, Kokusai Kohki Kellner, Orion DeepView, Orion E-Series, Russell Optics (2″ 52 and 60 mm only), Sky-Watcher Kellner, Sky-Watcher Super MA Series, Telescope Service RK, Unitron Kellner
50–55°	Celestron Erfle, Kokusai Kohki Erfle, University Optics Super Erfle
50–60°	Orion Optiluxe
50–65°	GTO Wide Field
50–70°	Russell Optics 1.25″/2″ Series
52°	Antares Elite Plössl, Baader Eudiascopic, Bresser 60° Plössl, Celestron Ultima, Kasai Astroplan, Meade Series 4000 Super Plössl (pre-1994, smooth sided, 5-elements), Meade Series 5000 Super Plössl, Orion Ultrascopic, Parks Gold Series Plössl, Takahashi LE
55°	Agena ED, Celestron X-Cel, Orion Epic ED II (older version), Vixen Lanthanum (LV), Vixen NLV
55°	Astro-Professional Long Eye Relief Planetary, Astro-Tech Long Eye Relief, Long Perng Long Eye Relief, Orion Edge-On Planetary, Smart Astronomy SA Solar System Long Eye Relief, Stellarvue Planetary, William Optics SPL, Zhumell Z Series
55°	Astro-Tech Long Eye Relief, Astro-Professional Long Eye Relief Planetary, Long Perng Long Eye Relief, Orion Edge-On Planetary, Smart Astronomy SA Solar System Long Eye Relief, Stellarvue Planetary, William Optics SPL, Zhumell Z Series
55°	I.R. Poyser Plössl and Adapted Military
55–65°	Astro-Professional EF Flatfield, Astro-Tech Flat Field, BST Flat Field, Orion Edge-On Flat-Field, Sky-Watcher Extra Flat, Smart Astronomy Extra Flat Field, Telescope-Service Edge-On Flat Field
58°	APM UWA Planetary, Burgess/TMB Planetary, Olivon Wide Angled Plössl, Owl Astronomy High Resolution Planetary, Telescope Service Planetary HR, TMB Planetary II
60°	Astro-Tech Paradigm Dual ED, BST Explorer ED, Olivon 60° ED Wide Angle, Orion Epic II ED, Pentax XF, Telescope Service NED "ED" Flat Field
60°	Celestron X-Cel LX

(continued)

Performance class	Eyepieces
60°	Meade Series 5000 HD-60
60°	Rini Various Eyepieces
60°	Siebert Optics Performance Series
60°	Surplus Shed Erfles
60–65°	Agena Wide Angle, Astro-Tech Series 6 Economy Wide Field, Burgess Wide Angle, Owl Advanced Wide Angle
60–80°	University Optics König/König II/MK-70/MK-80
65°	Denkmeier D21/D14
65°	TAL Super Wide Angle
65°	Tele Vue Wide Field
65–70°	Antares Classic Erfle
66°	Agena Enhanced Wide Angle (EWA), Orion Expanse, Owl Enhanced Superwide, Sky-Watcher Ultra Wide Angle, Telescope Service SWM Wide Angle, William Optics WA 66
68°	Baader Planetarium Hyperion/Hyperion Aspheric, Orion Lanthanum Superwide, Orion Stratus, Vixen Lanthanum Superwide (LVW)
68°	Explore Scientific 68° Nitrogen Purged (ES68 N2), Meade 4000 SWA, Meade 5000 SWA, Nikon NAV-SW, Pentax XL, Pentax XW, Tele Vue Delos, and Tele Vue Panoptic
68°	Long Perng 68° Wide Angle, Orion Premium 68° Long Eye Relief
70°	Agena Super Wide Angle, Astro-Professional SWA, Astro-Tech Wide Field, Garrett SuperWide Angle, GSO Superview, Meade QX, Olivon 70° Wide Angle, Opt Super View, Orion Q70 Super Wide-Field, Sky-Watcher PanaView, Telescope-Service WA, University Optics 70°, William Optics SWAN
70°	Antares W70
70°	Astro-Tech AF Series 70° Field, Olivon 70° Ultra Wide Angle, Telescope Service Expanse ED
70°	Astro-Tech Titan Type II ED, Sky-Watcher Aero, TMB Paragon
70°	Celestron Axiom
70°	Celestron Ultima LX
70°	GTO Proxima
70°	Siebert Optics Observatory
70°	Siebert Optics Ultra
75–80°	Agena Ultra Wide Angle (UWA), Apogee Widescan (20 and 30 mm), Astrobuffet 1RPD (30 mm), BW Optik Ultrawide (30 mm), Kokusai Kohki Widescan I/II/III (20 and 30 mm), Moonfish Ultrawide (30 mm), Olivion 80° Ultra Wide Angle, Owl Astronomy Knight Owl Ultrawide Angle, Sky-Watcher Ultra Wide Angle, Surplus Shed Wollensak (30 mm), TAL Ultra Wide Angle, University Optics 80°, University Optics Widescan I/II/III (20 and 30 mm)
82°	Antares Speers-WALER Series (Wide Angle Long Eye Relief), Astro-Professional UWA, Celestron Axiom LX, Celestron Luminos, Docter UWA, Explore Scientific 82° Nitrogen Purged, Meade Series 4000 Ultra Wide Angle, Meade Series 5000 Ultra Wide Angle, Orion MegaView Ultra-Wide, Sky-Watcher Nirvana UWA, Sky-Watcher Sky Panorama, Tele Vue Nagler, Williams Optics UWAN

(continued)

Performance class	Eyepieces
82°	Astro-Professional 28 mm UWA, Leitz 30 mm Ultra Wide, Orion 28 mm MegaView Ultra-Wide, Sky-Watcher 28 mm Nirvana UWA, Tele Vue 31 mm Nagler, and Williams Optics 28 mm UWAN
90°	Takahashi UW
100°	Agena Mega Wide Angle (MWA), Orion GiantView, TMB 100, Zhumell Z100
100°	Explore Scientific 100° Nitrogen Purged (ES100 N2), Nikon NAV-HW, Tele Vue Ethos
100°	Meade XWA
120°	Explore Scientific 120° Nitrogen Purged (ES100 N2)

Appendix 3

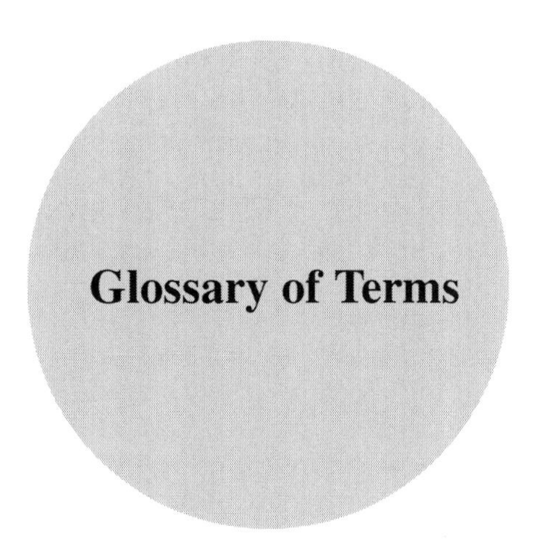

Glossary of Terms

Abbe (Ortho) eyepiece An eyepiece design invented in 1880.

Achromatic lens A lens designed to produce reduced color fringing.

Amplifier A type of eyepiece accessory, like a Barlow lens, that retains the original AFOV and eye relief of the eyepiece it is used with.

Angular magnification distortion (AMD) An optical aberration that occurs when the eyepiece's focal length is not constant across the field of view but varies slightly.

Aperture The diameter of the main objective of a telescope.

Apparent field of view (AFOV) The angular size of the field of view when observing through an eyepiece as measured from the furthest left to the furthest right.

Aspheric lens A lens where one of its surfaces is not spherical.

Astigmatism An optical aberration that deforms star points and causes extended objects to appear less sharp.

Axial chromatic aberration An optical aberration that causes an in-focus star point in the center of the field of view to separate into a small spectrum of colors around the fringe of the star point.

Axial longitudinal aberration See axial chromatic aberration.

Baffle A device within an eyepiece used to prevent unwanted light rays from entering the optical path; usually a small diaphragm with a circular opening placed at strategic points within the eyepiece's housing and barrel.

Barlow lens A negative (diverging) lens, usually a cemented doublet, which acts as a focal-length multiplier for the telescope. (e.g., a 2× Barlow will convert a telescope with a 500 mm focal length to 1,000 mm; effectively multiplies the magnification produced by the eyepiece by the magnification factor of the Barlow).

W. Paolini, *Choosing and Using Astronomical Eyepieces*, The Patrick Moore Practical Astronomy Series, DOI 10.1007/978-1-4614-7723-5,
© Springer Science+Business Media New York 2013

Barrel The lower portion of the eyepiece that is inserted into the telescope's focuser.

Barrel distortion An optical aberration that is a type of rectilinear distortion; the opposite direction of this distortion is called pincushion distortion.

Binoviewer A telescope accessory that uses two identical focal length eyepieces to allow the observer to view with two eyes similar to binoculars.

BK7 See crown glass.

Brandon eyepiece An eyepiece design invented in 1949.

Coated A marketing term that means that at least one optical surface of the lenses in an eyepiece is coated with a single anti-reflection coating to improve transmission.

Coma An optical aberration that occurs in the off-axis of the field of view of an eyepiece and makes a star appear as a small comet shape, with the tail always pointing directly away from the center of the field of view.

Crown glass A type of optical glass used in eyepieces, telescopes, and other optical instruments that has a relatively low refractive index and low dispersion.

Delos eyepiece An eyepiece design invented in 2011.

Double-concave A lens type where both surfaces are concave.

Double-convex A lens type where both surfaces are convex.

Doublet A lens group where two individual lenses are together.

Element An individual glass lens within an eyepiece.

Erfle eyepiece An eyepiece design invented in 1921.

Ethos eyepiece An eyepiece design invented in 2007.

Exit pupil The physical position where the image is formed in a circle by the eyepiece above the eye lens; the diameter of the circle that the image is formed in conveys the relative brightness of the image (e.g., two images with the same exit pupil diameter will be equally bright).

Eyeguard A device found sometimes on the eyepiece to help block stray light from entering the observer's eye when using the eyepiece.

Eye lens The top lens of the eyepiece that is closest to the observer's eye when observing.

Eye relief (ER) The distance above the surface of the center of an eyepiece's eye lens where the image is formed by the eyepiece (e.g., how far above the eyepiece the eye needs to be placed to view the image).

Eyepiece A device that allows an observer to view the image produced by the telescope; the human interface into the telescope.

Field curvature (FC) An optical aberration that occurs when the image comes to focus on a curved surface instead of a flat plane. It appears in the eyepiece as a focused image in the center of the eyepiece's field of view and an out of focus image in the off-axis of the field of view.

Field lens The bottom of the lens in the eyepiece that is furthest away from the observer when viewing through the eyepiece.

Field of view (FOV) See apparent field of view (AFOV).

Field stop A mechanical diaphragm device in the eyepiece that limits light rays and defines the apparent field of view.

Filter A device that is typically affixed to the bottom of the barrel of the eyepiece that filters the incoming light for a specific purpose (e.g., reduce brightness; only allow certain spectra of light to pass; block certain spectra of light; polarize the light).

Flint glass A type of optical glass used in eyepieces, telescopes, and other optical instruments that have a relatively high refractive index and low dispersion.

Focal length The distance from the telescope's objective lens to where the image is formed for a very distant object; for an eyepiece this characteristic is defined by the optical prescription of the eyepiece's design and is used to calculate how much magnification the eyepiece will produce in a telescope.

Focal plane For a telescope it is the position where the objective lens forms the image; for an eyepiece it is the position of the field stop where the telescope's image must be placed for the eyepiece to magnify the image.

Focal ratio For a telescope this is the ratio of the focal length of the telescope divided by the diameter of the aperture (main objective) of the telescope.

Fully coated (FC) A marketing term that means that every air-to-glass optical surface of a lens in an eyepiece is coated with a single anti-reflection coating to improve transmission.

Fully multi-coated (FMC) A marketing term that means that every air-to-glass optical surface of a lens in an eyepiece is coated with a multiple coats of an anti-reflection coating to further improve transmission.

Galilean eyepiece See Lippershey/Galilean eyepiece.

Ghost image A reflection of an object being observed through an eyepiece that forms on the surface on one of the lenses in the eyepieces so that it appears as a dim second image.

Housing The upper portion of the eyepiece that holds the optical lenses.

Huygen eyepiece An eyepiece design invented in 1662.

Kaspereit eyepiece An eyepiece design invented around 1923.

Kellner eyepiece An eyepiece design invented in 1849.

Kepler eyepiece An eyepiece design invented in 1611.

Kidney bean effect A popular name for an issue with the way the eyepiece forms the image at its exit pupil that has the technical term of spherical aberration of the exit pupil.

König eyepiece An eyepiece design invented in 1915.

Lanthanum A type of optical glass; a chemical element with the symbol La, atomic number 57, added as a "rare Earth element" into glass to give it special optical properties.

Lateral chromatic aberration See lateral color.

Lateral color A type of aberration that cause the color spectrum to separate slightly when a bright object is viewed off-axis.

Lens See element.

Lens assembly The collection of optical lenses in the eyepiece.

LER Long eye relief.

Lippershey/Galilean eyepiece An eyepiece design invented in 1608; the first eyepiece.

MAK Abbreviation for a Maksutov telescope.

MCT Abbreviation for a Maksutov-Cassegrain telescope.

Monocentric eyepiece An eyepiece design invented about 1883.

Mount See housing.

Multi-coated (MC) A marketing term that means that every air-to-glass optical surface of a lens in an eyepiece is coated with a single layer of an anti-reflection coating to further improve transmission, and where the eye lens of the eyepiece is treated with multiple coatings of an anti-reflection coating to further reduce the possibility of external light sources reflecting off the eye lens into the observer's eye.

Nagler eyepiece An eyepiece design invented in 1979.

Negative group A group of lenses that together diverge light rays; a Barlow lens is typically a doublet group of lenses that diverge light rays.

Negative meniscus =A lens type where one surface is convex and the other surface is concave and the lens diverges light rays.

Negative lens A lens type where it diverges the light rays; e.g., a plano-concave or double-concave lens.

Newtonian A type of telescope that uses a single spherical or parabolic primary mirror as a main objective and a flat mirror to redirect the light to the eyepiece.

Objective lens The primary lens of a telescope that gathers the light and forms the image; also called the main objective. The diameter of the objective defines the aperture of the telescope.

Ocular An alternate name for an eyepiece.

Off-axis The region of the field of view in an eyepiece that is not the central area of the field of view; generally refers to the central 50 % or less of the field of view.

Ortho See Abbe (Ortho) eyepiece.

Orthoscopic A characteristic of the field of view through an eyepiece where the objects being observed appear as they are, without distortion, across the field of view; linear features do not bend or bow, and angles do not change.

Panoptic eyepiece An eyepiece design invented in 1992.

Plano-concave A lens type where one surface is flat and the other surface is concave.

Plano-convex A lens type where one surface is flat and the other surface is convex.

Plössl eyepiece An eyepiece design invented in 1860.

Positive meniscus A lens type where one surface is convex and the other surface is concave and the lens converges light rays.

Public outreach eyepiece An eyepiece chosen by amateur astronomers for use at public observing events. These are typically inexpensive in case they are accidentally damaged or stolen at the public event. Ease of use and fair-to-good performance are also typical criteria for a public outreach eyepiece.

Radial distortion See rectilinear distortion.

Ramsden eyepiece An eyepiece design invented in 1782.

Rectilinear distortion (RD) An optical aberration that causes straight lines to appear bowed/bent inward (pincushion) or outward (barrel).

Refractor A type of telescope that only uses clear optical glass lenses, that does not require any mirrored surfaces, and has no obstructions in the light path.

RKE eyepiece An eyepiece design invented in 1977; the acronym stands for Rank-Kaspereit-Erfle.

SCT Abbreviation for a Schmidt-Cassegrain telescope.

Shoulder Where the housing and barrel of the eyepiece meet; where the eyepiece rests when inserted into the focuser.

Singlet A single-element lens.

Spherical aberration (SA) An optical aberration where the lenses do not focus all the light rays to a single precise point; can cause the image to be less than sharp when it is severe.

Symmetrical eyepiece An eyepiece design invented around 1860; today commonly marketed under the name Plössl.

Telecentric An alternate name for an amplifier.

Transverse chromatic aberration See lateral color.

Triplet A lens group where three individual lenses are together.

True field of view (TFOV) Sometimes referred to as the actual field of view; the angular measure, expressed in degrees, of the maximum amount of sky visible in the eyepiece from the furthest left to the furthest right of the field of view.

White water glass A type of optical glass used in specialized eyepieces and other optical instruments; this glass type is low iron, ultra clear, and transmits 98–99 % of light.

About the Author

William Paolini has been actively involved in optics and amateur astronomy for 45 years, has published numerous product reviews on major online amateur astronomy boards, and volunteers with public tours at a famous vintage Clark refractor site.

William Paolini's professional background is as an officer in the U.S. Air Force and as a computer scientist, holding a Bachelor's degree in Computer Science and a Master of Science in Education. He has worked for the U.S. Department of Defense, the U.S. Department of Commerce, the Federal Trade Commission, the Federal Reserve, the World Bank, and a variety of commercial corporations in the information technology, information technology security, and telecommunications industries.

Bill has been observing as an amateur astronomer since the mid-1960s, grinding mirrors for homemade Newtonian telescopes during the 1970s and eventually owning, using, and testing several hundreds of eyepieces in a wide variety of telescopes from achromatic and apochromat refractors to Newtonian, Maksutov-Cassegrain, and Schmidt-Cassegarian designs. Today he enjoys observing from his suburban home west of Washington, DC, where his primary amateur astronomy pursuits are lunar, planetary, bright nebula, open cluster, and globular cluster observing.

W. Paolini, *Choosing and Using Astronomical Eyepieces*, The Patrick Moore Practical Astronomy Series, DOI 10.1007/978-1-4614-7723-5,
© Springer Science+Business Media New York 2013

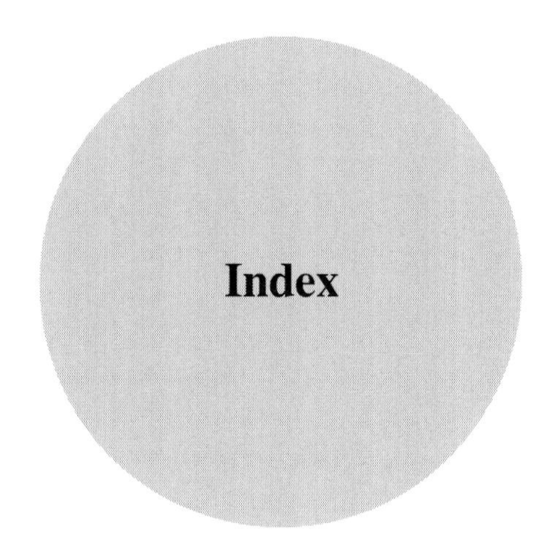

Index

A

AAVSO. *See* American Association of Variable Star Observers (AAVSO)

Abbe, 14, 18, 21, 22, 55, 67, 70, 80, 82, 109–112, 114–118, 126, 128, 134, 157, 163, 164, 166, 199, 204–205, 218, 220, 228, 229, 237, 238, 247, 254, 271–272, 275–276, 282–283, 286, 300, 341–343, 358, 359, 364, 385–388, 390, 392, 394, 409, 416, 419, 423, 426

Abell, 166

Aberration, 3, 7, 11, 16, 19, 23, 25–32, 43, 55–56, 67, 68, 71, 95, 111, 113, 115, 120, 127, 130, 131, 134, 146, 148, 151, 153, 176, 247, 340, 341, 360, 423–427

Absolute, 48, 49, 118, 179, 415

Academy, 158

Accessories, 34, 73–107

Acclimated, 163

Acclimation, 67

Accommodate, 16, 22, 28, 34, 42, 65, 128, 135, 157, 158

Accuracy, 43, 49, 50, 53, 56, 94, 413, 414, 416

Achromat, 126, 150, 284, 384

Achromatic, 95, 119, 120, 248–249, 284, 420, 429, 421

Acrtangent, 152

Adaptation, 155

AERO, 131, 135, 215, 344, 381, 421

AFOV. *See* Apparent field of view (AFOV)

Agena, 125, 127, 131, 136, 144, 146, 191–251, 264, 266, 275, 288, 297, 302, 303, 306, 310, 311, 317, 321, 323, 324, 326, 347, 349, 354, 362, 365, 368, 369, 379, 385, 386, 393, 397, 401, 402, 406, 411, 420–422

Air-to-glass, 69, 112, 115, 199, 220, 222, 372, 425, 426

Aluminum, 9, 39, 178, 250, 270, 338

AMD. *See* Angular magnification distortion (AMD)

American Association of Variable Star Observers (AAVSO), 149

Amplifier, 84–93, 124, 161, 338, 423, 427

Andromeda, 173

Angular, 25, 45, 46, 75, 423
  magnification, 25, 29–31, 48, 79, 131, 372, 378, 423

Angular magnification distortion (AMD), 25, 29–31, 48, 79, 131, 372, 378, 423

Anodized, 9, 35, 36, 39, 41, 178, 250

Antares, 110, 119, 125, 131, 136, 163, 197–202, 204, 208, 209, 211, 217–220, 224, 227, 229, 230, 232, 235–237, 239, 241, 245, 246, 251, 254, 255, 259, 263, 265, 267, 271, 272, 276, 280, 281, 282, 283, 289, 292, 294, 296, 300, 303, 305, 310, 315, 319, 321, 322, 326, 327, 343, 346–348, 352, 358, 361, 364, 367, 368, 373, 376, 385, 387, 388, 390, 391, 395, 399, 403, 419–421

Anti-reflection, 24, 63, 69, 117, 154, 161, 172, 270, 279, 370, 372–374, 378, 424–426

Anti-reflective, 40

W. Paolini, *Choosing and Using Astronomical Eyepieces*, The Patrick Moore Practical Astronomy Series, DOI 10.1007/978-1-4614-7723-5,

© Springer Science+Business Media New York 2013

APM, 89, 91, 127, 129, 203, 225, 304, 324, 366, 382, 384, 408, 420
Apochromat, 114, 421
Apogee, 110, 136, 196, 199, 204, 205, 218, 220, 226, 229, 237, 254, 272, 275, 276, 282, 283, 286, 297, 300, 303, 324, 343, 349, 354, 358, 362, 364, 385–388, 390, 391, 393, 395, 411, 419, 421
Apparent field of view (AFOV), 11, 43, 73, 109, 151, 189, 192, 254, 301, 354
AP-SPL. *See* Astro-Physics Super Planetary (AP-SPL)
Arctangent, 152
Aries, 113
Aristarcus, 115
AS80, 166
AS80/1200, 166
ASAPH, 6
ASPH, 10, 411
Aspheric, 10, 70, 80, 111, 206, 221–222, 328, 331, 380, 384, 404, 405, 408, 409, 420, 421, 423
Assyria, 4
Asterism, 78
Astigmatism, 25, 26, 34, 35, 68, 113, 120, 121, 131, 135, 145, 250, 372, 375, 423
Astrobuffet, 136, 196, 205, 226, 275, 297, 303, 324, 349, 354, 362, 386, 393, 421
Astrograph, 11, 19, 20, 23, 24, 29, 30, 32, 44, 46, 47, 51, 75, 76, 78, 79, 145
Astronomer, 3, 5–7, 9, 16, 17, 20, 56, 58, 66, 68, 83, 86, 95, 111, 122, 125, 128, 133, 134, 136, 145, 146, 149–184, 220, 247, 256, 371, 419, 426, 421
Astro-Physics (AP), 39, 70, 82, 91, 110–113, 118, 206, 328, 332, 338, 380, 384, 404, 405, 408, 420
Astro-Physics Super Planetary (AP-SPL), 70, 82, 91, 110, 112–114, 206, 328, 332, 380, 384, 404, 405, 420
Astro-Professional, 119, 125, 127, 131, 136, 143, 195, 200, 202, 207–209, 211, 212, 217, 223, 224, 227, 230, 232, 235, 236, 239, 245, 246, 251, 255, 259, 263–267, 277, 278, 280, 287–289, 292, 296, 299, 302, 303, 305, 306, 308, 309, 312, 314, 315, 317, 319, 322, 327, 345–348, 350, 352, 353, 361, 363, 367–368, 373, 376, 399, 400, 401, 407, 420–422
Astro-Tech, 119, 124, 125, 127, 129–131, 135, 154, 195, 196, 200, 207–217, 223, 224, 226, 227, 232, 236, 239, 245, 246, 255, 263–267, 278, 279, 287–289, 301–303, 305, 306, 308–310, 310, 314, 317, 319, 321, 322, 325, 327, 329, 344, 345, 347, 348, 350–353, 361, 364, 367–369, 376, 378, 381, 385, 399, 400, 401, 420, 421
Astrozoom, 103–104
Asymmetrical, 118, 119
ATC, 166
Athens, 139, 150
Axial, 25, 27, 423
Axiom, 131, 135, 136, 202, 209, 229–230, 235, 251, 259, 292, 296, 315, 346, 347, 373, 403, 421

B

Baader, 34, 91, 98, 111, 117, 118, 125, 131, 135, 154, 162–164, 174, 198, 199, 204, 218–221, 224, 229, 237, 241, 254, 271, 272, 276, 281, 282, 286, 289, 294, 300, 313, 320, 321, 326, 343, 358, 364, 385, 387, 388, 390, 391, 395, 398, 405, 408–411, 419–421
Baader Planetarium, 135, 199, 204, 218–221, 229, 237, 254, 272, 276, 281, 282, 286, 300, 343, 358, 364, 385, 387, 388, 390, 391, 395, 398, 405, 419, 421
Babylonian, 149
Baffle(s)/baffled, 16, 24, 40, 81, 115, 370, 372–374, 376, 423
Ball, 99, 111, 112, 155–161, 247, 341, 379, 419
Bandwidth, 98, 101
Barlow, 12, 17, 49, 57, 59, 60, 75, 84–93, 103, 104, 124, 141, 161, 164, 165, 169, 173, 180, 181, 184, 218, 262, 337, 397, 408–410, 423, 426
Barrel, 8–10, 12, 16, 24, 29, 30, 35, 39–41, 45, 49, 70, 77, 94, 104, 111, 113, 114, 118, 121, 123, 132, 145, 154, 155, 157, 159, 170, 177–180, 184, 187, 189, 192–238, 241–250, 254–256, 258–286, 288, 289, 291–300, 301–322, 324–342, 344–351, 353, 354, 358–394, 396–407, 411, 423–427
Bausch and Lomb, 187
BGO, 163, 174
Binoviewer, 38, 42, 43, 130, 132, 152, 255, 401, 424
BK7, 156, 424
Blackout, 23, 128, 140, 145, 163, 174
Blue Flash Nebula, 99
Bow Tie Nebula, 99
Box Nebula, 99
Brandon, 14, 18, 21, 67, 82, 110, 111, 117, 118, 126, 150, 153, 163, 164, 169, 174,

188, 199, 205, 218, 220, 229, 238, 254, 272, 276, 282, 283, 286, 300, 343, 352, 359, 364, 385, 387, 388, 391, 394, 395, 416, 419, 424
Braunschweig, 180
Bresser, 119, 125, 127, 198, 200, 208, 211, 217, 219, 223–224, 227, 232, 236, 239, 242, 245, 246, 255, 263, 265, 267, 271, 280, 281, 287, 289, 294, 303, 305, 312, 319, 321, 322, 326, 327, 348, 352, 358, 361, 367, 368, 376, 399, 420
BST, 127, 129, 130, 207, 211, 213, 222–223, 301, 308, 310, 329, 345, 350, 363, 364, 420
Bubble Nebula, 99
Burgess, 127, 196, 214, 225–227, 304, 321, 421
Burgess/TMB, 129, 203, 304, 324, 366, 382, 420
Butterfly Nebula, 99
BW Optik/BW Optic, 136, 196, 205, 226, 275, 297, 303, 324, 349, 354, 362, 386, 393, 421

C
C8, 152–154
Caliper, 49, 50, 176
Canon, 188
Carbon, 77, 114, 145
Carl Zeiss Jena (CZJ), 70, 82, 111, 113, 132, 163, 169, 206, 328, 331, 380, 384, 403–404, 405, 420
Carton, 119, 188, 200, 208, 211, 217, 224, 227, 232, 236, 239, 245, 246, 255, 263, 265, 267, 280, 287, 289, 304, 305, 312, 319, 322, 327, 348, 352, 361, 367, 368, 376, 399, 420
Cassegrain, 173, 426, 427
Cassini, 5
Cat's Eye Nebula, 99
Cave, 111, 199, 204, 218, 220, 228–229, 237, 254, 272, 276, 282, 283, 286, 289, 300, 343, 358, 364, 385, 387, 388, 390, 391, 395, 419
CCD, 175, 409
Celestron, 96, 110, 111, 118, 119, 122, 123, 125–126, 130, 131, 135, 136, 183, 188, 192, 198–200, 202, 204, 208, 209, 211, 217–220, 224, 227, 229–239, 241–246249, 251, 254, 255, 257, 259, 263–265, 267, 271–276, 280, 281, 283, 286, 287, 289, 292, 294, 296, 300, 304–307, 310, 312, 315, 319, 321, 322, 326, 327, 337, 343, 345–349, 352, 358, 361, 364, 366–368, 373, 376, 385, 387, 388, 390–392, 395, 397, 399, 403, 411, 419–421
Cemax, 119, 200, 208, 211, 217, 224, 227, 232, 236, 239, 245, 246, 255, 263, 265, 267, 280, 287, 289, 304, 305, 312, 319, 322, 327, 348, 352, 361, 367, 368, 376, 399, 420
Cemented, 11, 13, 85, 114, 115, 119, 270, 423
Chromatic, 25, 27, 32, 95, 423, 425, 427
Chromosphere, 95
Circle-NJ, 121, 240
Circle-T, 111
Circle-V, 122
Circumstance, 54, 56, 67, 68, 71, 87, 142
Clark, 6
Clavé, 110, 119, 120, 200, 208, 211, 217, 224, 227, 232, 236, 239, 244–246, 255, 263, 265, 267, 280, 287, 289, 304, 305, 212, 319, 322, 327, 348, 352, 361, 367, 368, 376, 399, 420
Cluster, 50, 51, 55, 65, 67, 71, 73, 74, 76–77, 97, 104, 116, 118, 130, 145, 153–155, 166, 173, 250, 405
Coated/coating, 5, 8, 35, 36, 39, 81, 94, 102, 105, 106, 113, 114, 137, 150, 153, 154, 187, 189, 192, 254, 270, 279, 301, 328, 351, 358, 360–362, 424–426
Cocoon Nebula, 94, 100
Collimated/collimation, 163
Color-corrected, 5
Coma, 25, 27, 28, 120, 131, 175, 360, 375, 424
Comet, 6, 27, 149, 170, 424
Compound, 5, 90
Compression, 9, 40
Concave, 4, 13, 120, 121, 124, 211, 289, 351, 352, 376, 424, 426
Cone Nebula, 99
Contrast, 43, 53–55, 62, 63, 65–71, 74, 79, 80, 84, 94, 96–98, 100, 101, 109, 112–115, 118, 124, 125, 129, 134, 135, 139, 140, 143, 145, 148, 150, 153, 154, 157, 162–166, 170, 171, 246, 247, 281, 333, 382, 384, 410
Contrast booster, 98
Convex, 4, 5, 118, 424, 426
Coronado, 100, 119, 200, 208, 211, 217, 224, 227, 232, 236, 239, 245, 246, 255, 263, 265, 267, 280, 287, 289, 304, 305, 312, 319, 322, 327, 348, 352, 361, 367, 368, 376, 399, 420
Couture, 111, 112, 155–161, 247, 341, 419

Crab Nebula, 98
Crescent Nebula, 99
Criterion, 67, 118, 232, 248–249, 257, 264, 274, 306, 307, 337, 345, 349, 366, 385, 420
Crown, 120, 262, 424
Crystal Ball Nebula, 99
CZJ. *See* Carl Zeiss Jena (CZJ)

D

D14, 131, 133, 249–250, 421
D21, 131, 133, 249–250, 421
Dakin,
Dark-adapted, 60, 64, 65, 175
Darkening, 54, 60, 65, 96
Dawes limit, 66
Declination, 48, 49, 53, 415
Deep sky object (DSO), 6, 59, 60, 66, 70, 73, 98, 100, 101, 129, 153, 162, 167
Deep view, 118, 125, 232, 234, 249, 257, 264, 274, 306, 307, 337, 345, 349, 366, 385, 420
Defocus/defocused, 25, 26, 77
Degradation, 87, 115
Degrade/degraded, 80, 86, 94, 105, 171
Dellechiaie, 134, 144
Delos, 14, 19, 22, 38, 55, 70, 89, 131, 133, 134, 152, 258, 291, 295, 299, 330, 333, 370–371, 417, 421, 424
Delrin, 8, 35, 36, 113, 159–161, 206, 337, 338
Denatured, 105
Denkmeier, 131, 133, 134, 152, 154, 249, 421
Density, 101, 102
Deposition, 114, 258
Dewing, 140
Diagonal, 94, 150, 155, 157, 261, 312
Dialsight, 119
Diaphragm, 10, 24, 423, 424
Diascope, 166
Dioptrx, 34, 35, 135, 145
Dispersion, 115, 192, 213, 215, 223, 243, 261, 301, 328, 330, 343–345, 358, 364, 365, 424, 425
Distortion, 4, 7, 16, 19, 25–32, 48, 67, 68, 71, 79, 109, 111, 126, 128, 130, 131, 151, 250, 370, 372, 377, 378, 405, 414, 423, 424, 426
Dobsonian, 7, 43–45, 58, 66, 68, 128, 135, 142, 150, 167, 179, 250, 292, 339
Docter, 70, 136, 143, 191–251
Double-convex, 5, 424
Doublet, 13, 63, 85, 114, 118–120, 124, 250, 423, 424, 426
DSO. *See* Deep sky object (DSO)
Dust-free, 114
Dutch, 4, 5

E

Eagle Nebula, 20
Earth, 5, 49, 151, 262, 397–399, 425
eBAY, 168
Eclipse, 149
Edge-of-field, 93
Edge-On, 125, 127, 207, 211, 212, 223, 278, 308, 309, 345, 350, 363, 400, 407, 420
Edmund, 6, 111, 118, 119, 153, 156, 164, 188, 199, 200, 204, 208, 211, 217, 218, 220, 224, 227, 229, 232, 234, 236, 237, 239, 245, 246, 249, 253–300, 304–307, 312, 319, 322, 326, 337, 343, 345, 348, 349, 352, 358, 361, 364, 366–368, 376, 385, 387, 388, 390, 391, 395, 399, 419, 420
Edmund Scientific, 111, 118, 119, 153, 156, 188, 199, 200, 204, 208, 211, 217, 218, 220, 224, 227, 229, 232, 234, 236, 237, 239, 245, 246, 249, 253–300, 304–307, 312, 319, 322, 326, 337, 343, 345, 348, 349, 352, 358, 361, 364, 366–368, 376, 385, 387, 388, 390, 395, 395, 399, 419, 420
Egyptian, 3, 4
Ejecta, 31, 102, 115
E-Lux, 118, 231–232, 234, 249, 257, 264, 274, 306, 307, 337, 345, 349, 366, 385, 420
Epic, 127, 129, 130, 174, 192, 213, 223, 243, 301, 310, 329, 364, 397, 420
Erfle, 197, 232–233, 385
ES-68, 126
ES-82, 88, 136, 142, 143, 164, 202, 209, 230, 235, 251, 259–260, 292, 296, 346, 347, 373, 403, 421
ES-100, 163
ES-120, 14, 19, 22, 147, 261–262, 416, 422
Eskimo Nebula, 99
Ethos, 14, 19, 22, 45, 55, 66, 70, 80, 134, 135, 144–146, 152, 163, 176–180, 261, 298, 371–372, 417, 422, 424
Ethos-SX, 144
Eudiascopic, 125, 198, 219, 224, 241, 271, 281, 289, 294, 321, 326, 358, 420
Excelsior, 133
Exhale/exhaling, 107
Exotic, 5, 153, 373, 377, 409
Expanse, 131, 135, 193, 210, 302, 310–311, 323, 349, 363, 368, 402, 421
Exploded, 7

Explorer, 127, 129, 130, 213, 222–223, 301, 310, 329, 364, 411, 420
Explore Scientific, 14, 19, 22, 55, 65, 88, 126, 131, 133, 136, 142–144, 146, 147, 162, 164, 188, 202, 209, 230, 235, 251, 258–262, 291, 292, 295, 296, 298, 299, 315, 330, 343, 346, 347, 371–374, 403, 416, 421, 422
External-facing, 124
Eyeglass(es), 21, 34, 114, 115, 128, 132, 135, 145, 161, 162
Eyelashes, 104, 140

F

Fall-off, 114
Festoons, 96–98
Fetus Nebula, 99
Field curvature, 25, 28, 29, 424
Field of view (FOV), 32, 150, 154, 168, 170, 171, 177, 424
Filaments, 170
Filter, 94–102, 425
Fine-tuning, 181, 221, 222
Finicky, 163
Fish-bowl, 29
Fish-eye, 29
Flaming Nebula, 100
Flare, 24, 41, 80, 94, 101, 102, 178 (Found as Glare)
Flat-field, 127, 207, 211, 223, 308, 345, 350, 363, 420
Flat/plano, 120, 121
Flint, 425
Floaters, 151
Fogging, 113, 146, 206
Formula/formulae, 22, 47–53, 61–63, 86, 87, 91, 151, 152, 413–417
FOV. *See* Field of view (FOV)
Fringe/fringing, 27, 32
Full-aperture, 183
Full-surface, 115, 329, 331, 333

G

Gailand, 262–263, 420
Galaxy, 55, 65, 68, 71, 74, 77–79, 118, 129, 134, 143, 154, 166, 167
Galaxy(ies)/nebula, 68, 77–79
Galileo, G., 4–6, 13, 411
Galland, 262–263, 420
Galoc, 262–263, 420
Garrett Optical, 119, 200, 208, 211, 217, 224, 227, 232, 236, 239, 239, 245, 255, 263–265, 267, 280, 265, 289, 304, 305, 312, 319, 322, 327, 348, 352, 361, 367, 368, 376, 399, 420
Gemini Nebula, 99
Genesis, 4
Ghost/ghosting, 16, 24, 80, 94, 98, 99, 129, 250, 339
Ghost Nebula, 98
Ghost of Jupiter Nebula, 99
Giantview, 144, 146, 194, 311, 379, 406, 422
Giovanni, 5
Globs, 4, 77
Globular, 55, 67, 71, 73, 74, 77, 104, 116, 130, 145, 155, 173
Glow, 54, 94, 142, 299
GRS, 96–98
GSO. *See* Guan Sheng Optical (GSO)
GTO, 119, 131, 200, 208, 211, 217, 224, 227, 232, 236, 239, 245, 246, 255, 263, 265, 267, 268, 280, 287, 289, 304, 305, 312, 319, 322, 327, 348, 352, 361, 367, 368, 376, 399, 411, 420, 421
Guan, 127, 264–266
Guan Sheng Optical (GSO), 85, 91, 118, 119, 127, 131, 195, 200, 208, 209, 211, 217, 224, 232, 234, 236, 239, 245, 246, 249, 255, 257, 263–267, 274, 280, 287–289, 302, 304–307, 312, 317, 319, 322, 327, 337, 345, 347–349, 352, 361, 366–369, 376, 385, 399, 401, 420, 421

H

Halley, 6
H-alpha. *See* Hydrogen-alpha (H-alpha)
Hands on optics, 96, 201, 253, 254, 268, 293, 397
Haze, 83, 96
H-beta. *See* Hydrogen-beta (H-beta)
HD, 111, 158, 163, 199, 205, 218, 220, 229, 238, 254, 272, 276, 282, 283, 286, 300, 343, 359, 364, 378, 385, 386, 388, 391, 395, 419
HD-60, 127, 130, 293, 421
Heart Nebula, 99
Helical Nebula, 99
Hieroglyphs, 4
Hioptic, 188
H/K, 95, 102
H-line, 95
Horsehead Nebula, 94, 100
Hue(s), 73, 77, 80, 81, 102
Humid, 140, 146
Humidity, 183

Huygen, 5, 6, 14, 18, 21, 29, 111, 166, 416, 425
Hydrogen, 79, 94, 99–102
Hydrogen-alpha (H-alpha), 99, 101, 102
Hydrogen-beta (H-beta), 79, 94, 98, 100, 101
Hyperion, 34, 131, 135, 154, 155, 164, 221, 222, 313, 320, 398, 409, 411, 421

I
IC 05, 100
IC417, 100
IC434, 100
IC1848, 99
IC5067-5070, 99
IC5146, 100
Imax, 165
Immersion, 135
Immersive, 165, 373
Imperfections, 11
In-focus, 25–27, 423
Inspect/inspecting, 11, 18, 40
Inspection, 105, 106, 328
Intensified, 162
Intensifying, 5
Intensity, 101
Interpupillary, 42
In-travel, 137
Inverse-square law, 53, 54
I. R. Poyser, 269–270, 420
Italian, 4, 5

J
Jinghua, 188
Jupiter, 5, 24, 62, 71, 73, 79, 81, 96–99, 145, 153, 162, 163, 171, 174, 250

K
Kasai, 198, 219, 224, 242, 270, 289, 294, 321, 326, 358, 420
Kaspereit, 14, 18, 21, 232, 234, 249, 255–257, 264, 274, 306, 307, 337, 345, 349, 366, 385, 416, 425, 427
Kellner, 12, 14, 18, 118, 127, 136, 150, 232, 234, 248–249, 256, 257, 264, 274, 306, 337, 345, 349, 366, 384–385, 416, 420, 425
Kepler, 5, 14, 18, 21, 416, 425
Kidney-beaning, 23, 140, 145, 174
Klee, 85, 89–91, 124
K-line, 95
Knife-edge, 50
Knight kwl, 136, 196, 205, 226, 275, 298, 303, 324, 349, 354, 362, 386, 393, 421
Kogaku, 188
Kokusai, 111, 118, 127, 136, 196, 199, 204, 205, 218, 220, 226, 229, 232–234, 237, 249, 254, 257, 264, 270–276, 282, 283, 286, 298, 300, 303, 306, 307, 324, 337, 343, 345, 349, 354, 359, 362, 364, 366,385–388, 390–393, 420, 421
Kokusai Kohki, 111, 118, 127, 136, 196, 199, 204, 205, 220, 226, 229, 232–234, 237, 249, 254, 257, 264, 270–276, 282, 283, 286, 298, 300, 303, 306, 307, 324, 337, 343, 345, 349, 354, 359, 362, 364, 366, 385–388, 390–393, 420, 421
König, 12, 14, 18, 21, 55, 67, 126, 127, 131, 136, 170, 335, 336, 388–390, 416, 420, 421, 425
Kson, 188, 199, 204, 218, 220, 229, 238, 275–276, 282, 283, 286, 300, 343, 359, 364, 385, 387, 388, 390, 391, 395, 419
Kumming, 188
Kupco, 166, 329, 374, 403

L
Lagoon Nebula, 98
Laminated, 115
Lanthanum, 115, 117, 222, 262, 313, 320, 329, 333, 397–399, 405, 421, 425
Lateral color, 27, 32, 68, 120, 128, 129, 134, 139, 140, 292, 372, 375, 425, 427
LE, 14, 18, 22, 125, 198, 219, 224, 242, 271, 281, 289, 294, 322, 326, 357–358, 416, 420
Leica, 10, 70, 80, 408, 409, 411
Leica/Leitz, 168
Leipzig, 170–172, 223, 369
Leitz, 276, 422
LER. *See* Long eye relief (LER)
LET, 125, 346, 420
Lightening, 96
Light-polluted, 54, 59, 60, 65, 97, 101, 150
Light-transmission, 88, 134, 174
Limb, 27, 32, 96
Lippershey, 5
Lippershey/LGlilean, 5, 13, 14, 18, 21, 425,
Long eye relief (LER), 125, 346, 420,425
Long Perng, 131, 188, 207, 212, 277–280, 309, 314, 317, 350, 353, 400, 407, 420, 421
Losmandy, 157
Lumicon, 98
Luna, 115
Lunar, 11, 19, 23, 29–32, 74, 75, 96, 97, 101, 105, 115, 118, 144, 163–164, 174
LV. *See* Vixen Lanthanum (LV)

LVW. *See* Vixen Lanthanum Superwide (LVW)
LX, 127, 130, 131, 135, 136, 202, 209, 230–231, 235, 243–244, 251, 259, 292, 296, 315, 346, 347, 373, 403, 420, 421

M
M1, 98
M8, 98, 101
M16, 20, 98
M17, 98
M20, 98, 100
M31, 77, 168, 173
M42, 44, 46, 47, 78, 98, 101, 153
M43, 100, 223, 409
M45, 51, 77
M57, 77, 98, 169
M76, 99
M97, 99
MA. *See* Modified achromat (MA)
Magnification, 69–72
MAK. *See* Maksutov telescope (MAK)
Maksutov-Cassegrain telescope (MCT), 426
Maksutov telescope (MAK), 426
Mask, 94, 101
Masuyama, 111, 125, 199, 204, 218, 220, 229, 238, 254, 271, 272, 276, 280–282, 283, 286, 300, 343, 359, 364, 385, 387, 388, 390, 391, 395, 419
MCT. *See* Maksutov-Cassegrain telescope (MCT)
Meade, 6, 42, 91, 110, 162, 188, 195, 254, 302, 354
Megaview, 136, 143, 202, 209, 231, 235, 251, 259, 277, 292, 296, 314–315, 346, 347, 373, 403, 421, 422
Meniscus, 426
Mercury, 97, 101, 171
Messier, 6, 46, 50, 51, 78, 126, 145, 352
Micrometric, 117, 405
Microscope/microscopic, 106, 111, 165, 168–169
Micro-stain, 107
Milky Way, 5, 174
Minimalist, 57, 58, 115, 180
Minimum glass, 70, 80, 81, 409
Misalignment, 27
MK-70, 127, 388–390, 421
MK-80, 127, 388–390, 421
Modified achromat (MA), 118, 232, 234, 249, 257, 264, 274, 306, 307, 337, 345, 348–349, 366, 385, 420
Mohr, 151, 176–179, 372
Monkey Head Nebula, 99
Mono, 164, 165
Monocentric, 14, 18, 21, 68, 70, 80, 109, 110, 114, 115, 337–338, 417, 419, 426
Moon, 5, 6, 27, 29–32, 41, 46, 54, 71, 74–75, 79, 95–98, 101, 102, 118, 142, 153, 155, 163, 164, 166, 169, 174, 196, 250, 352, 378
Moonfish, 136, 196, 205, 226, 275, 297–298, 303, 324, 349, 354, 362, 386, 393, 421
Museum, 3
Mustache, 30
Mylar, 102

N
Nagler, 3, 12, 14, 18, 22, 55, 65, 88, 120, 128, 131, 134, 136–145, 148, 150–152, 155, 163–165, 167–169, 175–179, 189, 202, 209, 231, 235, 250, 259, 277, 292, 296, 315, 346, 347, 372–374, 403, 408,417, 421,426
Narrowband/narrowbanded, 94, 98, 101
NAV, 55
NAV-HW, 144, 146, 261, 298, 372,422
NAV-SW, 131, 133, 134, 258, 291, 295, 299, 330, 333, 371, 417, 421
NAV-SW/HW, 14, 19, 22, 417
NEAF. *See* Northeast Astronomy Forum (NEAF)
Near-sightedness, 4
Nebula, 5, 55, 65, 68, 69, 74, 77–79, 94, 98, 100, 101, 118, 126, 129, 130, 134, 150, 153, 154, 170–171, 182
Nebulosity, 65
NED, 127, 129, 130, 213, 223, 301, 310, 329, 364, 420
Neptune, 6, 96
Neutral, 73, 76, 77, 81, 101, 102, 134, 153, 179, 292, 333
Newtonian, 7, 27, 63, 84, 150, 173, 175, 180, 426
NGC
  NGC281, 99
  NGC896, 99
  NGC1499, 100
  NGC1514, 99
  NGC 2070, 99
  NGC2174, 99
  NGC2237, 99
  NGC2264, 99
  NGC2327, 100
  NGC2359, 99
  NGC2371/2372, 99
  NGC2392, 99
  NGC2440, 99

NGC (*cont.*)
NGC3242, 99
NGC4666, 166
NGC4845, 166
NGC6210, 99
NGC6445, 99
NGC6543, 99
NGC6781, 99
NGC6888, 99
NGC6905, 99
NGC6960-6995, 99
NGC7000, 99
NGC7008, 99
NGC7009, 99
NGC7293, 99
NGC7635, 99
Nickel-plated, 9, 39
Nikon, 14, 19, 22, 55, 82, 110, 111, 131, 133, 134, 144, 146, 168, 188, 199, 204, 218, 220, 229, 238, 253–300, 330, 333, 343, 359, 364, 371, 372, 385, 387, 388, 391, 395, 411, 417, 419, 421, 422
Nimrud, 4
Nippon, 188
Nirvana, 136, 143, 202, 209, 231, 235, 251, 259, 277, 292, 296, 315, 346, 347, 373, 403, 421, 422
Nitrogen, 259, 261
Nitrogen-purged, 146, 202, 209, 230, 235, 251, 258–262, 292, 296, 315, 346, 347, 373, 403, 421, 422
NJ, 121–123, 158, 240
North American Nebula, 99
Northeast Astronomy Forum (NEAF), 144
NP-101, 152
NPL, 119, 124, 200, 208, 211, 217, 224, 227, 232, 236, 239, 245, 247, 255, 264, 266, 267, 280, 287, 289, 304, 305, 313, 319, 322, 327, 348, 352, 362, 367, 368, 376, 399–400, 420

O

Objective, 4, 7, 29, 48, 52, 53, 61, 67, 415
Observatory, 6, 131, 338–339, 421
Obsession, 168
Off-axis, 19, 25–30, 32, 48, 55, 67, 68, 71, 109, 112–115, 118, 120, 124, 126, 130, 131, 134, 135, 139, 142, 144, 146, 147, 153, 170, 194, 211, 250, 256, 289, 292, 299, 311, 338, 340, 351–353, 376, 378, 379, 384, 402, 406, 419, 424–426
O-II, 98, 111
OIII, 98, 111, 166
Oil, 105, 106
Olivion, 136, 196, 205, 226, 275, 298, 324, 349, 354, 362, 386, 393, 411, 421
Olympus, 168
Omcon, 125, 198, 219, 224, 242, 271, 281, 289, 294, 321, 326, 358
Omni, 119, 200, 208, 211, 217, 224, 227, 232, 235–236, 239, 245, 246, 255, 263, 265, 267, 280, 287, 289, 304, 305, 312, 319, 322, 327, 348, 352, 361, 367, 368, 376, 399, 420
On-axis, 19, 26, 28, 112–114, 124, 130, 140, 164, 247, 250, 338, 341, 405,
Opposition, 115
OPS, 111, 199, 205, 218, 220, 229, 238, 254, 272, 276, 282, 283, 286, 300, 343, 359, 364, 385, 387, 382, 390–391, 395, 419
Opt, 119, 131, 180, 195, 200, 208, 209, 211, 217, 224, 227, 232, 236, 239, 245, 247, 255, 264, 266, 267, 280, 287–289, 302, 304–305, 312, 317, 322, 327, 347, 348, 352, 362, 367–369, 376, 385, 400, 401, 420, 421
Optic/Optik, 3, 34, 81, 110, 162, 187, 193, 254, 302, 355
Optiluxe, 127, 315–316, 420
Orange, 77, 96, 97
Orange-yellow, 96
O-ring, 155, 262
Orion, 46, 58, 78, 101, 157, 274, 277, 287, 310, 315, 411
Orion Nebula, 44, 46, 65, 78, 98
Ortho, 6, 67, 82, 109, 150, 198, 253, 327, 354
Orthoscopic, 79, 109, 115, 128, 162, 166, 169, 174, 182, 377, 405, 426
Orthostar, 111, 199, 204, 218, 220, 228–229, 237, 254, 272, 276, 282, 283, 286, 300, 343, 358, 364, 385, 387, 388, 390, 391, 395, 419
Out-focus, 26
Owl, 275
Owl Astronomy, 127, 129, 131, 136, 196, 203, 205, 225, 226, 275, 298, 303, 304, 321–324, 349, 354, 362, 366, 382, 386, 393, 420, 421
Owl Nebula, 99
Oxygen-III, 79, 101

P

Pacman Nebula, 99
Palomar, 182
Panaview, 131, 135, 195, 209, 217, 264, 266, 288, 302, 306, 317, 347, 369, 385, 401, 421

Panoptic, 14, 16–18, 22, 46, 55, 88, 126, 131–134, 141, 142, 166, 169, 177, 258, 291, 295, 330, 333, 374, 378, 417, 421, 426
Panorama, 136, 176–180, 202, 209, 231, 235, 251, 259, 292, 296, 315, 346, 347, 373, 403, 421
Parabolic, 27, 426
Paracorr, 131, 179
Paradigm, 76
Paragon, 131, 135, 167, 168, 215, 344, 365, 381, 421
Parchment, 5
Parks, 119, 125, 198, 200, 208, 211, 217, 219, 224, 227, 232, 236, 239, 242, 245, 247, 255, 264, 266, 267, 269, 280, 281, 287, 289, 294, 304, 305, 313, 319, 322, 325–327, 348, 352, 358, 362, 367, 368, 400, 420
Patent/patented, 4, 120, 121, 131, 136, 137, 233, 245, 375, 405
Pelican Nebula, 99
Pentax, 14, 18, 22, 42, 51, 55, 57, 70, 82, 89, 91, 110–118, 127, 129, 131, 133–135, 150, 163, 166, 174, 175, 184, 206, 213, 223, 258, 291, 295, 299, 301, 310, 327–333, 364, 371, 374, 378, 380, 384, 404, 405, 408, 411, 417, 420, 421
Peripheral, 140, 147, 167, 175
Personal Solar Telescope (PST), 100, 246
Photon, 164
Pincushion, 30, 250, 424, 426
Planesphere, 111, 112, 247, 419
Planetary, 39, 73, 109, 150, 199, 253, 304, 355
Plano, 120, 121
Plano-concave, 4, 426
Plano-convex, 4, 5, 118, 426
Plastic, 8, 41, 140
Pliny, 4
Plössl, 12, 42, 89, 117, 150, 198, 254, 303, 354
Plössl/Symmetrical, 14, 18, 21, 120
Polar, 73, 81, 96–98, 157
Polarize/polarizing, 102, 425
Pollen, 106, 107
Polluted/pollution, 54, 59, 60, 65, 79, 94, 97, 98, 101, 150, 174, 182
Polymer, 8, 35
Proprietary, 81, 113, 115, 122, 188, 328
Protégé, 144
Prototype, 137, 138
Proxima, 131, 268, 421
Pseudo-Masuyama, 125
PST. *See* Personal Solar Telescope (PST)
Public outreach, 169, 346, 353, 426
Pupil, 16, 22–23, 42, 45, 56, 58–66, 74, 77, 78, 84, 100, 128, 134, 137, 140, 142, 151, 167, 170, 177, 179, 181, 372, 377, 403, 415, 424, 425
Push-pull, 38

Q

Q70, 131, 195, 209, 217, 264, 266, 288, 302, 306, 317, 347, 369, 385, 401, 421
Q-Tip, 103, 106
Queasy, 29
Questar, 164

R

Radian, 25, 55, 117, 127–129, 173, 377–378
Raleigh, 151, 176–179, 372
Ramsden, 5, 14, 18, 21, 29, 248–249, 417, 426
Rank, 114, 126, 256
Rank-Kaspereit-Erfle (RKE), 14, 18, 21, 110, 111, 118, 153, 164, 189, 232, 234, 249, 254–257, 264, 274, 306, 307, 336, 337, 345, 349, 366, 385, 417, 420, 427
Rayleigh, 67
Rectilinear distortion, 25, 30–31, 48, 79, 130, 131, 173, 424, 426
Reflections, 16, 24, 41, 94, 114, 128, 154, 180, 250, 330, 331, 333, 360–362, 370, 372–374, 376, 378, 379, 425
Reflective, 24, 40, 381
Reflector, 27, 167, 173, 179, 182, 183
Refraction, 13, 115, 329, 330, 333
Refractive, 4, 261, 424, 425
Refractor, 6, 7, 46, 50, 63, 64, 84, 95, 114, 150, 153, 173, 175, 180, 182, 426
Refurbished, 36
Residual Oil Remover (ROR), 105
Ring Nebula, 98, 99, 352
Rini, 127, 334, 421
RKE. *See* Rank-Kaspereit-Erfle (RKE)
ROR. *See* Residual Oil Remover (ROR)
Rosenstock, 112, 262
Rosette Nebula, 99
Russell Optics, 118, 131, 232, 234, 249, 257, 264, 274, 306, 307, 335–337, 345, 349, 366, 385, 420

S

Saturn, 5, 62, 96–99, 118, 153, 171, 174, 250
Saturn Nebula, 99
Scanning, 29
Scatter-free, 116

Scatter/scattered/scattering, 70, 77, 80, 112–116, 118, 124, 126, 134, 153, 170–172, 174, 394
Schmidt-Cassegrain, 173, 427
Schmidt-Cassegrain telescope (SCT), 46, 64, 84, 427
Schröter, 115
SCT. *See* Schmidt-Cassegrain telescope (SCT)
Seagull, 100
Seneca, 4
Sharp-to-the-edge, 134, 142, 333
Shorty, 89, 103
Shoulder, 8, 10, 12, 89–91, 427
Siebert, 85, 91, 340, 342
Siebert Optics, 110–112, 127, 131, 199, 204, 214, 220, 229, 238, 254, 272, 276, 282, 283, 286, 300, 336–344, 359, 364, 385, 387, 388, 391, 395, 419, 421
Silvertop, 239, 240
Single-coated, 137
Singlet, 4, 5, 13, 110–112, 114, 247, 250, 341, 419, 427
Sirius, 390
Sketcher, 154
Sketch/sketching, 67, 71, 173–174
Sky-Watcher, 118, 119, 125, 127, 131, 135, 137, 143, 193, 195, 197, 200, 202, 205, 207–209, 211, 215, 217, 219, 224, 226, 227, 231, 232, 234–236, 239, 245, 247, 249, 251, 255, 257, 259, 264, 266, 267, 274, 277, 280, 287–289, 292, 296, 298, 302–308, 311, 313, 315, 317, 319, 322–324, 327, 337, 344–350, 352, 354, 362, 363, 366–369, 373, 376, 381, 385, 386, 393, 400–403, 411, 420, 421
Smart Astronomy (SA), 125, 207, 212, 278, 309, 314, 350, 353, 400, 407, 420
SMC. *See* Super multi-coatings (SMC)
Smoothie, 162, 163
Smooth-sided, 123, 125, 198, 219, 224, 242, 271, 281, 286, 290, 294, 321, 326, 358, 376, 420
Smyth, 104
Solar, 95, 99, 101, 102, 125, 149, 167, 183, 207, 212, 278, 309, 314, 350, 353, 400, 407, 420
Soul Nebula, 99
Spacer, 159–161
Space-walk, 3, 139, 140, 142, 147, 165, 167, 292, 373
Spectra/spectrum, 23, 27, 94, 95, 98, 99, 360–362, 423, 425
Speers-WALER, 136, 201–202, 209, 230, 235, 251, 259, 292, 296, 315, 346, 347, 373, 403, 421
Sphere, 110, 112, 155, 156, 247, 341
Stellarvue, 125, 207, 212, 278, 309, 314, 350, 352–353, 407, 420
Sterling, 117, 126, 352
Strategy/strategies, 33–72, 84, 87, 145, 169
Stratus, 131, 135, 222, 287, 313, 319–320, 398, 421
Sun, 5, 95, 102, 183
Supermono, 117, 118
Supermonocentric, 70, 71, 82, 111, 113, 115, 116, 147, 157, 166, 206, 261, 328, 331, 380, 383–384, 404, 405, 420
Super multi-coatings (SMC), 114, 115, 328–330, 333
Superview, 131, 195, 209, 217, 264, 266, 288, 302, 306, 317, 347, 369, 385, 401, 421
Superwide (SW), 131, 163, 193, 195, 209, 217, 222, 264, 266, 288, 302, 306, 311, 313, 317, 320, 322–323, 347, 349, 368, 369, 385, 398, 401, 402, 421
Surplus Shed, 127, 131, 136, 157, 196, 205, 226, 275, 298, 301–355, 362, 386, 393, 421
Swa, 126, 131, 133, 163, 208–209, 258, 290–291, 295, 299, 330, 333, 371, 374, 416, 421
Swan Nebula, 98
SWA/UWA, 14, 18, 22, 55, 416

T

T1, 163
T2, 163
T4, 128, 140, 145, 151, 152, 163, 165, 167, 373
T5, 143, 145, 148, 163, 167, 177, 178, 373, 403
T6, 88, 167, 177, 179, 292, 373
Takahashi, 14, 18, 22, 110, 111, 125, 136, 143, 198, 199, 205, 218–220, 224, 229, 238, 242, 254, 271, 272, 276, 281, 283, 286, 289, 294, 300, 321, 326, 343, 357–411, 416–420, 422
Takahashi-UW, 14, 19, 22, 136, 143, 359–360, 417, 422
TAL, 131, 196, 205, 226, 298, 303, 354, 360–362, 386, 393, 421
Tarantula Nebula, 99
TEC140, 157
Telescope service (TS), 118, 119, 127, 129–131, 135, 171, 193, 195, 199, 200, 203, 205, 207–211, 213, 217, 218, 220, 223–225, 227, 229, 232, 234, 236, 238, 239, 245, 247, 249, 254, 255, 257, 264, 266, 267, 272, 274, 276, 280, 282, 283, 286–289, 300, 301, 302, 304–308, 310–313, 317, 319, 322–324, 327, 329, 337, 343, 345, 347–349, 352, 359,

362–369, 376, 378, 382, 385, 387, 388, 391, 395, 400–402, 419–421
Tele Vue, 3, 34, 80, 108, 151, 188, 200, 255, 304, 358
TFOV. *See* True field of view (TFOV)
Thor's Helmet Nebula, 99
Thousand oaks, 102
TMB, 10, 55, 68, 70, 71, 82, 88, 110, 111, 113, 115–118, 127, 129–131, 135, 144, 146, 147, 157, 163, 166, 167, 194, 203, 206, 215, 225, 261, 304, 311, 324, 328, 331, 338, 344, 366, 378–384, 404, 405, 420, 421
Tonal, 73, 76, 77, 81, 292
Tone-neutral, 73
Tone/toned, 73, 76, 77, 81, 84, 153, 179, 333
Topographical, 5
Transmission, 63, 64, 67–69, 80, 88, 96–99, 101, 113–115, 124, 130, 134, 135, 145, 154, 162, 174, 279, 351, 381, 384, 405, 424–426
Transparency, 67, 74, 117
Transparent, 100, 405
Transverse, 25, 27, 32, 427
TRAP-E, 78
Trapezium, 78
Trap-F, 78
Trifid Nebula, 98, 100
Triplet, 13, 111, 427
True field of view (TFOV), 43–53, 58, 59, 65–68, 75, 77, 78, 132, 141–143, 146, 151, 152, 389, 403, 409, 410, 413–415, 427
Turret, 166, 218
Turtle Nebula, 99
Tuthill, 125, 198, 219, 224, 242, 271, 281, 289, 294, 321, 326, 358
TV, 91, 126, 163, 176–180
Type-1, 136, 169
Type-2, 137, 138
Type-4, 137, 139–141
Type-5, 137, 139–143
Type-6, 137, 139, 140

U

UHC, 166
Ukraine, 113
Ultima, 125, 131, 135, 198, 219, 224, 241–243, 271, 281, 289, 294, 321, 326, 358, 420, 421
Ultra-high, 6, 98, 101, 162
Ultra-narrow, 99
Ultrascopic, 125, 198, 219, 224, 242, 271, 281, 289, 294, 320, 326, 353, 420
Ultra wide angle (UWA), 14, 18, 22, 55, 70, 104, 127, 129, 136, 143, 163, 169, 195–196, 202, 203, 209, 225, 230, 231, 235, 250–251, 259, 277, 291–292, 296, 304, 315, 324, 346, 347, 366, 373, 382, 403, 416, 420, 421
Uncorrected, 30, 131
Undercut, 9, 39, 40, 180, 312
Undriven, 151
United-Optics, 188
Unitron, 111, 118, 199, 205, 218, 220, 229, 232, 234, 238, 249, 254, 257, 264, 272, 274, 276, 282, 283, 286, 300, 306, 307, 337, 343, 345, 349, 359, 364, 366, 384–385, 387, 388, 391, 395, 419, 420
University optics (UO), 89, 90, 110, 111, 117, 124, 126, 127, 131, 136, 174, 195, 196, 199, 205, 209, 217, 218, 220, 226, 229, 233, 238, 254, 264, 266, 272–276, 282, 283, 286, 288, 298, 300, 302, 303, 306, 317, 324, 343, 347, 349, 354, 359, 362, 364, 369, 385–393, 401, 419–421
Uranus, 6, 96
UW, 14, 19, 22, 136, 143, 359–360, 417, 422
UWA. *See* Ultra wide angle (UWA)
UWAN, 14, 18, 22, 65, 136, 143, 163, 168, 202, 209, 231, 235, 251, 259, 277, 292, 296, 315, 346, 347, 373, 402–403, 417, 421

V

Variable, 83, 102, 139, 149, 180
Veil Nebula, 99
Venus, 5, 96, 97, 149, 171, 174
VERNONscope, 82, 111, 150, 153, 163, 164, 174, 188, 199, 205, 218, 220, 229, 238, 254, 272, 276, 282, 283, 286, 300, 343, 359, 364, 385, 387, 388, 391, 394–395, 419
Vignette, 88, 93, 103, 124
Vixen, 34, 119, 122, 124, 125, 131, 135, 163, 188, 192, 200, 208, 211, 217, 222, 224, 227, 232, 236, 239, 243, 245, 247, 255, 264, 266, 267, 280, 287, 289, 304, 305, 310, 313, 319, 320, 322, 327, 348, 352, 362, 365, 367, 368, 376, 396–400, 411, 420, 421
Vixen Lanthanum (LV), 125, 192, 243, 310, 365, 396–398, 420
Vixen Lanthanum Superwide (LVW), 34, 131, 135, 163, 222, 313, 320, 397–398, 421

Volcano, 111, 121, 123, 199, 205, 218, 220, 229, 238, 254, 272, 276, 282, 283, 286, 300, 343, 359, 364, 385, 387–388, 391, 395, 419
VR-1, 95

W

W70, 131, 202, 421
WA, 131, 171, 193, 195, 209, 217, 264, 266, 288, 302, 306, 311, 317, 323, 347, 349, 368, 369, 385, 401–402, 421
Waterproof, 146, 251, 258, 259, 261, 262
Wavelength, 94, 95, 98, 101
Webster/Lockwood, 45, 135
White water, 222, 384, 427
Widescan, 136, 196, 205, 226, 275, 297, 298, 303, 324, 349, 354, 362, 386, 393, 421
William Optics, 43, 65, 95, 131, 143, 163, 193, 195, 207, 209, 212, 217, 264, 266, 278, 288, 302, 306, 309, 311, 314, 317, 323, 347, 349, 350, 353, 368, 369, 385, 400–403, 407, 411, 420
WO, 163
Wollensak, 127, 131, 136, 196, 205, 226, 275, 298, 303, 324, 349, 354, 355, 362, 386, 393, 421

X

X-Cel, 127, 130, 192, 243–244, 310, 397, 420
XF, 127, 129, 213, 223, 301, 310, 329, 364, 378, 411, 420
XL, 14, 18, 22, 55, 131, 133, 134, 258, 291, 295, 299, 330, 333, 337, 371, 374, 417, 421
XO, 70, 82, 110, 112, 114, 115, 163, 174, 206, 328, 331, 380, 384, 404, 405, 409, 420
XP, 110, 113, 114, 206, 328, 331–333, 380, 384, 404, 405, 420
XW, 14, 18, 22, 42, 51, 55, 57, 70, 89, 91, 131, 133–135, 144, 163, 174, 175, 184, 258, 291, 295, 299, 330–333, 371, 374, 408, 417, 421
XWA, 144, 297, 422

Z

Z100, 16, 144, 194, 311, 379, 406, 422
ZAO. *See* Zeiss abbe orthos (ZAO)
ZAO-I, 111, 113, 117, 118, 206, 328, 331, 380, 384, 404–405, 420
ZAO-II, 117, 118, 405
Zeiss, 70, 82, 110–113, 115–117, 126, 134, 155, 157, 165, 166, 168, 169, 206, 220, 271, 328, 331, 380, 384, 394, 403–405, 408, 409, 420
Zeiss abbe orthos (ZAO), 70, 82, 112, 115, 116, 126, 134, 163, 220, 394, 409
Zhumell, 144, 146, 194, 207, 212, 278, 309, 311, 314, 350, 353, 379, 400, 406–407, 411, 420, 422
Zoom, 10, 70, 80, 103, 104, 154–155, 163, 180–181, 408–410
Zoomset, 85, 103, 104